AF552445

SHEEP PRODUCTION

P.N. Bhat

Centre for Integrated Animal Husbandry Dairy Development,
Flat No. 205, Block No. F – 64/C9,
Sector – 40, Noida – 201 301

C.L. Arora

Former Assistant Director General
(Animal Production and Breeding)
ICAR, New Delhi

2009

Studium Press (India) Pvt. Ltd.

SHEEP PRODUCTION

ISBN: 978-93-80012-02-5

Published by:

Studium Press (India) Pvt. Ltd.
4735/22, 2nd Floor, Prakash Deep Building
(Near Delhi Medical Association),
Ansari Road, Darya Ganj, New Delhi-110 002
Tel.: 23240257, 65150447; Fax: 91-11-23240273;
jngovil@studiumpress.com; shray@studiumpress.com

Printed at:

Salasar Imaging Systems
C-7/5, Lawrence Road Indl. Area, Delhi-110035

ABOUT THE SERIES

According to the 2003 Census data, the country had 485 Million (M) livestock population and 489 M poultry population, having the second highest number of cattle 185 M, the highest number of buffaloes 97 M, the third highest number of sheep 61 M, the second highest number of goats 124 M, the sixth highest number of camels 632 M, the fifth highest number of chickens 489 M and the fourth highest number of ducks 33 M in the world.

Livestock Sector has been playing an important role in Indian economy and is an important sub-sector of Indian agriculture. The contribution of livestock to GDP was 4.36% in 2004–05 at current prices. According to CSO estimates, gross domestic product from livestock sector at current prices was about Rs. 935 billion during 1990–2000, (about 22.51% of agriculture and allied GDP). This rose to Rs.1239 billion during 2004–05 with 24.72% share in agriculture and allied GDP. But the share of livestock sector in the plan allocation hovered at around 7% of the agricultural out lay.

This sector plays an important and vital role in providing nutritive food, rich in animal protein to the general public and in supplementing family incomes and generating gainful employment in the rural India, particularly among the landless, small, marginal farmers and women. Distribution of livestock wealth in India is more egalitarian, compared to land. Hence, from the equity and livelihood perspectives, it is an important component in poverty alleviation programmes. This fact however has not been appreciated by Policy planners and implementers.

The development of animal husbandry has been envisaged as an integral part of system of diversified agriculture. With its large livestock population, India has vast potential for meeting the growing need of millions, in respect of livestock products such as milk, eggs, meat and wool. This sector has the greatest potential in creating new self sustaining jobs in villages, if the knowledge base in veterinary and animal husbandry technology is based on real time data and is used in transforming rural India and other similarly placed developing countries.

Livestock production systems are based on low cost agro by-products as nutritional inputs, using traditional technologies. The spectacular growth of livestock products especially in milk, meat, eggs and poultry meat is attributable to the initi-atives taken by Government and the organized private sector and has primarily been driven by horizontal increase in numbers. It has been observed that with increasing income, demand for cereals is decreasing, which is causing a livestock revolution, which is demand driven.

With the livestock sector assuming an important role in the national economy, there is a requirement to improve the present state of knowledge gathering and information dissemination. Although considerable resources have been directed towards collecting and disseminating information on basic crops, little attention has been given to collecting, analyzing and disseminating information on livestock.

It is necessary that livestock units are made financially viable through generating a service provider industry which becomes a technology catalyser in a low educated farming community. This requires several initiatives, one of the major initiative is to upgrade the knowledge base and make the information available to students, teachers, planners and farmers.

The **Studium Press (India) Pvt. Ltd.** has decided in association with **Centre for Integrated Animal Husbandry and Dairy Development (CIAH&DD)** to bring out a series of books for under graduate and post graduate scholars in Animal and Veterinary Sciences in several volumes under the chief editorship of Professor (Dr) P.N. Bhat, Former Vice Chancellor and Director of Indian Veterinary Research Institute, Izatnagar–243122 (U.P) and former Animal Husbandry Commissioner of India and Deputy Director General (Animal Science) of Indian Council of Agricultural Research, Ministry of Agriculture, Krishi Bhawan, New Delhi. The basic idea of the series is provide first rate text books to students and scholars in developing countries based on the experiences of developing countries themselves with special focus on Southern Asia in conformity with standards laid out by regulatory agencies in India (VCI, ICAR, UGC, AICTE) and similar agencies in other developing countries.

The titles to be brought out in present series are given below.

1. Goat Production
2. Dairy co-operatives in India
3. Sheep Production
4. Buffalo Production
5. Dairy cattle production
6. Pig Production
7. Poultry Production

8. Cross breeding of cattle for improved milk production in tropics
9. Camel Production
10. Yak Production
11. Mithun Production
12. Rabbit Production
13. Laboratory Animal Production
14. Dog Management, Breeding and Health
15. Animal Biodiversity
16. Livestock Statistics
17. Livestock Economics
18. Livestock Extension
19. Breeding and Health of Equine
20. Animal Nutrition
21. Animal Physiology

It is hoped that in the next two years, these books will be available for the benefit of student, teachers and professionals in the area and fill the gap which is currently wide.

Prof (Dr) Pushkar Nath Bhat

ABOUT THE AUTHORS

Prof (Dr) P.N. Bhat got his Ph.D in population genetics from Institute of Population Genetics Purdue University, West Laffayette, Indiana, USA followed by several post doctoral assignments and visiting professorship in Genetics, biotechnology, livestock production systems. He returned to India and joined Punjab Agricultural University at Hisar–Ludhiana followed by several professional assignments in India and abroad. He joined as Project co-ordinator (Animal Breeding) and subsequently joined as Head, Division of Animal Genetics at Indian Veterinary Research Institute in 1971. He was responsible for establishing coordinated projects on cattle, sheep, goat, pigs and poultry during 1970–74. He was founder Director of Central Institute for Research on Goats. In 1984 he became Vice-Chancellor and Director of IVRI. He joined as Deputy Director General (Animal Science) in May, 1992 and Animal Husbandry Commissioner in December, 1992. He is fellow of several National and International Science Academies.

Late (Dr) C.L. Arora obtained his Bachelor's degree in veterinary Sciences in 1958 from Punjab University followed by M.V.Sc. in Animal Genetics and Breeding from the University of Madras and Ph.D from Haryana Agricultural University. He has been involved in research in the field of Animal Genetics and Breeding for nearly 30 years and over 21 years in teaching post-graduate and under-graduate students. He has been the Director of Central Sheep and Wool Research Institute at Avikanagar, Rajasthan from 1981–85 and joined as Assistant Director General (Animal Production and Breeding) at Indian Council of Agricultural Research Headquarters from 1985–90. He has published 140 Research papers, 36 Research abstracts, 13 Popular articles, 2 Books and bulletins. He has also handled a number of projects on sheep, goat and cattle improvement.

PREFACE

Sheep with its multi-facet utility (for meat, wool, skin, manure and to some extent milk and transport) play an important role in the Indian agrarian economy. They are better adapted to arid and semi-arid tropics with marginal and sub-marginal lands, other wise unfit for crop production. They are perhaps the most suitable small ruminants to utilize the sparse vegetation available in dry-land areas through range land management and reseeded pastures. They have an excellent ability to survive over a prolonged period of drought and semi starvation and are less prone to extreme weather conditions, ectoparasites as well as other diseases. They are unique for their fibre which allows ventilation and also protects the skin from the hot sun, rain and abrasions. Sheep can also constrict or relax blood vessels in face, legs and ears for control of heat loss. Because of their hardiness and adaptability to dry conditions, the North-Western and the Southern Peninsular regions of the country have a large concentration of sheep. In the tropics they are non-seasonal breeders and can be made to lamb throughout the year. The visual sense is exceedingly well developed in sheep that is how they can discern movement far better than humans but cannot distinguish shapes as well as the human beings.

The subject of sheep production and management was introduced in many Agricultural Universities at graduate and post graduate levels. Many books have been written by foreign authors and we have been using them. The climate of this region being different from that of those countries, these are far from the realities under our conditions. Very few books have been written on sheep management under tropical and subtropical conditions. These are not according to syllabi being followed in Agricultural Universities. This book has been written, keeping in view the requirements of undergraduate and post graduate students of sheep production which is rare combination.

While sheep's place among the livestock species is an important one, the research and development in this species is meagre and information available among the species is small. The editorial board decided that the second project should be a text book on **Sheep**, so that a basic text is available to the students of Veterinary and Animal Science Colleges as supplement to their college notes.

A number of personal have contributed to the development of this text. The basic draft was prepared by Late Dr Chaman Lal Arora, former Assistant Director General (Animal Production and Breeding) at Indian Council of Agricultural Research

Headquarters before his tragic demise. Which was then worked over by Dr P.N. Bhat, *Former* Director and Vice-Chancellor of Indian Veterinary Research Institute, Izatnagar.

We are grateful to Mr R.C. Garg of Project Directorate on Cattle for carefully going through the manuscript and for making several new suggestions which have improved the text. The advantage of having outstanding colleagues and friends like Dr R.M. Acharya and Dr N.K. Bhattacharyya for referral discussion is acknowledged.

Dr J.N. Govil, Publishing-Director and Managing Editor, Researchco Books & Periodicals Pvt. Ltd., Daryaganj, New Delhi, who is the brain behind this initiative deserves special thanks for making it possible to see that this volume is brought out in time and to the expected standards. Mr Anil Jain and Mr Shray Jain, proprietors of Studium Press (India) Pvt. Ltd. need to be complemented for sustained support.

The hard work put up by Dr H.P.S. Arya and Dr Stella Esther for proof reading the manuscript and preparation of content list is gratefully acknowledged. The coordination work of Mr G.P. Gangadharan Pillai, Executive Assistant to Chairman, in preparation of the draft manuscript and typing by Mr S.K. Chowdhry and Mr Pius Joseph is gratefully acknowledged.

New Delhi
10th March, 2008

Prof (Dr) Pushkar Nath Bhat
Author

CONTENTS

List of Colour Figures

List of Tables

CHAPTER 1

INTRODUCTION

Sheep with its multi-facet utility (for meat, wool, skin, manure and to some extent milk and transport) play an important role in the Indian agrarian economy. They are better adapted to arid and semi-arid tropics with marginal and sub-marginal lands, other wise unfit for crop production. They are perhaps the most suitable small ruminants to utilize the sparse vegetation available in dryland areas through rangeland management and reseeded pastures. They have an excellent ability to survive over a prolonged period of drought and semi-starvation and are less prone to extreme weather conditions, ectoparasites as well as other diseases. They are unique for their fibre which allows ventilation and also protects the skin from the hot sun, rain and abrasions. Sheep can also constrict or relax blood vessels in face, legs and ears for control of heat loss. Because of their hardiness and adaptability to dry conditions, the North-western and the Southern peninsular regions of the country have a large concentration of sheep. In the tropics they are non-seasonal breeders and can be made to lamb throughout the year. The visual sense is exceedingly well developed in sheep, that is how they can discern movement far better than humans but cannot distinguish shapes as well as the human beings.

Sheep in India are mostly maintained on natural vegetation on common grazing lands, wastelands and uncultivated (fallow) lands, stubbles of cultivated crops and top feeds (tree loppings). Rarely they are kept on grain, cultivated fodder or crop residue.

The sheep are mostly reared for wool and meat. Sheep skins and manure constitute important sources of earning, the latter particularly in southern India. Milk from sheep is of limited importance, limited to areas of Jammu and Kashmir,

Rajasthan, parts of Uttar Pradesh and Gujarat. Indian sheep are not regarded as dairy sheep.

Because of their close grazing nature and ability to utilize very low set vegetation which no other animal can utilize and their capacity to cover long distance in search of forage and water, they have often been associated with creation of desertic conditions. In reality, it is not the sheep but the man who owns the sheep, who is to be blamed for misconceived management of the grazing lands leading to desertification. A controlled and judicious grazing on the non-cultivable land could prevent soil erosion and make it fertile and suitable for crop production through sheep droppings collected over the years.

CHAPTER 2

CLASSIFICATION, ORIGIN AND DOMESTICATION

2.1 Taxonomy

As a member of Animal Kingdom, they belong to the Phylum–Chordata (back bone), Class–Mammalia (suckle their young ones), Order–Artiodactyla (hooved, even footed), Family–Bovine (ruminants), Genus–*Ovis* (domesticated and wild sheep) and Species–*aries*. Within this species many different breeds exist. All the pure breds and crossbreds have their place.

Class	**Mammalia**
Subclass III	Eutheria (females possessing placenta)
Order IV	Ungulata (hoofed mammals)

Suborder 1. Artiodactyles
(even-Suborder toed animals)
A. Sunina
Family C: Suidae
(true pigs)
Genus : *Sus*
Sus domesticus (domestic pig)
C. Tylopoda
Family : Camilidae
Genus : *Camellus* Linn
Camelus dromedarius Linn
(Arabian camel)

2. Perissodactyla
(uneven-toed animals)

Family C: Equidae
(horses, donkeys and zebras)
Genus : *Equus*
1. *Equus caballus* (horse)
2. *Equus asinus* (donkey)
(**Note:** Mule is a cross between male ass and female horse. Jennet is a cross between female ass and male horse. Both are sterile)

2. *Camelus bactrianus* Linn.
(two-humped bactrian camel)
C. Pecora (true ruminants)
Family D: Bovidae (hollow horned)
Genus : *Bos*

1. Taurine group:

Bos taurus (humpless cattle)
Bos indicus (humped or zebu cattle)

2. Bubaline group:

Bos (Bubalus) *bubalis* (Indian water buffalo)
Genus : *Ovis*
Ovis aries (domestic sheep)
Genus : *Capra*
Capra hircus (domestic goat)

Sheep in India as well as in Arabia have originated from their wild ancestor *Ovis orientale* Vignei. They appear to have been domesticated in the mountain of Iran, Turkestan and Baluchistan.

2.2 History of Domestication

Sheep along with goats were perhaps the first ruminants to be domesticated by man. They appear to have been firstly domesticated in the mountains of Iran, Turkestan and Baluchistan. Reference to their role in the economy of mankind is found in the history of civilization of Mesopotamia, Mohenjodaro and Harappa in northern India.

2.3 Theories of Domestication

The earliest evidence of sheep domestication seems to point to the near East, around 9000 B.C. Unlike our modern sheep, completely reliant on humanity, these early sheep had hair rather than wool. It took about 3,000 years until wooly sheep appeared and another 2,000 years before wool was used for garment making.

2.4 Wild Species of Sheep in India

There are three different varieties of wild sheep, the three geographical races are distinguished by colour, size, and horns.

These are:

a) The Shapoo or Urial of Ladakh and Punjab.

b) The Nyan of Tibet

c) Bharal of Trans-Himalayan Mountains

d) The Marchopolo Sheep

a) The Shapoo or Urial

The Shapoo or Urial is known as Shapoo in Ladakh or Urial in Punjab, this wild sheep lives under the varied climatic conditions. The Ladakh shapoo stands about 3 feet at shoulders. During summer season, its coat is grey or fawn which changes into mixtures of grey and brown during winters. The sheep inhabits in open valleys, hillside and grassy mountain slope at moderate elevations. It is steady, active and is always alert. The old male lives separately from group except during the breeding season. This wild sheep freely crosses with the domestic sheep. The Ladakh Shapoo sheep are found from Astor to Zanskar range in Ladakh and in Tibet region. The Punjab range of this wild sheep is found in Jhelum salt range and Northern parts of Pakistan.

b) The Nyan or Tibetan Sheep

It is the largest of old wild sheep and is found in the Himalayas and highlands of Central Asia. The variety of sheep varies in different areas in form of its horn, colour and size. The sheep stands up to 4' at shoulders and seems like antelope in built. The lambs are of light brown colour whereas the adult have white head and a white ruff developed around the neck which is shed with the age. The sheep ascends to higher level during the summer at the height of 15000', and in winters descend at the shelter of lower valleys. During the spring, male and female live separately and rejoin in the late autumn which is the beginning of mating season. The sheep are mainly found in the desolate plains and sand hills which can be extremely hot in summer and freezing cold in winters.

The sheep are plentiful in Ladakh and Tibet and are occasionally seen in Spiti valley of Himachal, Kumaun, Nepal and extreme north of Sikkim.

c) Bharal

Bharal is supposed to have the place between sheep and goats. Bharal is not bearded and doesn't have goat-like smell. It has rounded and smooth horns with curves backward. It stands 3' at shoulders and its powerfully built. Bharal weighs 120 to

150 pounds which varies to different time of the years. It has got a brown coloured head suffused with salty blue. It has got a dense coat. The face and chest of old Bharal is black and the black stripes run along the middle of each fringes and down the front of legs. During summer, Bharals' are found at 16000' and rarely below than 12000' during the winters.

They are found between freeze and snow lines with abundance of grass. The Bharal are found in most inaccessible places. They are normally seen in large flock of almost 200 in numbers. Bharal is found in Tibet, Ladakh, Sikkim and Nepal.

d) The Marchopolo Sheep

The Marchopolo sheep has the magnificent horns deeply wrinkled with colour of gold-ivory. Adult sheep can stand upto 12 hands and weigh up to 250 pounds. During the winter, this sheep has got creamy white head, legs and belly. This colour makes the animal almost invisible except while they are in movement. Its choice of food is limited to bunches of wiry grasses and later to the species of wild onion. After hard winter conditions, many die due to starvation. The breeding habits are quite similar to other sheep.

CHAPTER 3

ECOLOGY, HABITAT AND DISTRIBUTION

3.1 Geographical Distribution

India can be divided, on the basis of the agro-climatic conditions and the type of sheep into 4 regions, *viz.* (i) the Northern-temperate region, (ii) the North-western, Central and Semi-arid regions, (iii) the Southern region, and (iv) the Eastern region.

3.1.1 Northern temperate region

The Northern region comprises Jammu and Kashmir, Himachal Pradesh and hilly regions of Uttar Pradesh. About 7.68 million hectares (Mha) of land is available for natural grazing. *Andropogon*, *Apluda*, *Aristida* and *Setaria* species among annual and *Dichanthium*, *Eragrostis*, *Heteropogon* and *Themeda* species are among the main perennial grasses found in the region except in the Alpine pastures where temperate grasses and legumes, *viz.* red and white clover and rye grass are found. The important fodder trees and shrubs are *Acacia*, *Bauhinia*, *Morus*, *Terminalia* and *Zizyphus* species. This region has 4.5 M sheep and accounts for 7.3% of the total population (2003). It produces about 9.4 Mkg wool (20.6%). Of this about 1.2 Mkg is of 36s to 48s quality and 4.3 Mkg of 48s and above quality. The largest population of crossbred sheep primarily developed for apparel wool is in this region. In Jammu and Kashmir the crossbred sheep, mostly Merino crosses with variable level of Merino inheritance known as Kashmir Merino, constitute almost 70% of the population.

3.1.2 North-western, central arid and semi-arid region

This region comprises the states of Punjab, Haryana, Rajasthan, Gujarat, plains of Uttar Pradesh and Madhya Pradesh. About 45.63 Mha of land is under natural vegetation and is available for animal grazing. The major annual grasses of the region are *Andropogon* and *Aristida* species. The important perennial grasses are *Cenchrus*, *Lasiurus*, *Panicum*, *Sehima*, *Andropogon* and *Heteropogon* species. The major shrubs and trees are *Acacia, Albizia, Dicrostachys, Prosopis* and *Hardwickia* species. This region has the second largest sheep population (14.9 M; 24.4% of the total population; 2003 estimates). The annual wool production is 22.2 Mkg (about 49.3% of the total wool produced in the country). This region is most important for carpet wool production. Chokla and Patanwadi produce good apparel quality carpet/medium apparel quality wool. Malpura and Sonadi have extremely coarse and hairy fleeces and are to some extent used for milk. The other breeds produce medium to good quality carpet wool.

3.1.3 Southern region

This region is semi-arid in the Central peninsular area and hot and humid along the coast. It comprises the states of Maharashtra, Andhra Pradesh, Karnataka, Tamil Nadu and Kerala. About 34.90 Mha area is available for natural grazing. The major annual grasses of the area are *Aristida* species and the perennial grasses are *Andropogon*, *Cenchrus*, *Chrysopogon*, *Dicanthium*, *Heteropogon*, *Sehima* and *Themeda* species. Important fodder trees and shrubs are *Acacia*, *Hardwickia* and *Bauhinia* species. This region has the largest sheep population (37.3 M; 62.2% of total population). Almost 50% of the sheep in the region do not produce any wool. The rest produce extremely coarse, hairy and coloured fleece. Most of the sheep in the region are primarily maintained for meat and manure except Nilgiri. Nilgiri sheep produce fleece of good quality. Their number is extremely small and is restricted to Nilgiri hills. Their rearing poses problems as shearing facilities are not available and disposal of wool is also a limiting factor. Nearly 12 Mkg wool, about 26.6% of the total wool in the country, is produced in this region. The wool is of below 36s quality, excepting for a small quantity produced by Nilgiri sheep which is above 48s. The wool is thus suitable for extremely coarse carpets, burrack blankets and handspun woven kumblies. The Nellore is the tallest breed and Mandya, also known as Bandur or Bannur, is the shortest and has the best meaty conformation among the Indian breeds. The majority of breeds in Tamil Nadu, Andhra Pradesh and Karnataka, except Bellary or Deccani, fall within the range of height and colour of Nellore and Mandya.

3.1.4 Eastern region

This region comprises the states of Bihar, West Bengal, Orissa, Assam, Meghalaya, Arunachal Pradesh, Mizoram, Manipur, Tripura, Nagaland and Sikkim. It is mostly hot and humid, except for some parts of Eastern states which are sub-temperate and humid. About 30.48 Mha of land with natural vegetation is available for grazing. The predominant annual grasses are *Andropogon*, *Apluda* and *Aristida* species. The major perennial grasses are *Dichanthium*, *Heteropogon*, *Sehima* and *Themeda* species. The important trees and shrubs are *Acacia*, *Albizia*, *Bauhinia* and *Terminalia* species. This region has about 3.0 M sheep representing 4.9% of the total population. They produce about 13 Mkg of wool, *i.e.* about 2.9% of India's total wool production. Most of the wool is of below 36s quality except in the Tibetan sheep found in Arunachal Pradesh which produce good carpet quality wool. The majority of the wool produced in the region is suitable only for extremely coarse carpets, blankets and kumblies. The breeds of these regions are presented in Table 3.1.

Table 3.1. Breeds of Sheep in Different Agro-Ecological Regions in India

Northern temperate	North-west arid and semi-arid	Southern peninsular	Eastern region
Gaddi	Chokla	Deccani	Chottanagpuri
Rampur Bushair	Nali	Nellore	Balangir
Bhakarwal	Marwari	Bellary	Ganjam
Poonchi	Magra	Hassan	Tibetan sheep
Karnah	Jaisalmeri	Mandya	Bonpala
Gurez	Pugal	Mecheri	Garole
Kashmir Merino	Malpura	Kilakarsal	Shahabadi
Changthangi	Sonadi	Vembur	
	Patanwadi	Coimbatore	
	Muzaffarnagri	Nilgiri	
	Jalauni	Ramnad white	
	Hissardale	Madras Red	
	Munjal	Tiruchy Black	
	Avivastra	Kenguri	
	Bharat Merino		

The country has more recently been divided into 15 agro-climatic regions from the view point of crop production. The predominant breeds of the region have been presented in Table 3.2.

There are 40 descript breeds of sheep and others are being identified. However, the majority (about 75%) of the animals are non-descript. There is a large intermixture amongst the breeds in regions where 2 or more breeds exist. There are no Breed Societies or Agencies to register the animals of a particular breed, maintain flock herd-book that ensure the preservation of the purity of a breed or type. Little systematic effort has been made to describe, evaluate and manage (conserve and improve) these breeds. Central and State Government breeding farms, which maintain flocks of indigenous breeds, mostly for production of studs for distribution to the farmers, have been established.

Table 3.2. Agro-Climatic Region-wise Breeds of Sheep

S.No.	Agro-climatic region	State/Union territory	Sheep breeds
1.	Western Himalayan	J and K, H.P., Northern Punjab, Uttaranchal and hill districts of U.P.	Changthangi, Gaddi Gurez, Karnah, Poonchi, Kashmir Merino Rampur Bushair, Bhakarwal
2.	Eastern Himalayan	Sikkim, Arunachal Pradesh Meghalaya, Nagaland, Manipur, Tripur, Mizoram, Assam and Jalpaiguri, Coochbihar district of West Bengal.	Bonpala, Tibetan
3.	Lower Gangetic Plain	West Bengal Plains	Garole
4.	Middle Gangetic Plain	Uttar Pradesh excluding hill districts and Eastern U.P.	Shahabadi
5.	Upper Gangetic Plain	Uttar Pradesh excluding hill districts and Eastern U.P.	Muzaffarnagri
6.	Trans-Gangetic Plain	Punjab, Haryana, Chandigarh Delhi and Sriganganagar district of Rajasthan	Lohi, Magra, Nali Pugal, Munjal
7.	Eastern Plateau and Hills	South Bihar, Chattisgarh and Baster Plateau area of M.P. Eastern Vidarbha zone of Maharashtra and Plateau zone, Eastern Ghat, Western Undulating zone and West and Mid Cenmol Table land area of Orissa.	Balangir, Ganjam Chottanagpuri
8.	Central Plateau and Hills	Kymare Plateau, Satpuda Hill and Plateau Vindhya Plateau, Central Narmada Valley, Grid and Bundelkhand area of MP, Semi arid Eastern plain zone, Flood prone Eastern plain sub-humid zone and Aravali Hills zone of Rajasthan and Bundelkhand area of Maharashtra.	Malpura, Sonadi and non-descript
9.	Western Plateau and Hills	Malwa Plateau Ninar Valley and Jhabua Hill MP's area Western Ghat, Sub-Montane zone, Plain and Plateau zone, Central Vidharbha zone of Maharashtra.	Non-descript
10.	Southern Plateau and Hills	Telangana and Scarce Rainfall zone of Rayalaseema of A.P. Karnataka and North Western Zone of Tamil Nadu	Bellary, Coimbatore Deccani, Hassan Kilakarasal, Mecheri Nellore, Nilgiri.

Table 3.2. (*Contd.*)

S.No.	Agro-climatic region	State/Union territory	Sheep breeds
11.	East Coast Plains and Ghats	Krishna-Godavari zone, Hilly and tribal zone of Eastern of Orissa, North Eastern zone, Cauvery Delta zone and Southern zone of Tamil Nadu and Pondicherry	Coastal Kilakarsal Kenguri, Vembur Madras, Red Ramnad White Tiruchi Black
12.	West Coast Plains and Ghats	Coastal zone of Karnataka, Konkon Coastal zone of Maharashtra High Rainfall North Konkon Coastal zone of Tamil Nadu	Non-descript
13.	Gujrat Plains and Hills	Gujarat viz. Saurashtra and Caostal zone	Marwari, Patanwadi
14.	Western dry	Arid Western Plain, Transitional Plain zone of Inland Dronoge and Luni Basin	Chokla, Jaisalmeri Magra, Nali Marwari Pugal
15.	Island	Andman and Nicobar Island Lakshadweep, Minicoy and Amindivi	Non-descript

Most of the breeds of sheep in India have evolved through natural selection for adaptation to specific agro-ecological conditions. Very little concerted effort for developing these breeds through artificial selection has been made. Most of the breeds are very well adapted to harsh climate, long migration, tropical diseases, poor nutrition, shortages of drinking water and poor water quality.

Of the breeds of sheep, Marwari in North-West and Deccani and Bellary in Southern Central Pennisula are numerically the most important and make the largest contribution to carpet wool and meat production in the country.

3.2 Population Distribution

The current population of sheep in the world is 1,059.8 M (FAOSTAT, 2006). Of this, around 61.0 M (according to 2003 Census) are in India. Based on the figures available from different states it is estimated that sheep population during 2003 has been about 61 M. The country stands 3rd in sheep population in the world. These sheep produce 141 Mkg mutton, 35 Mkg of greasy wool and 39.78 M pieces of skin annually (FAO, 1993). Of the total wool produced in the country roughtly 10.49% falls in the range of above 56s, 22.31% between 44s and 56s 34.0% between 36s and 43s and the rest 33.18% is below 36s in quality. The wools of 54s and above quality are suitable for worsted sector and hosiery, of 36s to 53s quality for good quality carpets and those below 36s only for coarse quality carpets, druggets, felts and rugs. About 80 Mkg of fine wool is imported for the worsted sector for preparing woollen cloth for suitings, hosiery and other apparel, as only 4.0 Mkg of indigenous wool is combs worthy. The country requires 15 Mkg of 64s quality wool for the manufacture of fine apparels, 10 Mkg of 56s quality for medium quality for medium quality apparels and 20 Mkg of 36s to 48s quality for carpet manufacture. The availability is thus short by 15 Mkg of fine wool and about 7.2 M kg of medium quality apparel wool. The woollen industry provides employment to about 3 M people and earns more than 2,000 M of foreign exchange by way for exports. State-wise sheep population from 1977 to 2003 quinquinally is presented in Table 3.3. In Rajasthan, the largest sheep rearing state, the sheep population increased by about 34.61% as against 18.00% in the country during 1977–92.

Table 3.3. Sheep Population in Different States of India (in "000)

State	1977	1982	1987	1992	2003
Andhra Pradeh	7064	7507	6872	7768	21376
Assam	59	100	67	67	170
Bihar	1121	1337	1520	1520	382
Gujarat	1592	2357	1559	2037	2062
Haryana	541	783	890	1044	633
Himachal Pradesh	1055	1090	1112	1074	926
Jammu & Kashmir	1210	1861	2493	2917	3411
Karnataka	4536	4615	4727	4727	7256
Kerala	3	7	30	30	4
Madhya Pradesh	968	958	834	836	546
Maharashtra	2636	2770	2872	3075	3094
Meghalaya	20	26	15	15	18
Orissa	1432	1990	1840	1840	1620
Punjab	493	525	508	508	220
Rajasthan	9938	13890	9933	12168	10054
Sikkim	16	11	11	11	6
Tamil Nadu	5289	5475	5881	5881	5593
Uttar Pradesh	2059	2232	2181	2181	1437
West Bengal	793	1034	2312	1460	1525
Total	**40830**	**48568**	**45657**	**49159**	**60333**

3.2.1 Population trends in the states

According to 2003 census, there are 61.1 M sheep in India and ranks 3rd in the world. Sheep population in Indian is given in Table 3.3.

The sheep population recorded an increase of 22.0% in 1982 over 1972. The population of sheep in 1977 over 1966 had shown a decline of 2.6% whereas there has been 18% increase in 1977–82. In 1987, it recorded a decrease of

6.3% as compared to 1982. The overall rate of growth of sheep population from 1972 to 1987 has been 14.9%. This has been possible because of its substantial increase in the states of Maharashtra, Tamil Nadu, West Bengal, Rajasthan and Punjab. The increase of population in 1992 over 1987 is also substantial *i.e.* 49.2 M as against 48.7 M in 1987 (7.6%) due to increase in population in the states of Gujrat, Haryana, Jammu and Kashmir, Maharastra and Rajasthan.

The productivity of Indian sheep is lower than of those in agriculturally more advanced countries. Yet considering their nutritional and physical environment, their productivity cannot be considered as inefficient. The major reasons for low productivity are inadequate grazing resources, diseases, causing high mortality, morbidity and consequent reduced production and serious lack of organized effort for bringing genetic improvement.

3.2.2 Sheep population in the world

There are 1086 M sheep in the world, out of which, 13 and 14% are found in Australia and USSR, respectively. 268 M sheep (26% of the world) are found in Asia. They are found in larger numbers in China, Turkey, India and Iran. 70% of the total sheep population of Asia is found in these countries. Their number increased by 13% between 1955 and 1975.

Sheep population increased by 3% in Europe during this period. Largest number of sheep in Europe are found in Bulgaria, France, Romania and Spain.

Of the total 113 M sheep found in South America, 23 and 32% alone are found in Brazil and Argentina, respectively. Their number decreased by 4%. During 1955 and 1975, the sheep number decreased by 44% in North and Central America whereas it increased by 38% in USSR during the same period.

There were 155 M sheep in Africa according to 1975 census. The sheep population registered an increase of 35% between 1955 and 1975. 50% of the total sheep population of this continent is found in South Africa, Ethiopia, Morrocco and Sudan alone. In 2005, there were 1053 M sheep, with China having 171 M sheep followed by Australia having 106 M sheep and India having 63 M sheep (FAOSTAT-Website year 2006).

Australia is the biggest wool producing and exporting country of the world. About half of the world Merino wool is produced here. Australia produces about one third of the world wool (Table 3.4).

Table 3.4. Sheep Population in the World

Country	Numbers in '000		Percentage increase or decrease 1979–1994	As % of world total in 1994
	1979–81	1994		
Africa	180465	208845	15.5	19.2
Algeria	13111	17850 F	35.5	1.6
Angola	225	225 F	0.0	0.02
Benin	972	940 F	–3.2	0.08
Botswana	147	344 F	134.0	0.03
Burkina Fasco	3200	5686	77.5	0.05
Burundi	301	350 F	16.2	0.03
Cameroon	2167	3770 F	74.5	0.34
Cape Verde	1	7 F	600.0	0.0006
Cent Afr Rep	84	152 F	90.9	0.013
Chad	2620	2152	–17.9	0.198
Comoros	8	15 F	87.5	0.00
Congo	69	111 F	60.8	0.01
Cote Divoire	1020	1251	22.6	0.11
Djibouti	417	470 F	12.7	0.04
Egypt	1791	3382 *	98.8	0.31
Eq Guinea	33	36 F	9.0	0.00
Ethiopia PDR	23250	–	NA	–
Eritrea	1510	1510	Nil	0.13
Ethiopia	21700	21700 F	Nil	1.90
Gabon	105	170 F	61.9	0.01
Gambia	136	121	F	–11.0
0.01				
Ghana	1942	3288 *	59.3	0.30
Guinea	436	435 F	–0.2	0.04
Guineabissau	177	263 F	48.5	0.02
Kenya	5100	5500 F	7.8	0.50
Lesotho	1062	1691 F	59.2	0.15
Liberia	200	210 F	5.0	0.01
Libya	5380	3500 F	–34.9	0.32
Madagascar	695	740 F	6.4	0.06
Malawi	84	196 F	133.0	0.01
Mali	6247	5173	–17.0	0.47
Mauritania	5166	4800 F	–7.0	0.44
Mauritius	10	7 F	30.0	0.00
Morocco	15228	15594	2.4	1.43
Mozambique	106	119 F	12.2	0.01
Namibia	4084	2620	–35.8	0.24
Niger	3007	3700 *	23.0	0.34

Table 3.4. (*Contd.*)

Country	Numbers in '000		Percentage increase or decrease 1979–1994	As % of world total in 1994
	1979–81	1994		
Nigeria	8022	144455 F	80.0	13.29
Reunion	2	2 F	Nil	0.00
Rwanda	303	400 F	32.0	0.03
St Helena	2	1	–50.0	0.00
Sao Tome Prn	2	2 F	Nil	0.00
Senegal	1966	4600 *	133.0	0.42
Sierra Leone	268	302 F	12.6	0.02
Somalia	10467	13000	24.2	1.19
South Africa	31625	29134	–7.8	0.26
Sudan	17628	22870 F	29.7	21.04
Swaziland	32	27 F	–15.6	0.00
Tanzania	3754	3955 *	5.3	0.36
Togo	592	1250 F	111.1	0.11
Tunisia	4651	7100 F	52.6	0.65
Uganda	1319	1980 F	50.1	0.18
Zaire	726	1012 F	39.3	0.09
Zambia	29	69 F	137.9	0.00
Zimbabwe	481	550 F	14.3	0.05
N C America	21073	17529	–16.8	1.61
Antigua Barb	12	13 F	8.3	0.00
Bahamas	35	40 F	14.2	0.00
Barbados	50	41 F	–18.0	0.00
Belize	3	4 F	33.0	0.00
Br Virgin is	8	6 F	–25.0	0.00
Canada	480	691	43.9	0.06
Costa Rica	2	3 F	50.0	0.00
Cuba	356	310 F	–12.9	0.02
Dominica	6	8 F	33.3	0.00
Dominican Rp	65	134 F	106.1	0.01
El Salvador	4	5 F	25.0	0.00
Greenland	20	22 F	10.0	0.00
Grenada	14	12 F	–14.2	0.00
Guadeloupe	3	4 F	33.0	0.00
Guatemala	615	440 F	–28.4	0.04
Haiti	88	85 F	–3.4	0.00
Honduras	5	14 F	180.0	0.00
Jamaica	4	2 F	–50.0	0.00
Martinique	77	110 F	42.8	0.01
Mexico	6484	5905 *	–8.9	0.54

Table 3.4. (*Contd.*)

Country	**Numbers in '000**		**Percentage increase or decrease**	**As % of world**
	1979–81	**1994**	**1979–1994**	**total in 1994**
Montserrat	4	5 F	25.0	0.00
Neth Antille	8	7 F	–12.5	0.00
Nicaragua	3	4 F	33.3	0.00
Puerto Rico	6	8 F	33.3	0.00
St Kitts Nev	14	14 F	Nil	0.00
Saint Lucia	13	16 F	23.0	0.00
St Vincent	13	12 F	–7.6	0.00
Trinidad Tob	10	14 F	40.0	0.00
USA	12670	9600	–24.2	0.88
US Virgin IS	4	3 F	–25.0	0.00
South Americ	102944	94054	–8.6	8.60
Argentina	31473	20000	–36.4	1.84
Bolivia	8967	7789	–13.1	0.71
Brazil	18414	20500 F	11.3	1.88
Chile	6059	4649	–23.1	0.42
Colombia	2399	2540 F	5.8	0.23
Ecuador	1148	1728	50.5	0.15
Falkland IS	658	727	10.4	0.06
Fr Guiana	3	4 F	25.0	0.00
Guyana	115	131 F	13.9	0.01
Paraguay	387	386	–0.2	0.03
Peru	13767	11600 *	–15.7	1.06
Suriname	3	9 F	200.0	0.00
Uruguay	19219	23441 *	21.9	2.15
Venezuela	333	550 F	65.1	0.05
Asia	316162	340102	7.5	31.29
Afghanistan	18667	14200 F	–23.9	1.30
Armenia	–	720 *	–	0.06
Azerbaijan	–	4339 *	–	0.39
Bahrain	7	29 F	314.2	0.00
Bangladesh	750	1070 *	42.6	0.09
Bhutan	10	59 F	490.0	0.00
China	101864	111649	9.6	1.02
Cyprus	29	285 F	882.7	0.02
Gaza Strip	15	24 F	60.0	0.00
Georgia	–	1300 *	–	0.11
India	**44987**	**44809**	**–0.3**	**4.12**
Indonesia	4124	6411 *	55.4	0.58
Iran	31672	45400 F	43.3	4.17

Table 3.4. (*Contd.*)

Country	Numbers in '000		Percentage increase or decrease 1979–1994	As % of world total in 1994
	1979–81	1994		
Iraq	10842	6320F	–41.7	0.58
Israe	0.001	243	330 F	35.8
Japan	13	0.00	25	92.3
Jordan	950	2100 F	0.19	121.0
Kazakhstan	–	33524 *	–	3.08
Korea D P RP	292	396 F	35.6	0.00
Korea REP	6	4 F	-33.3	0.00
Kuwait	250	150 F	-40.0	0.00
Kyrgyzstan	–	7077 *	–	0.65
Lebanon	152	258 F	69.7	0.00
Malaysia	65	336 F	416.9	0.00
Mongolia	14261	14392	0.9	1.32
Myanmar	235	304	29.3	0.00
Nepal	730	914	25.2	0.08
Oman	114	149 F	30.7	0.00
Pakistan	22580	28975 *	28.3	2.66
Philippines	30	30 F	Nil	0.00
Qatar	48	170 F	254.1	0.00
Saudi Arabia	4040	7257 *	79.6	0.00
Sri Lanka	27	19 F	-29.6	0.00
Syria	9311	12000 F	28.8	1.10
Tajikistan	–	2000 F	–	0.01
Thailand	25	98	292.0	0.00
Turkey	46199	37541	-18.7	3.45
Turkmenistan	–	6000 *	–	0.55
Untd Arab EM	132	333	152.2	0.00
Uzbekistan	8600F	–	–	0.79
Yemen	3002	3715	23.7	0.34
Europe	123288	130692	6.0	12.00
Albania	1232	1900 F	54.0	0.00
Austria	193	324 F	67.8	0.00
Belarus	–	289 *	–	0.00
Bel-Lux	110	160	45.4	0.00
Bosnia Herzg	–	600 F	–	0.05
Bulgaria	10358	3763	-63.6	0.34
Croatia	–	444	–	0.04
Czechoslovak	883	–	–	0.08
Czechrep	–	196	–	0.00
Denmark	55	82	49.0	0.00

Table 3.4. (*Contd.*)

Country	Numbers in '000		Percentage increase or decrease 1979–1994	As % of world total in 1994
	1979–81	1994		
Estonia	–	83	–	0.00
Faeroeis	67	69F	2.9	0.00
Finland	107	79	–26.0	0.00
France	12133	10452	–13.8	0.96
Germany	3148	2360 *	–25.0	0.21
Greece	8040	9604 *	19.4	0.88
Hungary	2960	1280	–56.7	0.11
Iceland	838	500F	–40.3	0.04
Ireland	2374	5991	152.3	0.55
Italy	9120	10370 *	13.7	0.95
Latvia	–	133 F	–	0.00
Liechtensten	2	3 F	50.0	0.00
Lithuania	–	48 *	–	0.00
Macedoniafyr	–	2444	–	0.22
Malta	5	6 *	20.0	0.00
Moldovarep	–	1373	–	0.12
Netherlands	856	2174 F	153.0	0.20
Norway	2033	2316 F	13.9	0.21
Poland	4105	870	–78.8	0.08
Portugal	4440	5991	34.9	0.55
Romania	15766	11499	–27.0	1.05
Russian Fed	–	41078 □	□□	3.78
Slovakia	–	397	–	0.03
Slovenia	–	21 F	–	0.00
Spain	14721	23838	58.5	2.19
Sweden	392	483	23.2	0.04
Switzerland	350	425 F	21.4	0.03
UK	21643	29300 *	35.3	2.69
Ukraine	–	6118	–	0.56
Yugoslav SFR	7359		–	0.67
Yugoslavia. FR		2752	–	0.25
Oceania	202272	182758	-9.6	16.81
Australia	134871	132609	1.6	12.20
Fiji	–	6	–	0.00
Fr Polynesia	2		–	0.00
Newcaledonia	4	4 F	Nil	0.00
New Zealand	67393	50135 *	–25.6	4.61
Papua n Guin	2	4 F	100	0.00
USSR	142591	–	13.12	–
World	**1088794**	**1086661**	**–0.19**	–

CHAPTER 4

TYPES AND BREEDS OF SHEEP

4.1 Classification of Sheep Breeds

The country has 40 breeds of sheep out of which 24 are distinct. Amongst them 5 can be classified into medium or fine wool, 14 into coarse carpet quality wool and rest into hairy meat type breeds. They vary from the non-woolly breeds of sheep in the Southern peninsular region mainly kept for mutton and manure to reasonably fair apparel wool breeds of Northern-temperate region. Although productivity from these sheep is of low order, they cannot altogether be considered inefficient in comparison to the physical environment and nutritional conditions they are reared in.

If we follow the breed classification in strict sense, there are no specific breeds, as majority of them lack characteristics of fixed nature. Neither there are Breed Societies or Agencies to register animals of particular breeds, maintain flock books and ensure purity of the breeds. The sheep from various states keep on migrating over long distances where they undergo lot of admixture and thus making it difficult to maintain purity of the breeds. Animals with distinct characters localized to a place and different from those of other place are termed as breeds and given some local name. There has been very little efforts to conserve and improve the native breeds except at some of the Central and State Government farms. Some important breeds of sheep are maintained for their purity and producing stud rams for distribution to the farmers. Most of the breeds of sheep in India have evolved through natural adaptation to agro-ecological conditions followed by some limited artificial selection for particular requirements. Most of the breeds have generally been named after their place of origin and on the basis of prominent characters. Amongst the most widely distributed native sheep breeds, Marwari and Deccani are the most prevalent; out of them Marwari covers the greater part of arid North-western region of Rajasthan and Gujarat. It is highly migratory following a trans-human system of management and has left greatest impact on other breeds, especially those with very coarse and hairy fleece like Malpura and Sonadi. Deccani occupies most of the Central part of the Southern peninsula being distributed over the states of Maharashtra, Andhra Pradesh and Karnataka.

Attempts have been made to define and document some of the important breeds of sheep under ad-hoc research scheme financed by the Indian Council of Agricultural Research. Such efforts were mostly based on the exterior phenotype, shape and length of ears, length and direction of horns, fleece type, body colour and tail length, etc. There is little serious consideration to body weight, body measurements, population and flock size and its structure, prevalent management practises, productivity status and problems associated with their conservation and further development.

4.1.1 Fine wool breeds

The fine wool breeds produce fine and crimpy wool. Their fleece is heavy, dense and of good quality. It contains a large amount of yolk which comprises of suint and grease. They have a strong banding instinct and the ability to graze on poor quality range. Of the exotic fine wool breeds imported in India, Rambouillet and Soviet Merinos have done well as purebreds. Their crosses with indigenous breeds have shown improvement in wool quality and also in production.

4.1.1.1 *Exotic*

i) ***Merino*** *(Plate I, A and B)*

The origin of various strains and breeds of fine wool sheep of the present time traces to sheep of Spain. Fine wool sheep represent Spain's chief contribution to livestock industry of the world. Spanish Merinos possessed most, if not all, the features now found in such sheep. Selection within the Merino group has resulted in large varieties of breeds and strains. In addition, Merino has been widely used in the development of many other crossbred wool breeds.

Merino is a thin tailed fine wool breed that is adapted to arid environment. Because of their banding instinct they are easy to herd. They are good graziers and able to forage over large areas of poor and sparsy rangelands. Merinos have strong constitution and known for hardiness and longevity. The face and legs are white and the skin is pink. Rams mostly have heavy spiral horns, whereas ewes are polled. The head and legs are generally covered with wool. They have long been bred for wool production and do not carry the straight line and compactness of mutton breeds. Mature rams weigh around 75 kg and the ewes about 65 kg. Height of ram and ewe is about 70 cm and 60 cm respectively. Fleece production varies widely depending on environmental conditions and time of breeding but average is 4–5 kg for rams and 3–4 kg for ewes. The staple length is 5–10 cm. Merino tends to be a seasonal breeder and this limitation results in low prolificacy and poor lamb crop.

India has imported Soviet Merinos from former USSR which are Stavropol and Grozny strains. These have more grease in the fleece, large skin folds and close face or larger quantity of wool on the face resulting in covering of eyes which are not desirable characters.

ii) *Rambouillet* *(Plate I, C and D)*

This breed has descended from the old Spanish Merino. It was developed as a reed in France. There are two types B and C depending upon the skin folds. The differences between the types is same as in the case of Merinos. B-type has lost much of its popularity and has largely disappeared. The C-type consistently improved both for meat and fleece, is enjoying the greatest popularity. Modern Rambouillet are large, rugged and fast growing sheep. They are hardy and apparently adapted to a wide range of climatic and soil conditions. They are almost free from skin folds, with acceptable mutton conformation and are good wool producers. The wool is of long staple, fair density, uniformity and moderate shrinkage. The rams may have large spiral horns or are polled. The ewes are polled. They have large head with white hair around nose and ears. The face and legs are white and the skin is pink. The ewes are good mothers, quite prolific and are unequalled for range qualities. Mature rams and ewe weight from 100–125 kg and 60–90 kg respectively. In twelve months, the fleece will attain length from 5.7–7.6 cm. The wool has good uniformity and fineness. India has imported the majority of Rambouillet from Texas (USA). The Rambouillet as purebreds and in crosses with Indian breeds have generally performed better than Soviet Merinos.

iii) *Polwarth* *(Plate I, E and F)*

The Polwarth breed originated at Tarndwarncoort in Victoria. It was evolved by mating first cross Lincoln Merino ewe with Saxon Merino rams. The backcross progeny of the mating was interbred and eventually the 'Polwarth' breed was established. The sheep resembles a plain bodied extra long stapled wool Merino. They have fairly level frame, clear eyes, soft face, pink nose but sometimes mottled and are free from skin folds. They may be horned or polled. Although the animals are bulky in appearance, yet they are neat and have symmetrical lines. Their fleece is of even quality of about 55 count and average length of the staple is not less than 10 cm. The value of fleece is almost the same as that of Merino and mutton produced is of desirable quality.

4.1.1.2 *Indigenous*

i) *Chokla (syn. Chapper and Shekhawati)* *(Plate I, G and H)*

It can be categorized as medium fine wool. It is distributed over Churu, Jhunjhunu, Sikar, and the border areas of Bikaner, Jaipur and Nagaur districts of Rajasthan. Animals true to the breed type are found in Sikar and Churu districts.

Chokla are light to medium-sized animals. Their face is generally devoid of wool and is reddish brown or dark brown in colour which may extend upto the middle of the neck. The skin is pink.

The ears are small to medium in length (7.77 ± 0.09 cm) and tubular. Both the sexes are polled. The coat is dense and relatively fine, covering the entire body including the belly and the greater part of the legs.

ii) *Hissardale* *(Plate I, I)*

The breed was synthesized in the earlier part of the century at the Government Livestock Farm, Hisar (Haryana) through crossing Australian Merino rams with Bikaneri (Magra) ewes by stabilizing the exotic inheritance to about 75%. The animals are small with short legs, giving them a low set appearance and leaf like medium sized ears. Most animals are polled. Colour is predominantly white, although some brown or black patches can also be observed. A small flock of Hissardale is located at the Government Livestock Farm, Hisar. The rams were earlier distributed primarily to the hilly regions of Kulu, Kangra etc.

iii) *Nilgiri* *(Plate II, A and B)*

Evolved during the 19th century, the breed has originated from a crossbred base and contains an unknown level of inheritance of Coimbatore (the local hairy breed) and Tasmanian Merino, Cheviot and Southdown exotic breeds. It is localized to Nilgiri hills of Tamil Nadu. The animals are medium sized. Their body colour is generally white with exceptionally brown patches on face and body. Face line is convex, giving a typical Roman nose. Ears are broad, flat and drooping. Males have horn buds and scars but the females are polled. The tail is medium and thin. The fleece is fine and dense. According to the 1972 census the Nilgiri breed totalled 8,000 and as per 1977 census, 7,677. Well adapted to the conditions of the Nilgiri hills, the breed produces fine fleece but there is little organized shearing and marketing of wool. The breed is mostly maintained for manure by tea planters and other flock owners.

iv) *Tibetan* *(Plate II, C and D)*

This breed is distributed in Northern Sikkim and Kameng districts of Arunachal Pradesh. These are medium-sized animals, mostly white with black or brown face and brown and white spots on the body. Both sexes are horned. The nose line is convex, giving a typical Roman nose. The ears are small, broad and drooping. The fleece is relatively fine and dense. The belly, legs and face are devoid of wool. The total number of Tibetan sheep in the area is about 30,000.

Tibetan sheep produce an excellent, lustrous carpet-quality wool, which was available in plenty to the Indian States bordering Tibet when the Indo-Tibetan border was open. However, after its closer in 1962, little Tibetan wool is available from Tibet.

v) *Gaddi* *(Plate II, E and F)*

Also known as Bhadarwah, the breed is distributed in Kistwar and Bhadarwah Tehsils in Jammu province of Jammu and Kashmir state, Hamirpur, Ramnagar, Udampur and Kulu and Kangra valleys of Himachal Pradesh and Dehradun, Nainital, Tehri garhwal and Chamoli districts of Uttrakhand. These are medium sized animals, usually white, although tan, brown and black and mixtures of these are also seen. Males are entirely horned but females to the extent of only 10–15% are horned. Tail is small and thin. The fleece is relatively fine and dense.

vi) *Rampur Bushair* *(Plate II, G and H)*

This is distributed in Simla, Kinnaur, Nahan, Bilaspur, Solan and Lahul and Spiti districts of Himachal Pradesh and Dehradun, Rishikaesh, Chakrota and Nainital districts of Uttrakhand. These are medium sized animals. The fleece colour is predominantly white, but brown, black and tan colour are also seen on the fleece in varying proportions. The ears are long and drooping. The face line is convex, giving a typical Roman nose. The males are horned but most of the females are polled. The fleece is of medium quality and dense. Legs, belly and face are devoid of wool.

vii) *Bhakarwal* *(Plate II, I and J)*

The breed has derived its name from a nomadic tribe which rears these sheep in the state of Jammu and Kashmir. Due to its entirely migratory nature, there is no distinct home tract of this breed. These are medium sized animals, with a typical Roman nose. The animals are generally white, although coloured fleeces are also occasionally observed. All animals are spotted, fawn or grey. Rams are horned and the ewes polled. Ears are long and drooping. Tail is small and thin. Fleece is coarse and open. The adult ewes weight between 29 and 36 kg, and rams as much as 55 kg. In adult females height at withers is around 62 cm, body length 65 cm and chest girth 82 cm. The sheep are shorn thrice a year and the total annual wool production ranges from 1–1.5 kg per animal. The wool quality is from 36 to 40 with fibre diameter varying from 36 to 38 μ.

viii) *Poonchi*

This breed is distributed in Poonch and part of Rajori districts of Jammu province. The animals are similar in appearance to Gaddi except being lighter in weight. Animals are predominantly white in colour, including the face but spotted sheep varying from brown to light black are also seen. Ears are medium long. Tail is short and thin. Legs are also short, giving a low-set conformation. The weight of the adult ram ranges from 35–40 kg and that of a ewe from 25–30 kg. Wool is of medium to fine quality, mostly white in colour. Sheep are shorn three times a year which produce between 0.9–1.3 kg greasy wool per sheep per year. Fibre length ranges between 15–18 cm and the fibre diameter between 22 and 30 μ.

ix) *Karnah* *(Plate III, A and B)*

This is distributed in Karnah, a mountainous Tehsil in North Kashmir. These are generally large animals. The flock size is normally 25 to 100 animals comprising all age groups. The Karnah animals are medium sized, with rectangular and compact body conformation. The predominant coat colour is white-creamy. The breeding rams are all white while females are majority with white coat (>90%) while brown and black animals are rare. The rams have large curved horns and prominent nose line. All females are polled. The horn is backward, forward and upward with pointed tips. Ear is pendulous and medium sized. The nose is typical roman shaped. The tail is thin, straight and of medium size. The avg. wt. at marketing is 30.75 ± 0.28 and 37.55 ± 1.50 kg in males and females respectively. The sheep are shorn twice a year, in spring and autumn which produce between 1 to 1.5 kg of wool per animal per year. Staple length ranges from 12–15 cm and the average fibre diameter between 29 and 32 μ.

x) *Gurez* *(Plate III, C and D)*

The breed is distributed in the Gurez area of Northern Kashmir, largest of the sheep breeds of Jammu and Kashmir, generally white in colour, although some animals are brown or black or have brown or black spots. A small proportion of the animals have small, pointed horns. Tail is thin and short. Fleece is generally coarse and hairy. Ears are long, thin and pointed. The annual greasy-fleece weight varies from 0.5–1 kg per animal.

xi) *Kashmir Merino* *(Plate III, E and F)*

This breed has originated from crosses of different Merino types (at 1st Delaine Merinos, and subsequently Rambouillet and Soviet Merinos) with predominantly migratory native sheep breeds, *viz.* Gaddi, Bhakarwal and Poonchi. The level of inheritance in the crossbred animals included in Kashmir Merino predominantly varies between 50–75% but may vary from very low to almost 100% Merino. As the animals are highly diversified because of a number of native breeds involved, no definite description of the breed can be given. Some flocks of Kashmir Merino are maintained at the state farms.

xii) *Changthangi* *(Plate III, G and H)*

This breed is distributed in the Changthang region of Ladakh. The animals are strongly built, large framed with good fleece cover of an extraordinarily long staple. Animals are usually shorn twice a year, generally in May/June and September/October but in some cases shearing is done only once a year, in July/August. Greasy-wool production ranges from 1 to 1.5 kg per animal per year. The wool is of a good carpet/medium apparel quality.

4.1.2 Carpet wool breeds

4.1.2.1 *Indigenous*

i) *Nali* *(Plate III, I and J)*

It is distributed in Ganganagar, Churu and Jhunjhunu districts of Rajasthan and southern parts of Hisar and Rohtak districts of Haryana.

Nali are medium sized animals. Their face colour is light brown and the skin colour pink. Both sexes are polled. Ears are large and leafy; tail is short to medium in length and thin. Fleece is white coarse, dense and long stapled. Forehead, belly and legs are covered with wool. The March and September clips are yellow but the September clip is golden yellow in colour.

ii) *Marwari* *(Plate IV, A and B)*

Deriving its name from its home tract in Marwar, the breed resembles black-headed Persian sheep but is smaller in size and has good fleece yield. It is distributed in

Jodhpur, Jalore, Nagaou, Pali and Barmer districts extending up to Ajmer and Udaipur districts of Rajasthan and the Heoria region of Gujarat. The animals are medium sized with black face, the colour extending to the lower part of neck, ears extremely small and tubular. Both sexes are polled. Tail is short, medium and thin. Fleece is white and not very dense. The total Marwari sheep population in Rajasthan and Gujarat as per 1972 census was 4.367 M in 1977 it was 5.018 M. The breed is being improved through selection for fleece weight and carpet quality.

Surplus male lambs not retained for breeding, are slaughtered between 6 and 8 months of age.

iii) *Magra* *(Plate IV, C and D)*

(Formerly known as Bikaneri, is also known as Bikaneri Chokla and Chakri)

Although the breed is largely found in Bikaner, Nagour, Jaisalmer and Churu districts of Rajasthan, animals true to the breed type are found only in the eastern and southern parts of Bikaner districts. The animals are medium to large in size. White face with light brown patches around the eyes are the characteristics of this breed. Skin colour is pink. Ears are small to medium and tubular. Both sexes are polled. Tail is medium in length and thin. Fleece is of medium carpet quality, extremely white and lustrous and not very dense.

The most important strain of Magra (Bikaneri Chokla) has flocks with extremely white and lustrous fleeces and found only in a few villages around Bikaner. Their fleece is of good carpet quality. The breeding programme involves improving this breed through selection; however, there is much crossing with other breeds in the vicinity.

iv) *Jaisalmeri* *(Plate IV, E and F)*

Deriving its name from its home tract at Jaisalmer, the breed is largely distributed in Jaisalmer, Barmer and Jodhpur districts of Rajasthan. Pure specimens are also found in south western Jaisalmer, extending up to north-western Barmer and southern and western Jodhpur. The animals are tall and well built with black or dark brown face, the colour extending up to the neck, typical Roman nose, long drooping ears, generally with a cartilaginous appendage. Both sexes are polled. The tail is medium to long. The fleece colour is white, of medium carpet quality and not very dense.

This is the largest breed in body size of Rajasthan which produce good quality carpet-wool. There is need for conserving this breed.

v) *Pugal* *(Plate IV, G and H)*

Its name has originated from the Pugal area of Bikaner district, its home tract. It is also distributed over Bikaner and Jaisalmer districts of Rajasthan, but pure specimens are available only in the North-western border area of the two districts. Fairly well built, animals have black face, with small light brown strips on either side above the eyes, lower jaws, of typically light brown colour. The black colour may extend to neck. Ears are short and tubular. Both sexes are polled. Tail is short to medium and thin. The fleece is of medium carpet quality but not very dense.

Considering these small numbers, there is need for conservation of this breed. The breeding policy involves improving this carpet wool breed through selection for greasy fleece weight and carpet quality.

vi) *Malpura* *(Plate IV, I and J)*

Through very similar to Sonadi, it is better in wool production and quality and in body size probably due to better environmental and nutritional conditions in the Malpura tract in comparison to that in the home tract of Sonadi at Jaipur, Tonk, Sawaimadhopur and adjacent areas of Ajmer, Bhilwara and Bundi districts in Rajasthan. The animals are fairly well built with long legs with face light brown. Ears are short and tubular, with a small cartilaginous appendage on the upper side. Both sexes are polled. Tail is medium to long and thin. Fleece is white, extremely coarse and hairy. Belly and legs are devoid of wool.

The milk production averaged 64.50 kg in a lactation period of 90 days. Tupping and lambing percentages in the spring and autumn seasons, are 61.54, 96.23% and 88.7, 32.7% respectively. The dressing percentage on the live weight basis at 6 months ranged from 40.90 to 49.49.

vii) *Sonadi* *(Plate V, A and B)*

Mainly found in Udaipur and Dungarpur districts of Rajasthan, also extends to northern Gujarat. The animals are fairly well built somewhat smaller to Malpura with long legs, light brown face with the colour extending to the middle of the neck, ears large, flat and drooping and generally have a cartilagenous appendage. Tail is long and thin. Both sexes are polled. Udder is fairly well developed. Fleece is white, extremely coarse and hairy. Belly and legs are devoid of wool.

viii) *Patanwadi (syn. Desi, Kutchi, Kathiawari, Vadhiyari and Charotari)* *(Plate V, C and D)*

The breed includes three distinct strains

i) Non-migratory, red faced animals with small bodies, yielding relatively finer fleeces. These are typical Patanwadis and are located in north-eastern Saurashtra.

ii) The migratory type, with larger body and long legs, typical Roman nose and long tubular ears. This variety, producing coarser fleeces, is found in western and northern Gujarat.

iii) The meat type, with big body, low stature ad coarser fleeces, found in south eastern areas around Palitana. It is distributed in coastal plain region of Saurashtra and Kutch districts and sandy loamy areas of Patan, Kadi Kalol, Sidhapur and Chanssama taluks of Mehsana district of Gujarat.

ix) *Muzaffarnagri* *(Plate V, E and F)*

Also known as Bulandshahri is vastly available at Muzaffarnagar, Bulandshar, Saharanpur, Meerut, Bijnor and districts of Uttrakhand and parts of Delhi and Haryana. Pure specimens are found in Muzaffarnagar district. The animals are medium to large in size, face lines slightly convex. Face and body are white with occasional patches of brown or black, ears and face occasionally black. Both sexes are polled. Males sometimes contain rudimentary horns. Ears are long and drooping. Tail is extremely long and reaches fetlock. Fleece is white, coarse and open. Belly and legs are devoid of wool.

As the breed is one of the heaviest, largest and very well adapted to irrigated areas, its gradual decline in number necessitates conservation.

x) *Jalauni* *(Plate V, G and H)*

It is distributed over Jalaun, Jhansi and Lalitpur districts of Uttar Pradesh. The animals are medium sized with straight nose line. Both sexes are polled. Ears are large, flat and drooping. Tail is thin and medium in length. Fleece is coarse, short-stapled and open, generally white. Belly and legs are devoid of wool. According to the 1972 census its population was 0.179 M. The Uttar Pradesh Government is presently engaged in upgrading Jalauni with Nali to improve its wool yield and quality. Although Jalauni does not very much differ from Nali in body size and in general appearance, Nali crosses show improvement in fleece production as reflected by fleece weight, staple length and quality.

xi) *Kheri* *(Plate V, I and J)*

Kheri, the carpet wool sheep of Rajasthan is considered to have originated from a crossbred base with unknown levels of inheritance of Marwari, Malpura and Jaisalmeri sheep breeds of Rajasthan. The Kheri sheep are distributed largely in Nagaur, Jodhpur and Tonk districts of Rajasthan state. The animals are maintained as medium to large sized flocks. These are medium to large sized animals and are relatively heavier than Marwari sheep. Kheri locally refers to colour pattern having mixture of black and brown. The animals have black brown face, sometimes extending up to neck and/or spots on the legs with white body. The other features of this breed are Roman nose, long drooping ears with ear ridge, long tail reaching up to hock. The males are horned but females are polled. The animals can walk long distances and are adapted to withstand the drought and heat conditions. In Kheri breed the age at first mating for male is 12–18 months while it is 20–24 months in females. Lambing interval is reported to be 7–8 months and twinning is uncommon in this breed. Kheri animals are reared for wool and meat. Kheri animals are shorn twice or thrice a year with about 1.5 kg greasy fleece yield.

4.1.3 Hairy type breeds

4.1.3.1 *Indigenous*

i) *Deccani* *(Plate VI, A and B)*

The breed is spread over the greater part of the Central peninsular region, comprising the semi-arid areas of Maharashtra, Andhra Pradesh and Karanataka. It covers the major part of Maharashtra especially the Pune Division, parts of Kurnool, Mehboobnagar, Nalgonda, Nizamabad, Anantpur, Warrangel and the entire districts of Madak and Hyderabad in Andhra Pradesh and Bidar, Bijapur, Gulbarga and Raichur districts in Karnataka. The animals are medium sized, predominantly black with white markings. White and brown/fawn animals are also seen. In a survey in Maharashtra, 54.92% animals were completely black, 21% black with white or brown spots on head and 24% black or white with brown spots. Rams are horned but ewes are polled. Ears are medium long, flat and drooping. Tail is short and thin. Fleece is extremely coarse, hairy and open. Belly and legs are devoid of wool.

ii) *Bellary* *(Plate VI, C and D)*

This breed is not very different from Deccani. Animals found to the north of the Tungabhadra River are called 'Deccani' and those to the south of it 'Bellary'. Mostly found in Bellary district of Karnataka. The animals are medium sized with body colour ranging from white through various combinations of white and black to black. One third of the males are horned, females are generally polled. Ears are medium long, flat and drooping. Fleece is extremely coarse, hairy and open. Belly and legs are devoid of wool.

4.1.4 Mutton producing breeds

Claims with respect to mutton production are based largely upon the appearance of the animal. Some breeds are superior to others in mutton production.

Specialized mutton breeds mature fast, have high prolificacy, higher body weight gains, high feed conversion efficiency, high carcass yield and produce good quality mutton. These are usually maintained under more intensive management as compared to wool breeds. In India, mutton breeds have been imported from time to time from U.K. Australia and other countries for bringing about improvement in mutton production in indigenous breeds. Brief description of these sheep breeds is given below.

4.1.4.1 *Exotic*

i) *Suffolk* *(Plate VI, E and F)*

It is a medium wool type with its native home in Suffolk and surrounding countries in England and has long reputation for the superior quality of meat. It is an alert and active sheep developed through the use of large-type dark faced Southdown rams on an old strain known as the Norfolk. They have straight legs and are large in size. It is black faced and ears and legs are the commanding characteristics. The ears are long and pointed and are generally carried at very slide droop or horizontally. The head and ears are entirely free from wool and black hair extend to a line or back base of the ears. There is no wool below the knees and hocks. Both rams and ewes are polled although the males frequently have scurs. The fleece is moderately short, dense and fine with 50s or 58s count. The greasy wool yield is 2.75 to 3.25 kg annually. The breed is the foremost mutton breed. The carcass is full of lean meat evenly marbled and with no waste fat and the flavour is excellent. Purebred or graded suffolks have the capacity for rapid growth because of abundant milk of their mothers. As grazer suffolk is among the best because of its active and rustling qualities. Suffolk ewes are prolific. The mature rams in good condition weigh from 100 to 135 kg and ewes from 70–100 kg. Suffolk imported in India have performed relatively poorly as producers, especially in reproduction and survivability, than Polled Dorsets. Males were in general heavier than females at all ages except at birth. The body weight were however observed to be lower than in their home tract because of change in the environment *i.e.* tropical Vs temperate. However the body weight improved over the years, lambing and fertility percentage also improved over years. Six monthly greasy fleece production in the spring was less than in autumn during 1976–1977. Males always clippied heavier fleece than females. Six monthly greasy fleece weight was found to be 1.459 kg. In general the greasy fleece production decreased over years which was indicative of some stress in the hot semi-arid climate.

ii) *Dorset* *(Plate VI, G and H)*

Dorsets are of medium size. The body is moderately lowest, compact and of a medium degree of smoothness and quality. The back is generally strong and the breed is generally deep bodied. The origin of Dorset is clouded in obscurity, but it is well known, that they developed largely through selection. The breed is native to southern England, especially to the counties of Dorset and Somerset. There are horned and polled strains of Dorsets named Dorset Horn and polled Dorset. Except for the presence or absence of horns, both are identical. In the horned strain both the rams and ewes are horned. The face, ears and legs are white in colour and practically free from wool. The ears are of medium size, thin silky and carried well forward. The nostrils, lips and skin are pink. The hooves are white. They produce a carcass of medium weight fine boned and of superior quality meat. The weight of the fleece is 2.75–3.25 kg and the wool is short, close fine in texure of 52s to 58s quality. A native ram and ewe in good condition weigh 80–110 kg and 50–80 kg respectively, the lambs weigh 18–22 kg at 4 months and 30–35 kg at 9 months of age. The breed is prolific, it is hardy and is capable of doing well under most conditions.

iii) *Southdown* *(Plate VI, I and J)*

Southdown is one of the oldest breeds of sheep. The best specimens now closely approach the ideal mutton type in form. The body is compactly made and there is exceptional fullness of the hind quarters together with a smoothness of outline, equal with specimens of other breeds. The body is oval or rounded on top, is wide and deep and is covered with firm flesh. The legs are short which with other features give the best specimens a symmetry that is unsurpassed. This is one of the smallest of the breeds. The native place of the breed is southdown chalk hills of Sussex in extreme south-eastern England. The face, ears and legs are mouse coloured or light brown and the skin is bright pink. The breed is polled although scurs are found sometimes on rams. The eyes are large bright and prominent and the ears are medium sizes and covered with short wool. The ewes are not too prolific with 125 to 150 lambs per 100 ewes and produce only average milk. The animals mature early. The fleece is short close, fairly dense and of fine quality. The annual greasy fleece weight is around 2.25–3.25 kg. Mature rams and ewes in good condition weigh around 80–100 kg and 55–70 kg respectively. The lambs weigh around 15–22 and 27 kg at 3–6 and 9 months of age.

iv) *Correidale* *(Plate VII, A and B)*

The dual-purpose breed combine meat production with wool production qualities. Corriedale is the only important dual-purpose breed imported in India. This breed had its origin in New Zealand and Australia where both mutton and wool production are sought for in a single animal to satisfy the producer's requirements, since neither of the breeds used in developing correidale *i.e.* Merinos and Lincolns, met producers need and yet both had some of the things desired, the two were crossed. By interbreeding and careful selection, a uniform type was established that produced a good balance of mutton and wool. These sheep were named after the correidale estate of Otage where the experimental crossbreeding was done. The animals inherited a good mutton conformation for its long wool ancestors and derived a dense fleece of good quality from its Merino percentage. Mature rams and ewes in good condition weigh 80–100 kg and 55–80 kg respectively. On an average they produce 4.5–5.5 kg greasy wool annually. The wool is of 56s 58s quality with a fibre diameter of 24.95 to 27.84 μ. It is bright and soft with a good length. Corriedales are known for outstanding efficiency to produce more kg of lamb and wool per kg of body weight as compared to other range breeds. The face, ears and legs of these animals are covered with white hair, although black spots are sometimes present. Both sexes are polled although rams sometimes have horns. The ewes are fair in prolificacy and milking ability. In India, Corriedales have mostly been imported from Australia.

4.1.4.2 *Indigenous*

i) *Nellore* *(Plate VII, C and D)*

Based on coat colour, three varieties of this breed are: 'Palla'-completely white or white with light brown spots on head, neck, back and legs. 'Jodipi' (also called Jodimpu) are white with black spots particularly around the lips, eyes and lower Jaw but also on belly and legs, and 'Dora' are completely brown.

The animals are relatively tall with little hair except at brisket, withers and breech. The rams are horned but the ewes are almost always polled. The ears are long and drooping. The tail is short and thin. 86% of the animals carry wattles. Nellore district and neighbouring areas of Prakasam and Ongole districts of Andhra Pradesh predominantly contain this breed population.

ii) *Mandya (syn. Bannur and Bandur)* *(Plate VII, E and F)*

This breed is found in Mandya district and bordering Mysore district of Karnataka. Relatively small, animals are white in colour but in some cases their face is light brown which may extend to the neck. Possess a compact body with a typical reversed U-shape wedge from the rear. Ears are long, leaf-like and drooping. Tail is short and thin. A large percentage of animals carry wattles. Slightly Roman nose. Both sexes are polled. Coat is extremely coarse and hairy. The actual number of the breed is too small. There is high incidence of cryptorchidism in Mandya, possibly due to selection of animals for meaty conformation. It calls for conservation as the numbers are dwindling. Its body weight, weight gain, feed conversion efficiency and carcass yield are not very superior to most other breeds of the region.

iii) *Hassan* *(Plate VII, G and H)*

This breed is localised to Hassan district of Karnataka. These are small sized animals with white body and light brown or black spots. Ears are medium-long and drooping. 39% of the males are horned, females are usually polled. Fleece is white, extremely coarse and open, legs and belly are generally devoid of wool.

iv) *Mecheri* *(Plate VII, I and J)*

Also known as Mainlambadi and Thuvaramchambali in Coimbatore district. Mostly found in Mecheri, Kolathoor, Nangavalli, Omalur and Tarmangalam Panchayat Union areas of Salem district and Bhavani taluk of Coimbatore district of Tamil Nadu. These are medium sized animals, light brown in colour. Ears are medium sized. Both sexes are polled. Tail is short and thin. Body is covered with very short hair which are not shorn.

v) *Kilakarsal (syn. Keezhakkaraisal, Karuvai, Keezha Karauvai, Ramnad Karuvi and Adikarasial)* *(Plate VIII, A and B)*

The breed is found in Ramnathpuram, Madurai, Thanjavur and Ramnad districts of Tamil Nadu. They are medium-sized animals. Coat is dark tan, with black spots on head (particularly the eyelids and lower jaw), belly and legs. Ears are medium sized, tail is small and thin. Males have thick twisted horns, most animals have wattles.

vi) *Vembur* *(Plate VIII, C and D)*

This is also called Karnadhai. The animals are tall. Their colour is white with irregular red and fawn patches all over the body. Ears are medium sized and drooping. Tail is thin and short. Males are horned but ewes polled. The body is covered with short hair which are not shorn. The breed is found in Vembur, Kavundhanpatty, Achangulam and some other villages of Pudur panchayat union and Vilathikulam panchayat areas of Tiruneveli district of Tamil Nadu.

vii) *Coimbatore* *(Plate VIII, E and F)*

This is also called Kurumbai. It is widely available in Coimbatore and Madurai districts of Tamil Nadu and bordering areas of Kerala and Karnataka. The animals are of medium size and white colour with black or brown spots. Ears are medium in size and directed outward and backward, Tail is small and thin. 38% of the males are horned but the females are polled. Fleece is white, coarse hairy and open.

viii) *Ramnad White* *(Plate VIII, G and H)*

This breed is distributed over Ramnad district and adjoining areas of Tirunelveli district of Tamil Nadu. The animals are medium sized, predominantly white though some animals have fawn or black markings over the body. The ears are medium sized and directed outward and downward. Males have twisted horns but females are polled. Tail is short and thin.

ix) ***Madras Red*** *(Plate VIII, I and J)*

It is distributed in Chengalpet and Madras districts of Tamil Nadu. The animals are medium sized. Their body colour is predominantly brown whose intensity varies from light tan to dark brown. Some animals have white markings on the forehead, inside the thighs and on the lower abdomen. Ears are medium long. Males have twisted horns and the ewes are polled. Their body is covered with short hairs which are not shorn.

x) ***Tiruchi Black*** *(Plate IX, A and B)*

Also known as Tiruchy Karungurmbai, the breed is largely found in Perambalur and Ariyalur taluks of Tiruchy districts, Kallakurichy taluk of South Arcot district, Tirupathur and Tiruvannamalai taluks of North Arcot district and Dharampuri and a portion of Krishnagiri taluk of Dharampuri district of Tamil Nadu. These are small sized animals with completely black body. Males are horned but ewes polled. Ears are short and directed downward and forward. Tail is short and thin. Their fleece is extremely coarse, hairy and open.

xi) ***Kenguri*** *(Plate IX, C and D)*

This is also known as Tenguri. It is found in hilly tracts of Raichur district (particularly Lingasagar, Sethanaur and Gangarti taluks) of Karnataka. These are medium sized animals. Their body colour is mostly dark brown, but colours ranging from white to black with spots of different shades are also not uncommon. Males are horned, while the females are generally polled. Although their exact number is not known, their population is too small.

xii) ***Chottanagpuri*** *(Plate IX, E and F)*

This is found in Chottanagpur, Ranchi, Palamau, Hazaribagh, Singhbhum, Dhanbad and Santhal Parganas of Bihar and Bankura district of West Bengal. These are small light weight animals, light grey and brown in colour which possess small ears parallel to the head. Tail is thin and short. Both sexes are polled. Fleece is coarse, hairy and open which is generally not clipped.

xiii) ***Shahabadi*** *(Plate IX, G and H)*

This is also known as 'plain type sheep'. It is distributed in Shahabad, Patna and

Gaya districts of Bihar. These are medium sized leggy animals. The fleece colour is mostly grey, sometimes with black spots. Ears are medium sized and drooping. Tail is extremely long and thin. Both sexes are polled. Fleece is extremely coarse, hairy and open; legs and belly are devoid of wool.

xiv) *Balangir* *(Plate IX, I and J)*

This is distributed over North western districts of Orissa, Balangir, Sambalpur and Sundargarh. These are medium sized animals of white or light brown or of mixed colours. A few animals are black also. The ears are small and stumpy. Males are horned and females polled. Tail is medium long and thin. Fleece is extremely coarse, hairy and open. Legs and belly are devoid of wool.

xv) *Ganjam* *(Plate X, A and B)*

This breed is distributed in Koraput, Phulbani and part of Puri district of Orissa. These are medium sized animals with coat colour ranging from brown to dark tan, some have white spots on the face and body. Ears are of medium size and drooping. Nose line is slightly convex. Tail is medium long and thin. Males are horned but females polled. Fleece is hairy and short which is not shorn.

xvi) *Bonpala* *(Plate X, C and D)*

This is found in southern Sikkim. Their number is about 16,000. The animals are tall, leggy and well-built. Fleece colour ranges from complete white to complete black with a number of intermediary tones. Ears are small and tubular. Both sexes are horned. Tail is thin and short. Fleece is coarse hairy and open. Belly and legs are devoid of wool.

xvii) *Garole* *(Plate X, E and F)*

The breed is known as Garole. The word 'Garole' is used by farmers to indicate not only the breed but also 'sheep' in general. The Garole sheep is a small sized breed with low-set body. The head is straight and well set but a little high up of the body and appears triangular from front. The nose bridge is straight and the muzzle is small, pale cream or black in colour. The eyes are black and well set in the long face, and the neck is long and fine but heavier in males. Legs are thin with black hooves. The chest and abdomen are barrel-like, and heavier posteriorly in females. The back is straight. The udder is not well developed even during lactation. The teats are small and placed on the ventro-lateral angle of the udder. The scrotum in adults is large. The breed has hairy type non-lustrous, straight fibre with more than

86% medulation. Average length and diameter of the fibre at 12 months of age are 4.99 ± 0.06 μ respectively.

4.2 Biometrical Measurements and Performance

The efforts of Acharya (1981) to define some of the existing breeds are based on consideration of agroclimatic region and the type of sheep found therein and adult body weights and linear biometry on representative sample of both sexes, and production performance, both published and unpublished and personal surveys are given in Tables 4.1 to 4.3.

4.3 Distribution of Types/Breeds in Different Agro-climatic Regions

4.3.1 North-western region

This region comprises the states of Punjab, Haryana, Rajasthan and Gujarat and the planes of Uttar Pradesh and Madhya Pradesh. Important breeds of sheep found in this region are chokla, Nali, Pugal, Marwari, Malpura, Sonadi, Jaisalmeri, Patanwadi, Muzaffarnagri, Jalauni and Hissardale. This region is the most important in the country for carpet wool production.

There was a slight decline in the total sheep population in the Chokla region 0.530 M, 1972 census and 0.511 M according to the 1977 census). Of these, adult males and females were 0.049 and 0.352 M respectively. Although Chokla is perhaps the finest carpet-wool breed, most Chokla wool is now being diverted to the worsted processing sector for scarcity of fine apparel wool in the country.

4.3.2 Southern peninsular region

This region (semi-arid in central peninsular and hot humid region along the coast) comprises of Maharashtra, Andhra Pradesh, Karnataka, Tamil Nadu and Kerala and has 19.53 M (47.78%) sheep, producing 9.68 Mkg (28.0%) wool of which 9.22 M kg is below 36s, mostly coloured and extremely coarse and hairy and only about 0.06 Mkg (Nilgiri wool) is above 48. The important breeds of this region are Deccani/Bellari, Nellore, Mandya, Mecheri, Ramnad, Madras Red, Coimbatore, Hassan, Tiruchi black and Nilgiri. About 50% of the population of sheep in this region does not produce any wool and the rest produce very coarse, hairy and coloured fleeces. Most of the sheep, except Nilgiri of this region, are of meat type.

Table 4.1. Body Weights (kg) at Different Stages of Life in Indian Sheep Breeds

Region/breed		Body weight				
		At birth	At weaning	6 months	9 months	12 months
A. North western region						
1.	Chokla	2.37 ± 0.02 (711)	11.13 ± 0.13 (499)	13.46 ± 0.18 (230)	15.40 ± 0.17 (171)	17.89 ± 0.215 (133)
2.	Nali	2.88 ± 0.02 (726)	10.19 ± 0.14 (263)	13.30 ± 0.20 (174)	14.54 ± 0.21 (91)	17.74 ± 0.31 (74)
3.	Marwari	2.98 ± 0.02 (617)	8.16 ± 0.84 (101)	9.40 ± 0.20 (101)	14.65 ± 0.25 (101)	21.06 ± 0.04 (536)
4.	Magra	2.98 ± 0.02 (660)	17.83 ± 0.05 (006)	20.14 ± 0.10 (812)	21.82 ± 0.19 (343)	27.99 ± 0.19 (445)
5.	Jaisalmeri	–	–	–	–	–
6.	Pugal	–	–	–	–	–
7.	Malpura	2.52 ± 0.01 (300)	9.14 ± 0.09 (998)	12.55 ± 0.13 (494)	17.26 ± 0.18 (336)	20.63 ± 0.29 (222)
8.	Sonadi	2.40 ± 0.02 (659)	9.29 ± 0.13 (428)	13.22 ± 0.21 (185)	16.19 ± 0.24 (152)	18.95 ± 0.26 (129)
9.	Patanwadi	2.99 ± 0.02 (2018)	13.68 ± 0.11 (727)	–	–	21.9 ± 0.25 (225)
10.	Muzaffarnagri	3.01 ± 0.63 (288)	10.76 ± 0.39 (226)	14.56 ± 0.59 (194)	18.39 ± 0.93 (143)	25.01 ± 2.16 (39)
11.	Jalauni	–	–	–	–	–
12.	Hissardale	–	–	–	–	–

Table 4.1. (*Contd.*)

Region/breed		Body weight				
		At birth	At weaning	6 months	9 months	12 months
B. Southern peninsular region						
1.	Deccani	2.82 ± 0.001 (289)	13.56 ± 0.09 (633)	20.86 ± 0.33 (107)	–	–
2.	Bellary	2.60 ± 0.02 (589)	11.09 ± 0.15 (394)	16.28 ± 0.02 (349)	–	18.68 ± 0.41 (115)
3.	Nellore	2.74 ± 0.03 (349)	11.98 ± 0.42 (335)	16.60 ± 0.54 (198)	–	22.72 ± 1.00 (90)
4.	Mandya	2.09 ± 0.03 (411)	9.71 ± 0.16 (822)	12.76 ± 0.29 (261)	–	21.02 ± 1.52 (107)
5.	Hassan	–	–	–	–	–
6.	Mecheri	2.24 ± 0.01 (967)	9.81 ± 0.04 (382)	11.04 ± 0.05 (356)	–	18.96 ± 0.09 (190)
7.	Kilakarsal	1.29 ± 0.01 (405)	8.53 ± 0.05 (383)	14.15 ± 0.07 (356)	–	27.26 ± 0.55 (62)
8.	Vembur	1.97 ± 0.04 (100)	8.42 ± 0.10 (100)	10.50 ± 0.12 (100)	–	16.50 ± 0.24 (100)
9.	Coimbatore	2.16 ± 0.02 (281)	7.50 ± 0.13 (207)	10.83 ± 0.21 (153)	12.90 ± 0.32 (111)	14.77 ± 0.26 (101)
10.	Nilgiri	2.96 ± 0.04 (371)	11.84 ± 0.20 (239)	15.04 ± 0.23 (215)	19.15 ± 0.37 (70)	19.77 ± 0.47 (122)
11.	Ramnad White	1.68 ± 0.02 (192)	7.31 ± 0.13 (184)	8.45 ± 0.11 (157)	14.50 ± 0.88 (9)	16.30 ± 0.09 (151)
12.	Madras Red	2.61 ± 0.02 (375)	13.50 ± 0.21 (376)	15.72 ± 0.12 (364)	21.70 ± 0.30 (90)	21.89 ± 0.21 (164)

Table 4.1. (*Contd.*)

Region/breed		Body weight				
		At birth	At weaning	6 months	9 months	12 months
13.	Tiruchi Black	2.13 ± 0.04 (55)	9.46 ± 0.28 (53)	10.73 ± 0.28 (48)	–	16.8 ± 0.40 (40)
14.	Kenguri	–	–	–	–	–
C. Eastern region						
1.	Chottanagpuri	–	–	–	–	–
2.	Shahabadi	–	–	–	–	–
3.	Balangir	–	–	–	–	–
4.	Ganjam	–	–	–	–	–
5.	Tibetan	–	–	–	–	–
6.	Bonpala	–	–	–	–	–
7.	Garole	1.915 ±0.07	4.12 ± 0.1	8.38 ± 0.11	10.63 ± 0.14	–
D. Northern temperate region						
1.	Gaddi	2.52 ± 0.05 (162)	7.4 ± 0.20 (109)	10.81 ± 0.35 (51)	14.29 ± 0.38 (38)	–
2.	Rampur Bushair	2.38 ± 0.04 (411)	12.69 ± 0.16 (302)	–	–	17.84 ± 0.46 (73)
3.	Bhakarwal	–	–	–	–	–
4.	Poonchi	–	–	–	–	–
5.	Karnah	–	–	–	–	–
6.	Gurez	–	–	–	–	–
7.	Kashmir Merino	–	–	–	–	–
8.	Changthangi	–	–	–	–	–
9.	Karnah	2.78 ± 0.02	7.55 ± 0.06	–	–	–

Table 4.2. Adult Body Weight (kg) and Linear Body Measurements (cm) of Indian Sheep Breeds

SI. No.	Breeds	Sex	Body weight (kg)	Body length (cm)	Height at withers (cm)	Chest girth (cm)
A. North western region						
1.	Chokla	M	34.98 ± 1.06 (42)	67.83 ± 0.72 (52)	66.89 ± 0.50 (52)	77.77 ± 0.78 (52)
		F	23.11 ± 0.01 (903)	60.18 ± 0.06 (901)	59.44 ± 0.17 (901)	69.08 ± 0.07 (901)
2.	Nali	M	34.61 ± 1.75 (20)	65.69 ± 0.96 (16)	65.25 ± 0.74 (16)	76.69 ± 1.08 (16)
		F	24.34 ± 0.50 (1420)	66.18 ± 0.09 (1423)	60.11 ± 0.13 (1423)	80.09 ± 0.36 (1423)
3.	Marwari	M	30.66 ± 0.46 (140)	72.11 ± 0.76 (140)	61.90 ± 0.36 (140)	73.34 ± 0.38 (140)
		F	26.18 ± 0.42 (791)	68.52 ± 0.44 (791)	58.85 ± 0.54 (791)	70.01 ± 0.45 (791)
4.	Magra	M	26.85 ± 0.48 (80)	65.02 ± 0.45 (80)	63.63 ± 0.42 (80)	73.28 ± 0.57 (80)
		F	24.36 ± 0.45 (74)	65.55 ± 0.31 (74)	62.30 ± 0.44 (74)	81.64 ± 0.51 (74)
5.	Jaisalmeri	M	27.78 ± 0.76 (32)	65.5 ± 0.49 (32)	66.20 ± 0.70 (32)	72.6 ± 1.02 (32)
		F	29.94 ± 0.50 (46)	60.55 ± 0.63 (44)	66.91 ± 0.52 (44)	82.49 ± 0.66 (44)
6.	Pugal	M	31.79 ± 0.63 (80)	68.62 ± 0.53 (80)	64.88 ± 0.46 (80)	77.49 ± 0.59 (80)
		F	26.96 ± 0.32 (60)	65.64 ± 0.51 (59)	62.07 ± 0.35 (59)	79.68 ± 0.57 (59)

Table 4.2. (*Contd.*)

Sl. No.	Breeds	Sex	Body weight (kg)	Body length (cm)	Height at withers (cm)	Chest girth (cm)
7.	Malpura	M	41.57 ± 1.14	72.12 ± 0.70	71.93 ± 0.68	83.62 ± 0.89
		F	24.28 ± 0.47	63.91 ± 0.14	64.04 ± 0.22	70.11 ± 0.21
			(816)	(872)	(872)	(872)
8.	Sonadi	M	38.59 ± 0.98	68.45 ± 1.14	69.73 ± 1.20	77.10 ± 1.07
			(32)	(11)	(11)	(11)
		F	21.10 ± 0.17	60.58 ± 0.22	61.28 ± 0.17	66.62 ± 0.18
			(743)	(811)	(811)	(811)
9.	Patanwadi	M	33.34 ± 0.62	59.34 ± 1.16	65.12 ± 0.95	78.46 ± 1.63
			(87)	(24)	(24)	(24)
		F	26.53 ± 0.23	59.28 ± 0.27	59.55 ± 0.16	70.42 ± 0.16
			(930)	(380)	(380)	(380)
10.	Muzaffarnagri	M	49.97 ± 1.15	78.53 ± 0.79	74.38 ± 0.32	85.29 ± 0.67
			(34)	(34)	(34)	(34)
		F	33.99 ± 0.36	69.39 ± 0.33	70.04 ± 0.28	76.78 ± 0.33
			(184)	(184)	(184)	(184)
11.	Jalauni	M	39.96 ± 2.8	72.1 ± 1.4	68.2 ± 1.7	82.5 ± 2.1
			(10)	(10)	(10)	(10)
		F	29.27 ± 0.51	64.74 ± 0.42	60.92 ± 0.47	74.03 ± 0.47
			(66)	(66)	(66)	(66)
12.	Hissardale	M	54.5–59	65	70.5	101.75
		F	34–36	57.5	58	90
13.	Kheri	M	41.2 ± 1.46	73.6 ± 1.21	77.8 ± 0.55	83.3 ± 0.57
		F	32.3 ± 0.14	66.8 ± 0.18	69.8 ± 0.09	77.0 ± 0.09

Table 4.2. (*Contd.*)

Sl. No.	Breeds	Sex	Body weight (kg)	Body length (cm)	Height at withers (cm)	Chest girth (cm)
B. Southern pennisular region						
1.	Deccani	M	38.48 ± 1.06 (47)	47.88 ± 0.69 (47)	67.44 ± 0.71 (47)	77.96 ± 0.87 (47)
		F	28.58 ± 0.11 (1070)	65.21 ± 0.10 (1070)	63.79 ± 0.09 (1070)	70.75 ± 0.10 (1070)
2.	Bellary	M	35.39 ± 0.99 (18)	70.83 ± 0.45 (18)	71.78 ± 0.73 (18)	77.78 ± 1.04 (18)
		F	27.42 ± 0.24 (167)	64.97 ± 0.24 (167)	67.01 ± 0.30 (167)	71.76 ± 0.27 (167)
3.	Nellore	M	36.69 ± 2.56 (13)	68.31 ± 0.63 (13)	76.46 ± 1.36 (13)	75.39 ± 1.74 (13)
		F	30.0 ± 0.27 (266)	67.05 ± 0.22 (266)	72.75 ± 0.24 (266)	72.78 ± 0.23 (266)
4.	Mandya	M	34.80 ± 1.55 (7)	63.68 ± 1.28 (7)	62.0 ± 0.65 (7)	78.57 ± 1.75 (7)
		F	23.50 ± 0.27 (140)	59.92 ± 0.22 (140)	56.71 ± 0.27 (140)	65.25 ± 0.31 (140)
5.	Hassan	M	25.78 ± 0.69 (18)	61.61 ± 0.73 (18)	61.83 ± 0.99 (18)	69.56 ± 0.82 (18)
		F	22.68 ± 0.18 (133)	59.0 ± 0.20 (133)	57.18 ± 0.27 (133)	65.38 ± 0.27 (133)
6.	Mecheri	M	35.44 ± 0.06 (31)	66.66 ± 0.45 (36)	72.22 ± 0.66 (36)	78.35 ± 0.62 (36)
		F	21.59 ± 0.18 (159)	58.32 ± 0.15 (293)	64.49 ± 0.17 (293)	67.74 ± 0.20 (293)

Table 4.2. (*Contd.*)

SI. No.	Breeds	Sex	Body weight (kg)	Body length (cm)	Height at withers (cm)	Chest girth (cm)
7.	Kilakarsal	M	29.68 ± 0.72 (41)	61.29 ± 0.58 (41)	69.64 ± 0.47 (41)	72.16 ± 0.71 (41)
		F	21.29 ± 0.16 (223)	58.29 ± 0.31 (223)	64.83 ± 0.23 (223)	64.83 ± 0.22 (223)
8.	Vembur	M	34.33 ± 1.71 (74)	69.43 ± 0.28 (74)	77.16 ± 0.92 (74)	81.57 ± 1.03 (74)
		F	27.93 ± 0.23 (174)	74.20 ± 0.15 (236)	72.68 ± 0.22 (236)	72.67 ± 0.24 (236)
9.	Coimbatore	M	24.97 ± 0.75 (25)	63.22 ± 0.60 (25)	62.38 ± 0.56 (25)	71.64 ± 0.80 (25)
		F	20.50 ± 0.18 (245)	59.20 ± 0.18 (245)	59.36 ± 0.18 (245)	67.05 ± 0.21 (245)
10.	Nilgiri	M	30.63 (82)	64.06 ± 0.42 (98)	64.97 ± 0.45 (16)	76.83 ± 0.73 (98)
		F	25.04 (82)	59.14 ± 0.43 (150)	57.95 ± 0.24 (68)	75.04 ± 0.36 (150)
11.	Ramnad White	M	31.20 ± 0.98 (27)	64.83 ± 0.69 (29)	72.91 ± 0.74 (29)	75.53 ± 0.89 (29)
		F	22.52 ± 0.18 (328)	60.01 ± 0.16 (388)	67.07 ± 0.18 (388)	67.71 ± 0.20 (388)
12.	Madras Red	M	35.54 ± 0.33 (11)	63.74 ± 1.04 (15)	72.73 ± 0.81 (15)	82.90 ± 1.22 (15)
		F	23.21 ± 0.59 (120)	58.12 ± 0.30 (217)	63.71 ± 0.21 (217)	70.97 ± 0.20 (217)
13.	Tiruchi Black	M	25.79 ± 1.10 (10)	58.60 ± 0.42 (10)	57.60 ± 0.07 (10)	66.20 ± 0.60 (10)
		F	18.51 ± 0.18 (108)	52.88 ± 0.21 (108)	54.96 ± 0.26 (108)	62.01 ± 0.27 (108)

Table 4.2. (*Contd.*)

Sl. No.	Breeds	Sex	Body weight (kg)	Body length (cm)	Height at withers (cm)	Chest girth (cm)
14.	Kenguri	M	32.33 ± 2.57 (12)	64.75 ± 1.39 (12)	69.42 ± 1.22 (12)	74.92 ± 2.07 (12)
		F	26.69 ± 0.37 (114)	64.11 ± 0.32 (114)	68.59 ± 0.38 (114)	73.51 ± 0.37 (114)
C. Eastern region						
1.	Chottanagpuri	M	19.48 ± 0.12 (21)	52.70 ± 0.23 (21)	54.51 ± 0.41 (21)	71.12 ± 0.61 (21)
		F	19.76 ± 0.12 (429)	52.31 ± 0.16 (429)	54.46 ± 0.78 (429)	75.03 ± 0.61 (429)
2.	Shahabadi	M	37.98 ± 0.62 (30)	51.38 ± 0.40 (30)	61.00 ± 0.12 (30)	85.00 ± 0.37 (30)
		F	21.01 ± 0.13 (420)	58.63 ± 0.45 (420)	58.73 ± 0.15 (420)	70.55 ± 0.45 (420)
3.	Balangir	M	23.60 ± 1.06 (4)	56.50 ± 2.12 (4)	63.00 ± 0.35 (4)	61.25 ± 1.14 (4)
		F	17.85 ± 0.19 (180)	53.24 ± 0.26 (180)	62.54 ± 0.28 (180)	57.66 ± 0.28 (180)
4.	Ganjam		24.18 ± 0.34	60.42 ± 0.40	66.41 ± 35	70.18 ± 0.42
5.	Tibetan		26.20 ± 0.45	69.12 ± 0.47	63.40 ± 40	79.92 ± 0.62
6.	Bonpala		–	–	–	–
7.	Garole	M	10.88 ± 0.14 (266)	53.5 ± 0.5	49.9 ± 0.5 (266)	65.4 ± 0.4
		F	10.37 ± 0.14 (843)	53.0 ± 0.3	48.7 ± 0.7 (3481)	65.2 ± 0.3

Table 4.2. (*Contd.*)

Sl. No.	Breeds	Sex	Body weight (kg)	Body length (cm)	Height at withers (cm)	Chest girth (cm)
D. Northern temperate region						
1.	Gaddi	M	2.52 ± 0.05 (162)	7.44 ± 0.20 (109)	10.81 ± 0.35 (51)	14.29 ± 0.38 (38)
		F	26.59 ± 1.9 (468)	57.45 ± 1.21 (574)	56.14 ± 1.10 (574)	70.42 ± 2.04 (574)
2.	Rampur Bushair	M	28.84 ± 0.21 (46)	62.33 ± 0.21 (46)	59.05 ± 0.22 (46)	67.97 ± 0.22 (46)
		F	25.38 ± 0.25 (162)	59.55 ± 0.23 (162)	57.83 ± 0.28 (162)	65.92 ± 0.31 (162)
3.	Bhakarwal	–	–	–	–	–
4.	Poonchi	–	–	–	–	–
5.	Karnah	M	40–48	72	70	102
		F	29–37	59–62	59–62	70–75
6.	Gurez	–	–	–	–	–
7.	Kashmir Merino		–	–	–	–
8.	Changthangi	M	38.64 ± 0.57 (51)	76.0 ± 0.77 (51)	69.0 ± 0.71 (51)	97.5 ± 1.28 (51)
		F	34.0 ± 0.62 (43)	75.2 ± 0.89 (43)	67.0 ± 0.65 (43)	89.0 ± 0.80 (43)

M = Male, F = Female

Table 4.3. Fleece Characters of Indian Breeds of Sheep

Sl. No.	Breeds	6 month grease fleece yield (kg)	Annual grease fleece yield (kg)	Staple length (cm)	Fibre diameter (μ)	Medullation (%)	Density (cm^2)
A. North western region							
1.	Chokla	2.79 (656)	24.07 ± 0.62 (655)	4.70 ± 0.07 (655)	28.22 ± 0.20 (720)	24.01 ± 0.62 (656)	1008.7 ± 46.5 (24)
2.	Nali	1.46 ± 0.16 (702)	–	8.31 ± 0.14 (604)	34.92 ± 0.69 (935)	30.74 ± 0.40 (374)	12.71 ± 37.28 (72)
3.	Marwari	0.89 ± 0.02 (1441)	–	6.56 ± 0.05 (1343)	36.93 ± 0.16 (1406)	65.18 ± 1.66 (1406)	17.14 ± 11.7 (294)
4.	Magra	1.09 ± 0.002 (1833)	–	5.81 ± 0.02 (2851)	32.45 ± 0.35 (2995)	48.29 ± 0.39 (2968)	–
5.	Jaisalmeri	0.77 ± 0.07 (28)	–	6.46	39.1 ± 2.76 (35)	64.1 ± 3.1 (35)	–
6.	Pugal	0.80 ± 0.02 (72)	–	5.71 ± 0.04 (513)	35.13 ± 1.00 (524)	61.86 ± 0.62 (524)	–
7.	Malpura	0.54 ± 0.004 (6322)	–	5.60 ± 0.0001 (1331)	41.95 ± 0.37 (453)	71.84 ± 0.17 (363)	626.25 ± 28.4 (24)
8.	Sonadi	0.45 ± 0.01 (5675)	–	4.58 ± 0.27 (60)	52.65 ± 1.77 (143)	88.15 ± 2.1 (60)	618.77 ± 50.33 (60)
9.	Patanwadi	0.628 ± 0.087 (195)	1.1 ± 0.015	8.51 ± 0.09 (452)	31.95 ± 0.26 (452)	29.88 ± 0.95 (452)	957.8 ± 17.0 (452)
10.	Muzaffarnagri	0.650 ± 0.08	–	3.72 ± 0.05	45.17 ± 0.37	69.92 ± 0.87	2811 ± 74.2
11.	Jalauni	–	0.860 – 0.950	–	41.1 ± 0.19	78	–
12.	Hissardale	–	2.27 – 2.72	6.15 ± 0.15 (6)	24.42 (6)	almost 0	–

Table 4.3. (*Contd.*)

Sl. No.	Breeds	6 month grease fleece yield (kg)	Annual grease fleece yield (kg)	Staple length (cm)	Fibre diameter (µ)	Medullation (%)	Density (cm^2)
B. Southern peninsular region							
1.	Deccani	0.359 ± 0.054 (724)	0.74 ± 0.02 (225)	8.58 ± 0.32 (331)	52.42 ± 1.86 (331)	73.75 ± 2.54 (331)	734.46 ± 28.90 (331)
2.	Bellary	0.300 (622)	–	–	59.03 ± 1.06 (36)	43.43 ± 1.42 (81)	346 ± 9.27 (36)
3.	Nellore	–	–	–	–	–	–
4.	Mandya	0.372 (1169)	–	–	–	–	–
5.	Hassan	–	0.3 – 0.4	–	–	–	–
6.	Mecheri	–	–	–	–	–	–
7.	Kilakarsal	–	–	–	–	–	–
8.	Vembur	–		–	–	–	–
9.	Coimbatore	0.365 ± 0.01 (119)	–	5.79 ± 0.39 (16)	41.05 ± 1.83 (16)	58.37 ± 3.05 (16)	376.0 ± 19.3 (16)
10.	Nilgiri	0.615 ± 0.028 (151)	–	–	27.34 ± 0.077 (63)	11.31 (63)	2199 ± 57 (27)
11.	Ramnad White	–		–	–	–	–
12.	Madras Red	–		–	–	–	–
13.	Tiruchi Black	–	0.400	–	–	–	–
14.	Kenguri	–		–	–	–	–

Table 4.3. (*Contd.*)

Sl. No.	Breeds	6 month grease fleece yield (kg)	Annual grease fleece yield (kg)	Staple length (cm)	Fibre diameter (µ)	Medullation (%)	Density (cm^2)
C. Eastern region							
1.	Chottanagpuri	⊟	0.184	–	52.54	83.61	–
2.	Shahabadi	–	0.240 (420)	–	49.83 ± 9.06	87.08 ± 6.81	
3.	Balangir	–		–	–	–	–
4.	Ganjam	–		–	–	–	–
5.	Tibetan		0.4 – 0.9 (226)	7.24 ± 0.11 (226)	13.22 ± 1.25 (226)	19.30 ± 0.64 (226)	–
6.	Bonpala	–	0.4	9.63 ± 0.47	66 ± 0.25	0.95 ± 1.4	–
7.	Garole	4.99 ± 0.06	–	53.02 ± 0.56	–	–	86
D. Northern Temperate region							
1.	Gaddi	0.78 ± 0.02 (2650)	–	5.70 (1245)	28.52 ± 0.07 (2736)	25.80 ± 0.33 (2697)	–
2.	Rampur Bushair	–	1.17 ± 0.06 (47)	7.70 ± 2.05 (185)	34.35 ± 2.70 (185)	23.81 ± 1.30 (185)	–
3.	Bhakarwal	–	1–1.5	–	36–38	–	–
4.	Poonchi	–	0.9–1.3	15–18	22–30	–	–
5.	Karnah	–	1–1.5	12–15	29–32	–	–
6.	Gurez	–	0.5–1	–	–	–	–
7.	Kashmir Merino	1.2 ± 0.02 (271)	2.8 ± 0.08 (38)	15.60 ± 0.07 (191)	20.4 ± 0.14 (197)	–	–
8.	Changthangi	–	1–1.5	–	–	–	–

4.3.3 Eastern region

This region (hot and humid) includes Bihar, West Bengal, Orissa, Assam and other eastern states and has 3.48 M (8.5%) sheep, producing 1.4 Mkg (2%) of wool, primarily of below 36s quality. This region has no distinguished breeds of its own except in the case of Bihar where Shahabadi and Chottanagpuri breeds are found. The sheep in this region are primarily of meat type but for Arunachal Pradesh, which has a small number of better wool sheep. The quality of wool produced by the sheep of this region in general is small and extremely coarse, coloured and of hairy quality.

4.3.4 Northern temperate region

This region comprises of Jammu and Kashmir, Himachal Pradesh and hilly parts of Uttar Pradesh and has approximately 3.69 M (9.04% of the country's sheep population) sheep. They produce about 3.08 Mkg wool (8.27% of the country's wool production) of which 0.77 Mkg is of 36s to 46s quality suitable for carpets. The rest 2.32 Mkg is of 48s and above standard and is suitable for apparel and superior quality carpets. The important breeds of this region are Rampur Bushair, Kashmir Merino, Gurez, Karnah and Gaddi. Around $^{1}/_{3}{}^{rd}$ of sheep in Jammu and Kashmir and about 15–20% in other parts of this region are expected to be the crosses of native breeds with exotic fine wool breeds.

Plate - 1, Fine Wool Breeds (Exotic)

♂ = Male, ♀ = Female

(A) MERINO ♂

(B) MERINO ♀

(C) RAMBOUILLET ♂

(D) RAMBOUILLET ♀

(E) POLWARTH ♂

(F) POLWARTH ♀

Fine Wool Breeds (Indigenous)

(G) CHOKLA ♂

(H) CHOKLA ♀

(I) HISSARDALE

Plate - 2, Fine Wool Breeds (Indigenous)
♂ = Male, ♀ = Female

Plate - 3, Fine Wool Breeds (Indigenous)

♂ = Male, ♀ = Female

(A) KARNAH ♂

(B) KARNAH ♀

(C) GUREZ ♂

(D) GUREZ ♀

(E) KASHMIR MERINO ♂

(F) KASHMIR MERINO ♀

(G) CHANGTHANGI ♂

(H) CHANGTHANGI ♀

Carpet Wool Breeds (Indigenous)

(I) NALI ♂

(J) NALI ♀

Plate - 4, Carpet Wool Breeds (Indigenous)
♂ = Male, ♀ = Female

Plate - 5, Carpet Wool Breeds (Indigenous)
♂ = Male, ♀ = Female

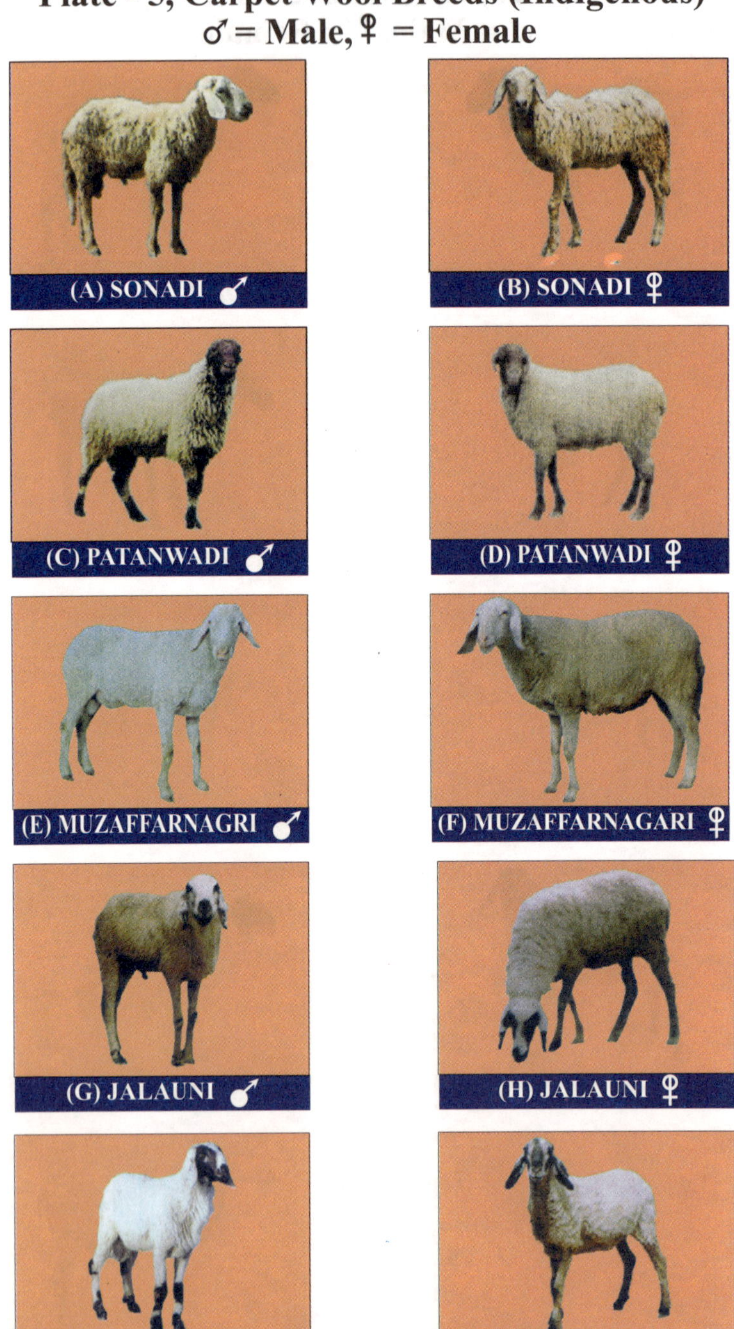

Plate - 6, Hairy Types Breeds (Indigenous)

♂ = Male, ♀ = Female

(A) DECCANI ♂

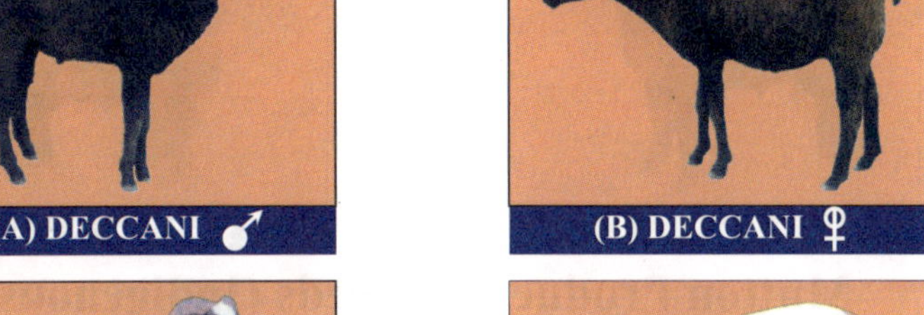

(B) DECCANI ♀

(C) BELLARY ♂

(D) BELLARY ♀

Mutton Producing Breeds (Exotic)

(E) SUFFOLK ♂

(F) SUFFOLK ♀

(G) DORSET ♂

(H) DORSET ♀

(I) SOUTHDOWN ♂

(J) SOUTHDOWN ♀

Plate - 7, Mutton Producing Breeds (Exotic)

♂ = Male, ♀ = Female

(A) CORRIEDALE ♂

(B) CORRIEDALE ♀

Mutton Producing Breeds (Indigenous)

(C) NELLORE ♂

(D) NELLORE ♀

(E) MANDYA ♂

(F) MANDYA ♀

(G) HASSAN ♂

(H) HASSAN ♀

(I) MECHERI ♂

(J) MECHERI ♀

Plate - 8, Mutton Producing Breeds (Indigenous)
♂ = Male, ♀ = Female

(A) KILAKARSAL ♂
(B) KILAKARSAL ♀
(C) VEMBUR ♂
(D) VEMBUR ♀
(E) COIMBATORE ♂
(F) COIMBATORE ♀
(G) RAMNAD WHITE ♂
(H) RAMNAD WHITE ♀
(I) MADRAS RED ♂
(J) MADRAS RED ♀

Plate - 9, Mutton Producing Breeds (Indigenous)

♂ = Male, ♀ = Female

Plate - 10, Mutton Producing Breeds (Indigenous)
♂ = Male, ♀ = Female

(A) GANJAM ♂

(A) GANJAM ♀

(C) BANPALA ♂

(D) BANPALA ♀

(E) GAROLE ♂

(F) GAROLE ♀

CHAPTER 5

GENETICS

5.1 Chromosomal Profile

All animal are made up of microscopic 'building blocks' called cells. The animal body may contain many millions of cells of different sizes and shapes. They are all alike in that they contain two major parts, the cytoplasm and the nucleus. The outer portion of the cell is the cell membrane, which serves as a framework and maintains the shape of the cell. By proper staining, cells can be seen to contain on oval-shaped body more or less near the center called the nucleus. The nucleus is said to be the heart and brain of the cell, and it is this part which is important in inheritance. The material between the nucleus and the cell membrane is called the cytoplasm. Within the cytoplasm are various bodies that play an important role in the function of the cell but in farm animals, have little or nothing to do with the transmission of inheritance.

When the cell is properly prepared and stained, a number of thread-like bodies appear within the nucleus. These microscopic threads are called chromosomes. The chromosomes are present in pairs in the body cells. The members of each pair are very similar in appearance and are called twins. In genetic terms these are called as 'homologous chromosomes', ('home' comes from the Greek work meaning equal or the same, and 'logous' from the word meaning proportion).

The different species of farm animals differ in the number of pairs of homologous chromosomes. Swine possess 19 pairs, cattle and horses 30 pairs, and sheep 27 pairs. Within a single species of farm animals, the number of pairs of chromosomes is constant; *e.g.*, sheep have 27 pairs of homologous chromosomes regardless of the breed. The same is true for the other species of farm animals.

Each pair of homologous chromosomes is distinct from the other pairs. An individual receives one member of each pair of homologous chromosomes from his/her father and the other member from his/her mother. Each individual, in turn, will pass on to each of its offspring one member of each pair of chromosomes it possesses. It is purely a chance, which one of the members of each pair, the individual will receive from his parent and there is no way of predicting. Both members of a pair are not passed to the offspring, but only one of each pair.

It is not known how many genes are there on any one of the 27 pairs of chromosomes in sheep. Suppose we assume that these are 10 and they do not interact, the 27 chromosomes in a sperm and the 27 in an ovum carry factors that determine 270 of the characteristics of the offspring of a ram and an ewe. The actual number of genes is undoubtedly much greater. If the chromosomes from the male and those from the female have within them genes for the same characteristics, the lamb will be pure or homozygous for those characteristics, when the chromosomes unite to form body cells with twice as many chromosomes as found in the reproductive cells. But, if one gene for colour, for *e.g.* on one chromosome is for white and the gene on the other chromosome from the other parent is for black, then the lamb is not pure for white colour, although it may appear white. In such cases the geneticist calls the lamb heterozygous for colour. Exactly the same thing may be true for all characteristics where it is possible to have one condition or another or even where there may not be such absolute differences as there are between white and black or horned and polled.

The factors for white and black, were received from the parents, one chromosome in the cells of the lamb will carry the factor for white, and the other chromosomes will carry the factor for black. In the subsequent formation of reproductive cells, at the point where the reduction of the number of chromosomes occurs one cell will receive the chromosome bearing the factor for white, and the other will receive the chromosome carrying the factor for black. Both, of course, come from the same parent. Hence, it becomes immediately apparent that identical parentage does not ensure identical inheritance. If both chromosomes carried the same factor, then inheritance would be identical for that character.

It is not easy to find two animals that are exactly alike. In the case of white and black, it will become still more evident as other factors are added. Suppose a ram and a ewe, both heterozygous (not pure for either white or black) for colour, are mated. Each of them is capable to produce two kinds of gametes (reproductive cells) with respect to colour. One gamete would carry the factor for white; and the other would carry the factor for black. These two kinds of gametes are capable of producing two kinds of lambs with respect to their colour appearance. They may be white or black. But there will be three kinds of lambs with respect to their genetic makeup. White is dominant over black, and there may be lambs that are pure or homozygous for white, heterozygous for white and black, and others that are pure for black. Since black is a recessive colour in most of our common sheep, only lambs that are pure for black have a black colour. Therefore, if a black lamb is horned, it gets that character equally from both parents.

The same processes are involved with respect to all characters that are due to single genes. If a ram was homozygous for all characters he possessed, his offspring from ewes that had dissimilar characters could not possibly be homozygous. Thus, the only way to maintain a homozygous condition once it has been developed is to restrict mating to individuals that are homozygous. Mating brother and sister does not in itself ensure progeny that is homozygous.

Even with a homozygous condition with respect to the genes, it would be improper to expect all progeny to be exactly alike, for some characters that are to be found in individuals are due to physiological reactions, which of course may in turn be related to the genes which have been transmitted from the parents. Horns are undoubtedly an inherited character, but certain physiological processes or substances must be present for the inheritance to become evident. It is possible that physiological processes are influenced by the interaction of various genes.

Some characters are due to single genes, but not all the characters. Some are due to two or more genes where the degree of interaction may vary. Dominance may not always be complete, and partial dominance introduces more complexity. Likewise, there are cases when a character may be dominant in one sex but not in the other.

5.2 Blood Groups

Systematic studies on blood groups of Indian sheep were conducted by Raina (1969). All antibodies were observed in normal sera of group of sheep and were used as reagents for detecting R blood group factor on sheep erythrocytes. The anti-O antibodies were found in normal sera of two out of 89 sheep tested and were used as reagents. The highest titres of anti-R and anti-O were found around September and lowest around March–May. The erythrocytes of newborn lambs lacked the R and O substances, though these antigens were present in the respective blood sera.

27 blood group factors belonged to 9 different blood group systems, of which A, B, C, M and R-O systems were previously described, and F, G, H and K systems were the new blood groups systems not reported earlier. 14 blood group factors were recognized in the B-system, where 51 alleles were recognized. Twelve C pheno-groups controlled by 12 different alleles were recognized in the C system. A-system was observed to have 1 factor with 2 phenotype and 2 allele system.

5.3 Biochemical Polymorphism

5.3.1 Haemoglobin

Indian sheep show normally three distinct haemoglobin (Hb) phenotypes, *viz.* AA, BB and AB resulting from two alleles A and B (Agar, 1968; Agar and Evans, 1969 a,b; Agar *et al.*, 1969; Kalla *et al.*, 1970; Arora, 1970; Arora *et al.*, 1970; Agar and Seth, 1971; Singh *et al.*, 1972 a,b; Singh *et al.*, 1976 a,b; Arora and Arora, 1978; Kandasamy, 1979 and Arora, 1979). Another type *i.e.* HbC has also been reported in a highly anaemic Nali lamb (Arora *et al.*, 1970b).

The gene frequencies of haemoglobin types in various breeds in India are given in Table 5.1. In Indian breeds, individuals with HbB predominated. Indian meat types breeds (Nellore and Mandya) and their crosses with Indian carpet wool breeds and exotic mutton breeds (Suffolk and Dorset) and their crosses with Malpura and Sonadi show a higher frequency of HbB type. In some breeds of hilly region the frequency of HbA is found more.

Haemoglobin types are inherited in simple Mendelian fashion. The two haemoglobin types, *viz*. HbA and HbB are controlled by two alleles co-dominant to each other, each responsible for the formation of one kind of haemoglobin although some deviation from the general rules has also been observed by Fesus (1972) and Singh *et al.* (1972b).

Haemoglobin polymorphism has been studied extensively in Indian sheep breeds. Arora and Acharya (1972b) found no significant effect of haemoglobin types on body weights at birth, and 3, 6, 9 and 12 months of age in Nali sheep, although the individuals with HbBB phenotypes were marginally better at birth, 6 months and yearling age than HbAB phenotypes. Seth *et al.* (1974) observed the influence on growth with HbAB showing significantly higher weights. Arora (1980) also found HbAB type to be superior at 9 month body weight. Trivedi and Bhat (1978) did not observe any relationship between haemoglobin types and growth.

Kalla *et al.* (1971) showed that animals of HbAA type were better as regards the quality of wool produced than the animals of HbAB and HbBB type. Kalla and Ghosh (1975) also observed Hb type to be better in fleece production per kg live weight.

Arora and Acharya (1972b) observed no significant difference of Hb types on first six monthly greasy-fleece weight in Nali sheep. However, the HbBB type individuals were only marginally superior to HbAB individuals. Similar non-significant results about the effects of Hb types on wool traits were found by Chopra and Chopra (1972) in Nali and Lohi sheep.

Table 5.1. Gene Frequencies of Haemoglobin Types in Indian Breeds of Sheep and their Crossbreds

Breed	Area	Haemoglobin gene		Reference
		HbA	HbB	
Chokla	Rajasthan	0.36	0.64	Ghosh *et al.* (1965)
Corriedale	Uttar Pradesh	0.10	0.90	Agar *et al.* (1969a)
Polwarth	-do-	0.05	0.95	Agar (1971)
Jaisalmeri	Rajasthan	0.20	0.80	Ghosh *et al.* (1965)
Magra	-do-	0.40	0.60	-do-
Malpura	-do-	0.15	0.85	-do-
Mandya	-do-	0.04	0.96	Agar *et al.* (1969a)
Marwari	-do-	0.16	0.84	Ghosh *et al.* (1965), Taneja and Ghosh (1965) Taneja and Abichandani (1967a)
Marwari	-do-	0.19	0.81	Agar *et al.* (1969a)
Rampur Bhushair	Uttar Pradesh	0.06	0.94	Agar and Seth (1971)
Nali	Hissar	0.15	0.85	Arora *et al.* (1970)
Lohi	-do-	0.15	0.85	-do-
Nellore × Nali	-do-	0.11	0.89	Arora and Acharya (1971a)
Nellore × Lohi	-do-	0.04	0.96	-do-
Pugal	Rajas	than 0.30	0.70	Kalla *et al.* (1971)
Marwari	-do-	0.25	0.75	-do-
Muzaffarnagari	Uttar Pradesh	0.04	0.96	Sharma *et al.* (1976)
Muzaffarnagari × Dorset	-do-	0.09	0.91	Trivedi and Bhat (1978)
Muzaffarnagari × Suffolk	-do-	0.07	0.93	-do-
Gaddi	Himachal Pradesh	0.05	0.95	Negi and Bhat (1980)
Gaddi × Gaddi	-do-	0.09	0.91	-do-
Rambouillet × Gaddi	-do-	0.01	0.99	-do-
Russian Merino × Gaddi	-do-	0.13	0.87	-do-

Bhat *et al.* (1981) observed that haemoglobin types had significant effect on average fibre diameter in Gaddi breed of sheep. HbAB type individuals had significantly finer fibre diameter than those of HbBB type individuals.

5.3.2 Transferrin

The available information about Indian breeds is meagre. Arora and Acharya (1971a) observed nine transferrin types Tf^{I}, Tf^{A}, Tf^{G}, Tf^{B}, Tf^{C}, Tf^{M}, Tf^{D}, Tf^{E} and Tf^{P} in Nali,

Lohi, Nali × Nellore 7, Lohi × Nellore and Nali × Mandya. Gene frequency was highest for Tf^C, Tf^D, Tf^B and Tf^M.

Sharma *et al.* (1976) and Trivedi and Bhat (1978) reported transferrin polymorphism in Muzaffarnagri breed of sheep and its cross with Corriedale and Suffolk and Dorset. Transferrin alleles Tf^A, Tf^B, Tf^C, $Tf^{D'}$, and Tf^E were observed. Frequency of Tf^B and Tf^D was highest.

Negi and Bhat (1980) observed five transferrin alleles Tf^A, Tf^B, Tf^C, Tf^D and Tf^E in Gaddi breed and its crosses with Russian Merino and Rambouillet. Frequency of Tf^B and Tf^D was highest in Gaddi while frequency of Tf^D and Tf^A was highest in crossbreds.

Arora and Acharya (1972a) observed significant variation between transferrin types in yearling weight of Nali and Lohi sheep and their three crossbreds (Nali × Nellore, Lohi × Nellore and Nali × Mandya). Sheep of TfMB type had the highest yearling weight followed closely by those of TfEC Birth weight was highest in lambs with TfEM, whereas 3 month weight was highest in sheep of TfEE type. Animals of TfEM type had the highest 6 month weight. Transferrin type explained about 5% of the variation in birth weight and 15% of that in yearling weight.

Sharma *et al.* (1978), Bhat *et al.* (1978, 1981) and Trivedi and Bhat (1978) also observed significant effect of transferrin types on body weights at different ages.

Bhat *et al.*(1981) observed that transferrin types had a significant effect on fleece yield in Gaddi breed of sheep. Animals of TfAD type had a significantly higher fleece weight than the other types.

5. 3.3 Albumin

Krishnamurty *et al.* (1974) observed a new albumin type slower than S in Mecheri breed of Indian sheep. Bhat (1978) did not observe any polymorphism in respect of serum albumin in Muzaffarnagari breed of sheep. Negi and Bhat (1980) reported similar results in Gaddi breed of sheep.

5.3.4 Alkaline phosphatase

Information pertaining to alkaline phosphatase polymorphism in Indian breeds of sheep is extremely rare. Bhat *et al.* (1973) observed two clearly marked zones, *viz.* a fast moving and a slow moving zone. A third zone near the starting end of the samples was also observed. Repeatability of this third zone was not very high.

Nainar *et al.* (1977) studied genetics of serum alkaline phosphatase and 'R-O-i' blood group system in Nilgiri sheep. They also observed the rare types 'X' and 'C' in addition to the 'A' and 'B' zones in Nilgiri sheep. 'C' band was observed as a thin band close to the point of insertion. These rare types were not reproduced by repeated electrophoresis. They found association between 'B' zone intensity and serum alkaline phosphatase level. They also found highly significant association between blood groups and serum alkaline phosphatase level.

5.3.5 Glutathione

Kalla and Ghosh (1975) observed glutathione polymorphism in Indian sheep breeds. High GSH animals were predominant in all the breeds. A significant correlation (0.13) was observed between wool production body weight with erythrocyte GSH level in the Marwari breed only.

5.3.6 Potassium

Two genetically determined types of potassium [(high, (HK) ,and low (LK)] exist in sheep. They are inherited in a simple Mendelian manner controlled by a single locus with a pair of major alleles. The LK is dominant over HK (Taneja and Abichandani, 1967; Arora and Acharya, 1971b; Taneja, 1972; Bhat *et al.*, 1980; Khan and Bhat, 1980. In addition to the major genes that place the animal into high groups, there are also minor genes controlling the concentration within the blood potassium types (Taneja *et al.*, 1969; Arora and Acharya, 1971b) and causing the bimodality of the potassium distribution to be less distinct.

Breed differences in potassium types were reported by Taneja (1972) and Arora (1980). However Arora and Acharya (1971b) did not find any breed differences. Taneja and Ghosh (1967) observed a significant relationship of blood potassium type with body weight and fleece weight in a flock of Marwari sheep. Wethers of LK type were significantly heavier than those of sub-types of HK < 33.16 kg Vs 28.39 kg). There were no other significant differences between different types with respect to body weight and fleece weight. Taneja (1968) reported significant difference in medullation between the two potassium types. LK animals showed 16.0 and 22.3% fewer medullated fibres respectively, as compared to HK animals. These finding have further been supported by a subsequent work (Taneja *et al.*, 1969). In the Russian Merino which produces fine wool, the frequency of LK was 100% whereas in Malpura and Pugal breeds producing coarse carpet quality wool it was only 12.2%. Within breed differences also revealed an association between LK gene and wool fineness. Arora (1970) observed that blood potassium type significantly affected body weight at 6 months and one year. The regression coefficient of weight on potassium concentration was positive. The potassium concentration however, contributed only 3.3, 2.5 and 1.7 to the variation in yearling weight, carcass yield and 6 month weight respectively. The intrasire mean comparisons using data on production traits adjusted for significant environmental effects (Arora and Acharya, 1972b) showed that individuals with HK phenotypes were slightly but not significantly heavier than those with LK phenotypes at birth, 6 months and one year of age but had lower first 6 monthly fleece weight. The differences between sire progeny groups for body weight were consistently in favor with HK phenotypes. Chopra and Chopra (1972) reported that animals of LK type had heavier, uniform and superior fleeces as compared to HK type animals. Similar results have been reported by Taneja (1968) and Taneja *et al.* (1969).

Taneja (1967) reported that on equal body weight basis, the LK type sheep drank relatively lesser quantity of water. On a large sample where potassium types were adjusted for variation in body weight, differences in water intake were however not statistically significant. The differences in water intake by the two types of animals varied from season to season (Taneja *et al.*,1971). This is suggestive of an evidence of genotype environment interaction. Taneja *et al.* (1967) further showed that after exercise there was greater rise in body temperature in LK animals than in HK animals. This was related to their ability to conserve body water as after the exercise, these animals did not drink as much water as the HK type animals. Physiologically, a rise in body temperature above normal limit is detrimental to an animal and natural situation may be in favor of animals which show a lower rise in body temperature during stress. The authors have further conjectured that in chronically water deficient areas the evolution and physiological adaptive mechanism might have been directed primarily towards saving of body water but the situation seems to be of very complex nature, as Purohit *et al.*(1973) working on effects of varying degree of water restriction on the distribution of body water in high and low potassium type in Marwari sheep have reported that there was no difference in the voluntary water intake of the HK and LK animals. In both types of animals the total body water, total blood and plasma volume and extracellular, intracellular and interstitial fluid volume started decreasing as the water intake was reduced below 25% of the daily requirement.

Summarizing there are two types, *viz.* HK (High Potassium) and LK (Low Potassium) distinguishable on the basis of whole blood potassium. These types are controlled by a single autosomal locus with two alleles, the allele for HK being recessive. In addition to the locus, a large number of other loci with minor effects also control the potassium inheritance. Most of the Indian breeds of sheep show a higher frequency of HK. The proportion of HK type varies from 62.91%. Although no specific reasons for the predominance of HK Indian sheep can be assigned but this is suggestive of adaptive advantage in hot arid climate. Genes controlling the potassium types are not completely neutral and contribute towards the variability in some production and adaptation characters and fitness need to be carried out. Potassium types are reported to be associated with fleece quality, the LK producing fleece with lower percentage of medullated fibres.

5.4 Inheritance of Qualitative Traits

Each chromosome carries hundreds and possibly even thousands of different genes. Some of these genes affect only one trait, whereas others affect several. Further, the same trait may be affected by many different pairs of genes. Therefore thorough knowledge of the different types of gene action is necessary to understand and apply genetics to devise satisfactory mating and selection systems for the improvement of sheep.

5.4.1 Dominance and recessiveness with one pair of genes involved-horn pattern

To study how genes, work together to bring about the expression of given characters, let us study the condition of horns in sheep. The symbol P will be used to designate the gene for absence of horns and to represent the gene for presence of horns. If a pure polled ram (PP) is mated to a group of horned ewes (pp). The lambs receive one-half of their genes from the ram and one half from the ewe. The lamb will receive the P gene from the sire and the p gene from the dam, for the sire carries only genes for polledness, and the dam carries only genes for horns.

Polled ram		Horned ewe
(genes Pp)	×	(genes Pp)
Genes in sperm		Gene in egg
P or p		P or p

Polled calves

Pp

Genes in the offspring

The polled gene (P) is said to be dominant. Therefore, the lambs will all be polled even though they also carry the gene for horns. The ability of the polled gene to cover the expression of the horned gene is called dominance. The gene for horns, then, is said to be recessive to the polled gene, because it is not expressed. When two genes affecting a character in different manners occur together in an individual, and occupy identical locations on each member of a pair of homologous chromosomes, this individual is said to heterozygous (in this, the lambs, Pp). For this reason, all of the lambs from the mating of a pure polled ram to pure horned cows will be heterozygous; *i.e.,* each will carry one gene for horns and one for polled. These two genes in the heterozygous animal are called alleles (allelomorphs) of one another, because they both affect the same characteristic in the individual but in different ways. When the two genes affect the character in the same ways, as is the case in the horned ewe (pp) and the polled ram (PP), the animal is said to be homozygous. This genetic structure is called the genotype. The expression of the genotype is called the phenotype.

These polled individuals, which are heterozygous, can produce two kinds of sex cells. One sperm or egg will carry the gene P and the other will carry the gene p. By a diagram that combines the genes from the eggs and the sperm in all possible ways, it can be seen that one-half of the offspring have the same genotype as that of the parents, one-fourth have the same genotype as that of the polled grandparent, and the other one-fourth have the same genotype as that of the horned grandparent.

Polled ram		Polled ewe
(genes Pp)	×	(genes Pp)
Genes in sperm		Genes in egg
P or p		P or p

Possible combinations of sperm and eggs would be: 1 PP Pure for polled (homozygous); 2 Pp Polled (heterozygous) and 1 pp Pure for horns (homozygous).

The same type of diagram can be used to demonstrate genes distribution for all traits where only one pair of genes is involved and where dominance is complete. This also assumes that the penetrance of the gene for polledness is 100% and that its expression is constant.

Three genotypes appear, in a definite ratio: 1 PP : 2 Pp : l pp. This is called a genotypic ratio. Two of the genotypes give rise to the same phenotype. The distribution of phenotypes in the offspring is called the 'phenotypic ratio'. All the lambs carrying Pp and PP are polled while the one carrying pp has horns. From the above expression, it can be seen that the phenotypic ratio in the offspring of this mating would be 3 polled : 1 horned.

The PP and the pp individuals are said to breed true, or to be pure, since they are homozygous and can pass only one kind of gene to their offspring. The Pp individuals, however, never breed true, *i.e.,* they do not have 100% Pp offspring, because they are heterozygous and pass two kinds of genes through their sex cells to their offspring.

For understanding the inheritance of a trait where only one pair of genes is involved, following facts (laws) can be stated (1) genes occur in pairs in the body cells of the individual, (2) one of each pair came from the father and one from the mother, (3) the genes of a pair separate during formation of sex cells, and (4) on fertilization, the genes are restored to the paired condition.

The law of chance determines which sperm and which egg will combine at the time of fertilization to form the new individual. The 1:2:1 genotypic ratio for the offspring of two heterozygous (Pp) individuals works well on paper, but under practical conditions the actual ratio may be far from the expected. Actually, when a small number of offspring are produced from such a mating, all might be of one genotype of either PP, Pp, or pp. The larger the number of offspring produced, the more likely it is that the expected ratio will occur. A similar condition is observed in the number of boys and girls that may occur in any one family. The chances of any one baby being a boy is one out of two, for it has an equal chance of being a boy or girl. Yet, many of us have seen even large families where all the children are of the same sex. This happens because of the law of chance.

6 different kinds of matings, with reference to parental genotype, are possible when one pair of genes is involved and where dominance is complete. These are listed as under:

Genotypic of parents	Genotypic ratio of offspring	Breeding ability of offspring
PP × Pp	{1 PP	Breeds true
	{1 Pp	Doesn't breed true
PP × PP	All PP	All breed true
PP × pp	All Pp	None breed true
Pp × Pp	{1 PP	Breeds true
	{2 Pp	None breed true
	{1 pp	Breeds true
Pp × pp	{1 Pp	None breed true
	{1 pp	All breed true
pp × pp	All pp	All breed true

5.4.2 Incomplete dominance – coat colour

Occasionally, modifications of this simple two-gene inheritance, are observed, for *e.g,* the case of coat colour in sheep. Three different coat colours appear in the sheep having red coat colour: red (RR), roan (RW), and white (WW). When a roan ram (RW) is mated to a group of roan ewes (RW), the offspring will be coloured in the ratio of one red (RR) to two roan (RW) to one white (WW). It should be noted that this is the same genotypic ratio that is obtained when dominance is complete, and as, in the example where polled was dominant to horned. But in this instance, the heterozygote expresses itself phenotypically in a different manner from either homozygote, and the genotypic and phenotypic ratios are the same, 1:2:1. This type of gene action is called **incomplete dominance**, or a blending type of inheritance. One practical point here is that, even though it is not possible to develop a pure breeding roan herd (RW), because the roan individuals are all heterozygous and will not breed true, it is always possible to produce roan sheep (RW) by mating a red parent (RR) to a white parent (WW).

Many cases of inheritance are known where dominance is incomplete and the genotypic and phenotypic ratios of the offspring of a mating between heterozygous parents are the same. For *e.g.* Palomino horses will not breed true, because they are heterozygous for genes for colour, but they can always be produced by mating Chestnut sorrels with Pseudo Albinos. Another *e.g.* is in Comprest Herefords, which are heterozygous for the Comprest gene. When mated, Comprests produce offspring in the ratio of 1 normal (cc) to two Comprest (Cc) to one dwarf (CC). Of course, in this instance it is impossible to mate the normals (cc) with the dwarfs (CC) to produce all Comprest (Cc) because the mortality rate of the Comprest dwarfs is so high that they do not survive to maturity and cannot be used for breeding purposes.

An important fact concerning selection when dominance is incomplete is that the genotype can be determined by inspection of the phenotype. Therefore, an undesirable gene can be eliminated from the herd by discarding all heterozygotes.

5.4.3 Conformation

Desirable type and conformation have also received attention in sheep as in other classes of farm animals. With this species, however, attention must also be paid to selection for wool production in addition to mutton quality and rate and efficiency of gains.

Animals possessing very obvious defects, such as overshot jaws, undershot jaws, black wool, wool-blindness, skin folds, shallow bodies, and poor mutton qualities, should be culled from the flock. If animals with these defects are eliminated from the breeding flock and selections are made on the basis of body weight and quantity and quality of wool, especially in the selection of rams, perhaps this will amount to proper attention to type.

Ewes which may be highly productive yet have some objectionable characteristic may be mated to a ram that is especially outstanding in the trait in which the ewes are inferior. Many times this corrects the fault in just one cross. This is a principle of mating that could be used to improve many traits and is effective, as shown by the history of the development of the many present day breeds of farm animals.

5.4.4 Wool and hair types

The main differences between hair and wool are: wool fibres are usually much smaller in diameter; the epidermal cells of hair are fastened to the cortex throughout their length; hair is always medullated; hair never is crimped like fine wool, although some is wavy like the coarser wools; both hair and wool may have pigment within the cortex, but most wool is free of pigment. Wool fibres grow much denser on the skin than hair grows. In very dense-fleeced, fine-wool sheep there may be more than 50,000 fibres per square inch of skin area. In the coarser, loose fleeced sheep the number may not exceed 5,000 per square inch.

Wool is one of the natural fibres of animal origin which is hygroscopic. It is a protein fibre and the protein is called Keratin because it contains sulphur bearing amino acids. The main sulphur containing amino acid cysteine forms the basic linkage in wool through disulphide bonds. This fibre has a small diameter of the order of 10^{-3} to 10^{-4} compared to its length going upto 10–12 cm. This property gives it an important place among textile fibres. Its chemical constituents are mainly about 18–20 amino acids linked through CONH bonds forming the polypeptides which give softness and resiliency, suitable for a textile fibre. Also its surface structure, the cuticle has scales which provide cohesion in the fibres when they are twisted into yarn. The chemistry of the fibre is most suitable for easy dyeing and other wet processes. Quality evaluation of textile materials is very important to utilize them for appropriate products. The properties of the yarn and fabric depend largely on the fibre arrangement and fibre properties. It is therefore important to know the fibre quality.

The fleece in wild or less developed sheep breeds contains 3 types of fibres, *viz.* wool, hair and kemps. The true wool fibres are relatively fine and 15 to 50 μ in diameter. Kemp fibres are very coarse (100 to 200 μ) brittle, tapered, heavily medullated and short because of their limited growth. Hairy or heterotype fibres have a network of hollow, air-filled cell walls (medulla). The medullated portion can be fragmented, interrupted or continuous. Medulla is commonly found in wools coarser than 30 μ. The carpet wools are evaluated in terms of average fibre diameter, staple length and percentage of medullated fibres, *viz.* kemp and heterotypes.

Variations in wool covering in different sheep are extremely large. Some sheep have wool fibres that are hard to distinguish from hair. In some, only the main portion of the body is covered, while in others the wool extends from just above the nostrils to the feet. Wool on any one sheep is never uniform in length or fineness. The finest and densest wool fibres are found on the shoulders, about midway between the top of the shoulders and floor of the chest; the coarsest wool is about the breech and dock, the shortest wool is usually on the belly. Sheep with a dense growth of relatively long wool on the main parts of the body yield the heaviest fleeces.

In the raw state, wool contains various adhering materials which may be considered as impurities. Some of these are not necessarily disadvantages as they may be essential to the proper preservation of the fibre while it is being developed to a length suitable for use. The amount of these materials vary widely; the chief influences are those inherent in the sheep and those associated with environment, such as climatic and nutritional factors. The adhering materials consist of yolk, suint, vegetable matter, dirt, and moisture. In some cases these materials make up a greater percentage of the entire weight than does the wool fibre. The percentages for different kinds of wool may show such wide variations as follows:

	Variations in percent of grease weight
Wool fibre	15 to 75
Yolk	7 to 40
Suint	5 to 15
Foreign materials	5 to 40
Moisture	3 to 20

These figures are approximations taken from various sources and are intended merely to show, what matters may influence the weight of raw wool obtained from a sheep. Furthermore, these amounts differ widely, depending upon what part of the fleece is selected for examination.

Yolk or wool grease seems to be of direct importance in the preservation of the fibre from detrimental action by the weather. Wool that is lacking in yolk may show considerable damage, especially near the tip. On the other hand there is no apparent advantage in an amount of yolk beyond the minimum amount needed to afford such protection. The amount of yolk secreted by fine-wooled, dense fleeced sheep is much greater than the amount found in the wool of loose, coarse-fleeced animals. Yolk is a mixture of a number of materials of which the principal one is cholesterol. This is not a true fat, and it does not form soap when combined with alkali. It is soluble in ether, alcohol and benzene. Much of it can be removed with water, with which it forms an emulsion. When purified, it has a number of uses in industry, such as in ointments, cosmetics, leather dressing, rope making, and rust preventive.

Suint cannot be distinguished from yolk, as the two are mixed together in the grease found in raw wool. Suint is very different from yolk, however, it is readily dissolved in water. Because of this and also because it has emulsifying and cleansing properties, some consideration is given to its presence in wool scouring, and purified suint may be added to the scouring bowls. It consists chiefly of potassium salts of various fatty acids and lesser amounts of sulphates, phosphates, and nitrogenous materials. There has been no relationship reported between the suint and wool production. It seems to be the source of the odour associated with sheep.

The dimensions of wool fibres vary from about 0.0003 to 0.002 inch for diameter and from one to twenty inches in length during a year's growth. Crimp has been studied, and efforts have been made to determine what causes this character in wool. To date, no completely acceptable explanation has been given. Various means of classifying wool on the basis of crimp have been proposed; for instance, (a) normally crimped wool, (b) deeply crimped wool, and (c) flat or wavy wool. Crimp is defined as that property which causes wool to assume its wavy appearance. It is much more pronounced in the wool of some breeds and some individuals than others. Variation in the kind of crimp in the fibre may indicate changes in the health of the sheep and, hence, differences in the strength and other features of the fibres. The number of crimps per inch of fibre ranges from about 10 to 36 per inch. The more numerous crimp are found in the finest wools, and this has given rise to the belief that numbers of crimp are definite indications of relative fineness. This is not strictly the case, although there is such a general relationship, there are numerous exceptions. Coarse wool, such as obtained from Cotswolds or Lincoln, may have only one or two waves per inch. Crimp makes for great differences in the staple length and the fibre length of a fleece. Fibre length is measured by stretching sufficiently to remove the crimp. Crimp may account for as much as one-third of the total length.

Much study on the size and shape of wool fibres has been made by means of various types of projection apparatus at both low and high magnifications. For some kinds of physical studies, this apparatus is well suited. Other apparatus has been developed for the study of length, strength, pliability and so on.

Wool fibres are seldom circular in cross-section. Most of them are irregularly shaped when viewed under a microscope or when projected onto a screen after being cross-sectioned. In studies of many wool samples at various wool laboratories, the general variations in the shape of the fibres and in their size have been universally noted. Fleeces sometimes show extreme differences in these respects, and none that have been found that are completely uniform, even though the fibres are from a very restricted area of the skin.

According to many, the wool fibre is hollow, and it is through this channel that the fibre is nourished by various "juices". There appears to be no basis for such statements, for once the fibre has been elaborated it is not further changed by any activity of the body. The yolk does serve to keep it in good condition, but this is in no sense a nourishment to the fibre. The fibres have no nerves or blood supply in them. There is also a common belief that dyes enter the wool fibres through the channel which is supposed to pass from end to end. This is not the case, as fibres are readily dyed in the center without immersing either end into the dye.

The chemical composition of wool is complex, and it is possible that there is considerable variation, depending upon the type of wool. The physical properties of wool are undoubtedly related to its chemical structure. Pure wool is composed of keratin, which is also the chief constituent of hair, feathers, horns, and hooves. Keratin is closely related in composition to some of the proteins, but it is not identical with them, as the keratin of wool is not, at least, readily digestible in gastric juices as many proteins are. It is probable that much of the material of which the wool fibre is composed is derived from the proteins in the feeds, but these are changed in the elaboration of the fibres. Studies have shown that sheep on a ration of alfa-alfa have deposited 6.8 kg protein daily in the production of the fleece for each 454 kg live weight.

While wool is represented as composed chiefly of keratin, cysteine is represented as the main constituent of keratin. It is in cysteine that sulphur is present in wool. The sulphur content of clean wool is about 3.4%. Cysteine occurs to a greater or lesser extent in most protein food stuffs and is an essential in body growth This has led to the investigation of the significance of sulphur in the diet, and its relation to wool quality and to wool growth. If sulphur is essential to wool production, then a deficiency would lead to a reduced weight of fleece if the percentage of sulphur was maintained, or it would lead to an abnormal fleece if the weight of fleece was maintained. Since keratin has a constant percentage of sulphur, investigators expected no change in chemical composition of wool as a result of feeding increased amounts of cysteine. Slight increases in weight and slight though noticeable differences in some features of wool (glossiness) have been reported as a result of such feeding.

The content of cysteine in wool is exceptionally high. Studies have shown that the cysteine content of the proteins of plants is less than the cysteine content of wool (13.1% of the dry wool fibre). Since it is unlikely that animals can manufacture cysteine, their only source is that contained in their feed. To produce 450 g of wool protein, a sheep must eat at least 3.2 kg of vegetable protein. It seems probable that, from a practical standpoint, cysteine may be in the same category as many other substances: namely, a certain liberal allowance is needed for normal production; a deficiency reduces production; an excess does not stimulate production to such an extent as to cover the increased cost of such feeding.

Since sulphur is found in wool, there have been recommendations that sulphur, even though not in the form of cysteine, may be the limiting factor in wool growth. The feeding of inorganic sulphur has, however, not been found to have any tendency to influence wool growth. Certain other minerals are also found in the ash of wool after it is burned. These are present in very minute amounts and it is possible that rations that are adequate for the general nourishment of sheep are also adequate for maximum wool growth. While there may be a measurable difference through the use of delicate laboratory apparatus in the amount of wool produced on rations which contain increased amounts of some minerals, the increase is below the added cost of such feeding at the present time.

Properties of wool such as elasticity, pliability, and softness may be influenced by environment, breed and nutrition, and individual inheritance. Studies have been made to determine how these features of the wool fibre are related to the external and internal structure of fibres. Further, it is possible that differences in the readiness with which various wools are dyed may be related to structural differences or to chemical differences.

In general, wool may be described as showing considerable variation in many of its properties. Many difficulties are encountered in research work on wool fibres because of the interrelationships, one factor may have to a host of other factors. Thus, the sulphur content of wool may have a relationship to its elasticity, but other items may have equally important influences on elasticity or on some other characteristic.

Wool absorbs and holds moisture so that it is released slowly. The absorption of moisture causes some changes in the fibre, especially in diameter. The swelling of wool from an air-dry to a saturated condition may amount to almost 15%. From a condition of complete dryness, the swelling would be considerably greater.

Most work reported on wool fibres has been done without completely controlled conditions, and this has caused difficulty in comparing reports of various investigators. In the absorption of water, wool evolves heat. A 45 gm thoroughly dry wool, in changing to a thoroughly wet condition, is reported as evolving 43 British Thermal Units. This is an unusual fibre quality. Wool releases its moisture slowly.

Electricity and heat transfer through wool is slow. Undoubtedly, part of the low conductivity of heat is due to the numerous air cells, which fabrics made of wool may contain. However, the conductivity of the fibres is also low.

Wool is not quickly inflammable, but it will burn and gives off a very disagreeable odour. It is very different from cotton, which burns readily. When wool is burnt, a charred bead remains where the burning has stopped. These differences serve as one easy means of distinguishing between wool and some other materials. If the material in question contains a mixture of several fibres, the so called burning test is of no value, and more accurate chemical or microscopic means are needed. Because wool is subject to severe damage when exposed to caustics, strong acids, and high temperatures, either moist or dry, care must be taken to preserve its original qualities during the scouring and other manufacturing processes and also after it has been completely fabricated. These are the reasons why carbonizing of greasy wool, for *e.g.* is a rather slow and costly process. Washing of woolen materials must be done with neutral soap that which contains no free alkali and the temperature of the bath must be in the neighbourhood of 120 ^{0}F. Violent agitation of the bath is apt to cause shrinkage of a fabric or a felting tendency of unmanufactured wool.

Because of its elasticity, the best quality wool gives to fabrics a striking ability to recover from crushing or compression or from temporary stretching. When the pressure used in bailing wool is released, this resiliency causes the wool to increase gradually in bulkiness. It is this property which causes wool to "drape" commonly on the human form and to retain the "shape" into which it is pressed when moist and is dried during the process of pressing.

5.5 Inheritance of Quantitative Traits

In quantitative inheritance, many pairs of genes are involved, and there is no sharp distinction between the different phenotypes, the differences being of one degree only. Many traits in farm animals which are of the greatest economic importance are good examples of this kind of inheritance, including fertility, rate of gain, efficiency of gain, milk production, and carcass quality. The expression of these traits is affected by many pairs of genes as well as by environment.

Since, in quantitative inheritance, the phenotypes are no distinct and separate but exhibit a series of variations between the extremes, mathematical methods have been devised for measuring and describing populations. Some of these methods are given below:

a) Mean

Everyone has calculated averages, or means. Nevertheless, to help the student to become familiar with the use of symbols and formulas, the mean may be stated as follows:

Many statisticians refer to the mean for a group of individual observations as bar X ($\overline{X}$) or X. The symbol X refers to each individual item or observation, and the Greek symbol $\sum$ means to add all items in the group. The letter n refers to the number of X items in the group of data to be summed.

$$X = \frac{\sum X}{n}$$

b) Range

The range is a very rough measure of the variation within a population. It is determined by finding the lowest and the highest values within a series or group of figures.

The chief disadvantages of the range as a measure of variation are that it is subject to chance fluctuations and that it becomes larger as the size of the sample increases.

c) Variance

No doubt we can use the average of the deviations from the mean as a measure of variation, but when we add all of the deviations and pay strict attention to the sign of each deviation, the sum is equal to zero. Therefore, the average of the deviations from the means would be zero also. This difficulty could be eliminated, however, by disregarding signs method is by the deviations and averaging them. Squaring the deviations and averaging them also makes it possible to perform other statistical measurements in a correct mathematical manner.

Once the sum of the squared deviations from the mean is determined, may be calculated easily by dividing by the number one less than the total number of observations in the sample. Instead of dividing by the number of individuals in a sample n we divide by the total minus one (n-l), at last for samples that include less than 25 observations.

Short-cuts have been devised by statisticians in which the variance may be determined by using a calculator. By using the following equation, the variance can be calculated:

$$\sigma^2 = \frac{\Sigma X^2 - \Sigma X^2 / n}{n-1}$$

where Σ X is the sum of all items, ΣX^2 is the sum of all items squared, and n is the number of observations in the sample.

By using the calculator, we could, of course, obtain the answer more quickly.

Another important point to mention here is that the numerator of the formula for the variance,

$$\Sigma X^2 - \frac{\Sigma X^2 / n}{n}$$

actually is the sum of the squared deviations from the mean and is often written as å X^2. The sum of little x squared divided by the number of observations minus one is the variance.

One of the most useful properties of the variance is that it can be separated by a special analysis into its various component parts. Special adaptations of the analysis of variance can be used to determine the percentage of the variation in a population that is due to inheritance and that due to environment.

d) Standard deviation

The standard deviation is a much more accurate measure of variation in a population than is the range, and can be used very effectively, together with the mean, to describe a population. Statisticians use various symbols to denote the standard deviation, but the one used here is S.D. The standard deviation is the square root of the variance.

e) Coefficient of variation

The coefficient of variation is another method of expressing the amount of variation within a particular population. When this co-efficient is multiplied by 100, it is expressed as a percentage. The coefficient of variation is the fraction or percentage that the standard deviation is of the mean. One important use of this statistic is that it can be used to compare the variations of two unrelated groups.

f) Standard deviation of the mean

In experimental work with livestock, we do not use an unlimited or indefinite number of animals. Actually, we use a very small sample of the entire population. This is true even if the sample includes the hundreds of animals for a given experiment, for ascertaining the closeness of the mean of the sample with the true mean of the entire population.

We use a method of calculating approximately the true mean of the population, and this statistic is called the Standard Deviation of the Mean (S.E.). The standard deviation of the mean may be determined by dividing the standard deviation of the distribution (S.D.) by the square root of the number of items in the population. The formula can thus be written:

$$S.E. = \frac{S.D.}{n}$$

We can use the standard deviation of the mean together with the mean of the distribution to describe the true mean of an infinite number of means drawn from a population. The mean of the distribution plus or minus one S.E. should include about 68% of the means. The mean of the distribution plus or minus two S.E. should include approximately 95% of the means. In other words, we can say that there are only 5 chances out of 100 that the true mean of an infinite number of means drawn from a population would fall outside the mean of a sample plus or minus two S.E. Quite often in scientific reports. The mean of a sample is reported together with plus or minus the S.E.

If the means of two large samples have been derived independently, the information can be used to determine the Standard Deviation of a Difference of means. The formula for this is:

$$(S.E._1)^2 + (S.E._2)^2$$

If a difference between the means of two samples is at least twice as large as the Standard Deviation of the Difference, we can accept this as a true difference at the 5% level of probability.

g) Coefficient of correlation

This statistic and the expansion of the idea are often used in animal breeding and livestock production research. The coefficient of correlation is referred to as r and gives a measure of how two variables tend to move together. They are said to be positively correlated if they tend to move together in the same direction; *i.e.,* when one increases the other increases, or when one decreases the other decreases. They are said to be negatively correlated if they tend to move together in the same direction, *i.e.*, when one increases the other increases, or when one decreases the other decreases. They are said to be negatively correlated if they tend to move in opposite directions *i.e.*, when one increases the other decreases. Thus, the co-efficient of correlation for two variables lies somewhere between 0 and ± 1.

Although the coefficient of correlation tells us how two variables tend to move together in like or in opposite directions, it does not necessarily mean that the movement of one is the cause or the effect of the movement of the other. The cause and effect relationship must be determined, if possible, from other known facts concerning these two variables.

A particular coefficient of correlation is usually said to be significant, highly significant, or nonsignificant, the degree of significance depending upon the size of the coefficient of correlation and the number of individual items used to calculate it.

The formula for calculating the simple co-efficient of correlation between two variables is:

$$r = \Sigma XY - \frac{\dfrac{(\Sigma X)(\Sigma Y)}{n}}{\sqrt{\Sigma X^2 - \dfrac{\Sigma X^2}{n} \Sigma Y^2 - \dfrac{\Sigma Y^2}{n}}}$$

where X is each individual observation for variable X, Y is each individual observation for variable Y, n is the number of observations for each variable, and the Greek symbol (å) means the summation of all items for each variable or pair of variables.

5.5.1 Growth rate/body weight

The age at which lambs are weaned varies under different conditions, but the age of 180 days is often used for selection purposes.

Lambs can be weighed as they reach 180 days, or can be corrected to this age, the correction is done by multiplying the average daily gain from birth by 180 and adding this product to the birth weight.

The age of the ewe may have considerable influence on the weaning weight of her lambs. Two-year-old ewes wean lambs that are from 2 to 4 kg lighter than those from mature ewes. Production of ewes usually increases to four or five years of age. Probably the most important adjustment is that for weaning weights of lambs from ewes that are two years of age. This adjustment can be made by comparing the production of the two-year-old ewes with that of mature ewes in the same herd and then adding the difference to the weaning weights of lambs from the younger ewes. Or, if this is not possible, an adjustment may be made by adding 3.1 kg to the weight of the lambs.

Postweaning Gain is also very highly repeatable. Heavier yearling ewes also wean heavier lambs and produce heavier fleeces, so selection for this trait would probably be effective and desirable.

5.5.2 Reproduction and fertility traits

Sheep in India, and other parts of tropics in the world breed through the year, but are seasonal breeders in European countries and in most areas of the United states the breeding season is in the fall. Some breeds, such as the Merino and Dorset Horn, may be bred under some conditions to produce two lamb crops per year. Most breeds produce just one crop per year, although rams produce sperm throughout the year. Some rams are susceptible to high summer temperatures, and may be infertile or of low fertility in late summer during the first part of the breeding season.

The number of lambs raised per ewe is one ofthe most important factors determining the efficiency of production. Lamb production varies a great deal under different conditions and with different breeds. Ewes under farm conditions are usually more prolific than those produced on the range. This is probably due to a higher level of nutrition generally found in farm flocks. Twinning in sheep is often desirable, because a ewe that weans twins produces from 10–15 kg more lamb than the ewe that weans a single lamb.

Fertility in sheep is lowly heritable, with an average heritability and repeatability estimate of 7–13%. These estimates are in agreement with those for other classes of livestock. This indicates that fertility in sheep is not greatly affected by additive gene action and could be improved very little by selection. Most of the phenotypic variation, therefore, is due to environmental factors and attention to these should improve the lamb crop.

5.5.3 Wool production/wool yield and wool quality

Wool production is largely dependent upon fleece weight and staple length. Greater gains are obtained with increase in staple length in both the fine-wool and the coarse-wool breeds. With some breeds, selection may have to be practised against coarseness of wool, but selection should be practised for the grade of wool most desirable in a particular area.

Heritability estimates show that most wool characteristics are highly heritable. The average heritability of fleece weight has been estimated as 40%, and of staple length as 45%. These heritability estimates are high enough that the mating together of the best individuals for these two traits should result in genetic improvement, and because of their economic value, these traits should be given attention in a breeding program.

Face covering in ewes is an important economic trait. Open-faced ewes produce more lambs and wean more pounds of lamb than those with wool-covered faces.

5.5.4 Carcass yield

Evaluation of carcass can be done when intact or partially cut or as wholesale cuts or retail cuts of meat.

Evaluation of an intact carcass depends on its conformation and finish. The most desirable carcass conformation is the one which will yield the greatest quantity of edible meat. Well-finished carcasses show greater depth of fat over the back and rumps, a covering of smooth white interior fat over the kidneys, much thinner covering over the inner surface of ribs and liberal deposits of fat between the muscles. Poorly finished carcasses show marked deficiency of external and internal fat with uneven distribution of soft, flabby and yellowish fat. Hot carcass weight, carcass weight/day of age and dressing percentage (carcass weight / live weight × 100) are three useful criteria which will help in comparing carcasses of sheep from different farms, breeds, sires, or production systems.

As consumers would hesitate to buy cuts with more bone and fat, their production is not economical. A carcass with higher lean to fat and lean to bone ratio should be the goal of any sheep husbandry system meant for meat production. In a country like ours where information on preferences for or against is not available, clubbing together meat and fat for meat to bone ratio estimation may be reasonable. However, in the long run due emphasis given for production, of optimum amount of fat in the carcasses will help.

There is very less information about carcass quality and the degree of heritability of the various carcass items. Studies to gain such information are in progress, and results should be available in the near future. Since nearly all carcass traits are highly heritable in swine and beef cattle, it seems very likely that the same will also be found true in sheep. Tenderness and a high proportion of lean to fat in mutton undoubtedly are needed to meet consumer desires.

At present, the proportion of leanness and tenderness of mutton can be measured only after the animal is slaughtered, making it impossible to measure this trait in animals to be used for breeding purposes. The next best measurement is probably that made in the close relatives of an individual.

Carcass information may also be obtained on the progeny of a ram, but the same disadvantage of progeny tests applies in sheep as in other classes of farm animals. It takes so long to get a good progeny test that the ram may be dead before his worth is fully known. Once he is progeny tested, however, and proved to be superior, whether he is dead or living, his offspring should be given preference when selecting breeding animals. If he is dead when the results of progeny tests are fully known, there is still the possibility of keeping the relationship of individuals in the flock closer to the outstanding ancestor through linebreeding.

5.5.5 Carcass quality

When a carcass is partially cut by ribbing at 12th and 13th ribs intersection, area of longissimus dorsi, its colour and marbling, fat thickness over the 12th rib, and per cent kidney and pelvic fat can be used to evaluate carcasses and grade them.

Sheep carcasses can be cut into 5 major wholesale cuts, *viz.* leg, loin, rack, neck and shoulder (with 4 ribs) and breast and shank. They can be further trimmed into retail cuts separating low-valued cuts like neck, breast, flank and shanks from higher valued cuts (also called as primal cuts) like square cut, shoulder, rib, leg and loin. Yield of higher valued cuts (either whole-sale or trimmed retail) can be used as a criterion in carcass evaluation. Other methods of evaluation include manual separation of whole or half carcass into lean, fat and bones.

Meat quality can be further assessed for tenderness using Warner-Bratzler shear on cooked muscles. Average shear force values of 4 kg or less on 1.3 cm cores of longissimus dorsi muscles indicate acceptable tenderness. Juiciness of cooked meat can be evaluated using the filter-paper press method. Colour, odour, flavour and proximate composition will also help in meat quality evaluation. Organoleptic tests using human volunteers can also help in assessing meat quality.

5.5.6 Lamb pelts

The lamb pelts are produced either through slaughter of lambs of pelt sheep, *e.g.* Karakul, within 24 to 48 hrs of birth or from unborn lambs removed through slaughter or through cesarean of pregnant ewes at about 130 to 140 days of pregnancy. The latter two pelts, known as broad (fat) tailed persian, are twice more valuable than the best type produced through slaughter after birth as they have better ornament, are thin and are more lustrous. The lamb pelts fetch 20 to 25 dollars per pelt in the international market depending upon size and quality. Good quality pelts of sur' and grey colour sell at about 40 to 50 dollars per pelt. Its lightness combined with a larger size of pelt, lustre combined with the richness of colouration, silky feel, beauty and grace permit it to be used for many kinds of highly priced fur garments such as overcoats, jackets and suits. The trimmings are used to make collars for gowns, head dresses etc. The quality of a pelt is generally determined by the ornament, *i.e.* colour, type of curls, size, tightness, colour, lustre and weight. The curls are usually formed as long as the length of the hair is more than 20 mm. The present trend is to get light skins with short hairs. The size of curls is completed during the foetal stage itself. The pelts are classified into various types and into quality grades within each type. The types are jacket, ribbed and caucasian.

CHAPTER 6

BREEDING

6.1 Components of Sheep Breeding

Sheep are bred for wool and mutton. In general, in western part of the world mutton is not as popular as pork and beef, although it is the preferred meat in some countries. In India, sheep have never been bred exclusively for mutton. The surplus males and females from wool production herd are available for mutton. Wool for use in the manufacture of clothing and other textiles is being replaced somewhat by various synthetic fibres, but there is still a good demand for it in many parts of the world.

Traits of economic importance in sheep are those related to the cost of production of the kind of wool and mutton demanded by the consumer. These important traits are fertility type and components of growth in terms of body weights at different ages, wool production and quality and carcass yield and quality.

6.2 Types of Sheep Breeding

Results from inbreeding in sheep are very similar to those with other farm animals. In breeding in some instances is followed by the appearance of defects that are due to the pairing of recessive genes.

Inbreeding is usually accompanied by a decline in vigour. Results show that weaning weight and yearling body weight are decreased by 0.5 to 1.5 kg for each 10% increase in inbreeding. Body score and condition score are not affected adversely.

The weight of wool produced decreased with inbreeding, as does the staple length. Possibly this is due to the decline in vigour usually associated with this system of breeding.

Crossbreeding has been used in the commercial production of sheep for many years. It has considerable merit. Data has to be generated for the different traits based on comparisons of measurements of the various traits in the pure breeds and their crossbred offspring to work out heterosis if any and for this purpose well-designed experiments have to be conducted and heterosis is to be estimated. Very few experiments have been reported in which the crossbreeds have maintained concurrently with the pure breeds used to produce the cross so that reciprocal crosses could be compared with the pure breeds.

Most experiments in crossbreeding of sheep compare the crossbreeds with only one of the parent breeds. These are usually comparisons between the offspring of rams of different breeds mated to ewes of a single pure breed and the offspring of purebred rams mated to purebred ewes of that same breed.

The overall performance of the crossbred ewes and lambs is determined to a considerable extent by the prolificacy and mothering ability of ewes from the breeds involved. Thus, crossing two breeds noted for a high percentage lamb crop will give more lambs at weaning than will crossing two breeds that are known for producing a low percentage lamb crop. Even though the degree of heterosis is the same in both instances, the level of fertility may be considerably different because of the average production of the pure breeds used in the crosses. This is probably also true for other traits.

Crossbreeding increases the weaning weights of lambs by about 6 to 7% and the mature weights of ewes by 10 to 15% over the average of the pure breeds used in the cross. On the same basis, it also increases the weight of wool produced, which could be closely related to increased growth rate and viability of the individuals involved.

6.2.1 Selection

The process in which certain individuals in a population are preferred to others for the production of that next generation is known as selection. Selection in general is of two types natural, due to natural forces, and artificial, due to the efforts of man.

No new genes are created by selection. Under selection pressure there is a tendency for the frequency of the undesirable genes to be reduced whereas the frequency of the more desirable ones is increased. Thus, the main genetic effect of selection is to change gene frequencies, although there may be a tendency also for an increase in homozygosity of the desirable genes in the population as progress is made in selection.

6.2.1.1 *Natural selection*

The main force responsible in nature for selection is the survival of the fittest in a particular environment. Natural selection is of interest because of its apparent effectiveness and because of the principles involved.

Natural selection can be illustrated by considering the ecology of some of our wild animal species.

Some of the most interesting cases of natural selection are those involving man himself. All races of man that now exist belong to the same species, because they are interfertile, or have been in all instances where matings have been made between them. All races of man now in existence had a common origin, and at one time probably all men had the same kind of skin pigmentation. As the number of generations of man increased, mutations occurred in the genes affecting pigmentation of the skin, causing genetic variations in this trait over a range from light to dark or black. Man began to migrate into the various parts of the world and lived under a wide variety of climatic conditions of temperature and sunshine. In Africa, it is supposed, the dark skinned individuals survived in larger numbers and reproduced their kind, because they were better able to cope with environmental conditions in that particular region than were individuals with a lighter skin. Likewise, in the northern regions of Europe, men with white skins survived in a greater proportion, because they were better adapted to that environment of less intense sunlight and lower temperatures. But Eskimos who lived in the polar regions of the North are also dark skinned. This is because, Eskimos are more recent migrants from Asia to the polar region and compared to the Negro in Africa and the whites in Europe, they have not lived so long in that region.

Further, evidence is available that there is a differential selection for survival among humans for the A, B and O blood groups. It has been found that members of blood group A have more gastric carcinoma (cancer) than other types and that members of type O have more peptic ulcers. This would suggest that natural selection is going on at the present time among these different blood groups, and the frequency of the A and O genes might be gradually decreasing unless, of course, there are other factors that have opposite effects and have brought the gene frequencies into equilibrium.

Natural selection is a very complicated process and many factors determine the proportion of individuals that will reproduce. Among these factors are differences in mortality of the individuals in the population, especially early in life; differences in the duration of the period of sexual activity; the degree of sexual activity itself; and differences in degrees of fertility of individuals in the population.

It is interesting to note that in the wild state, and even in domesticated animals to a certain extent, there is a tendency toward an elimination of the defective or detrimental genes that have arisen through mutations, through the survival of the fittest.

6.2.1.2 *Artificial selection*

Artificial selection is that which is practised by man. Under this man determines to a great extent which animals to be used to produce the next generation of offspring. Even in this, selection seems to have a part. Some research workers have divided selection in farm animals into two types one known as automatic and the other as deliberate selection.

Litter size in swine can be used as an illustration to define these two terms. Here, automatic selection would result from differences in litter size even if parents were chosen entirely at random from all individuals available at sexual maturity. Under these conditions, there would be twice as much chance of saving offspring for breeding purposes from a litter of eight than from a litter of four. Automatic selection here differs from natural selection only to the extent that the size of the litter in which an individual is reared influences the natural selective advantage of the individual for other traits. In deliberate selection, for *e.g.* it is the term applied to selection in swine for litter size above and beyond that which was automatic. In one study by Dickerson (1973) and coworkers involving selection in swine most of the selection for litter size at birth was automatic and very little was deliberate; the opportunity for deliberate selection however among pigs was utilized more fully for growth rate.

Definite differences between breeds and types of farm animals within a species proves that artificial selection has been effective in many instances. This is true, not only from the standpoint of colour patterns which exist in the various breeds, but also from the standpoint of differences in performance that involve certain quantitative traits. For instance, in dairy cattle there are definite breed differences in the amount of milk produced and in butterfat percentage of the milk.

i) *Basis of selection*

a) *Individuality*

Selection on the basis of individuality means that animals are kept for breeding purposes on the basis of their own phenotype. Selection may be made for several traits, such as coat colour, conformation, performance, or carcass quality. In the past, the emphasis in selection probably was based on coat colour and conformation, although performance and carcass quality have received more attention in recent years.

Most of the breeds of livestock are characterized by a particular coat colour or colour pattern, and this is one of the requirements for entry into the registry associations. Selection for coat colour has been practised because of its aesthetic value rather than its possible correlation with other important economic traits.

Attempts to relate variations of coat colour to performance within a breed have not met with success although many livestock men feel that there is a relationship. There is a strong belief of horse breeders that there is a strong relationship between colour and temperament which has no basis as per the evidence. There is however evidence that animals of some colours are better able to cope with certain environmental conditions, such as high temperatures and intense sunlight in some regions of the tropics or in the south and the south-western portions of the United States. Coat colour in some instances is closely related to lethal and undesirable genes in farm animals. Further, other species such as the mouse, dog, cat, mink, and fox, also show such relationships. Certain coat colours are the trademark of the some breeds of livestock. This is probably because this can be easily recognized. It is thus important that the breeder must conform to the breed requirements for this trait otherwise he will not be in the purebred business for long.

Type and conformation have been used as the basis of selection for many years throughout the world. Type may be defined as the ideal of body construction that makes an individual best suited for a particular purpose. This basis of selection has merit in some instances. The conformation of a draft horse is such that he is better suited to pulling heavy loads than he is to racing. On the other hand, the reverse is true of the thoroughbred.

The performance of individuals has also been given some attention in the development of some of our breeds of livestock. For many years thoroughbred horses have been selected for breeding purposes for their speed. Dairy cows have been selected for their ability to give large amounts of milk and butter fat. In beef cattle and swine, however, less attention has been paid to selection for performance and carcass quality until recently.

Increased emphasis is now being placed on selection for performance and carcass quality, because breeders realize that the type or conformation of an individual is not the best indicator of its potential performance or its carcass quality. Appropriate measures of these traits must be applied before progress can be made in selection for them.

The correlation between type and carcass quality is greater in some instances than is the correlation between type and performance. The meatiness of hogs by a visual inspection, can be assessed but this is not reliable. Better methods are backfat probes on live animals, actual weighings and measuring of lean meat in the carcass.

The fact that type and performance are not usually closely related indicates the importance of selecting separately for the important traits in livestock production. If the correlation between type and other traits is low, it means that they are inherited independently and that they can be improved only if selection is practised for each of them.

Individuality for certain traits should always be given some consideration in a selection program. However, it is more important in some instances than in others. It is most important as the basis of selection when the heritability of a trait is high, showing that the trait is greatly affected by additive gene action. High heritability estimates also suggest that the phenotype strongly reflects the genotype and that the individuals that are superior for a particular trait should also possess the desirable genes for that trait and should transmit them to their offspring.

The greatest disadvantage of selection on the basis of individuality is that environmental and genetic effects are sometimes difficult to distinguish. Much of the confusion may be avoided by growing or fattening of the offspring being compared for possible selection purposes under a standard environment. Even then, it is still possible to mistake some genetic effects for environmental effects. This is less likely to happen, however, in the outstanding individuals than in those that have a mediocre record. For instance, a bull calf placed on a performance test may make a poor record because of an injury or because of sickness while on test. But if he makes an outstanding record, it is certain that he possessed the proper genes and in the right combination as well as the proper environment to make the good record. It cannot always be certain, however, whether an individual with a mediocre record would have done better even if adverse environmental factors had not interfered. We can be certain that his record is poor and by culling on this basis, elimination of the genetically poor individuals is possible. This chance is worth taking, even though we may discard some genetically superior individuals occasionally.

Studies of selection on the basis of individuality within inbred lines of swine have shown that selection favored the less inbred litters. This is another way of saying that selection probably favored the more heterozygous individuals, and this may be true also in many cases where inbreeding is not involved to a great extent. Chance combinations of genes may make an individual outstanding, but his offspring may be inferior, because he cannot transmit his heterozygosity to his offspring. The breeder should avoid keeping superior individuals from very mediocre parents and ancestors. For breeding purposes, it would be much more desirable to keep superior individuals from parents and ancestors that themselves were outstanding.

b) ***Pedigrees***

A pedigree is a record of an individual and his ancestors that are related to him through his parents. Earlier, the information included in a pedigree has been simply the names and registration numbers of the ancestors, and little has been indicated as to the type and performance of the ancestors. Pedigrees now include information on the size of the litter at birth and weaning.

If full information is available on the ancestors as well as the collateral relatives, it may be of importance in detecting carriers of a recessive gene. Such information has been used to a great extent in combating dwarfism in beef cattle.

A disadvantage of the use of the pedigree in selection against a recessive gene is that there are often unintentional and unknown mistakes in pedigrees that may result in the condemnation of an entire line of breeding when actually the family may be free of such a defect. On the other hand, the frequency of a recessive gene in a family may be very low, and records may be incomplete. Then later, it will be found that the gene is present.

Another disadvantage of pedigree selection is that the individuals in the pedigree, especially the males, may have been selected from a very large group, and the pedigree tells us nothing about the merit of their relatives.

Still another disadvantage of pedigree selection is that a pedigree may often become popular because of fashion or fad and not because of the merit of the individuals it contains. The popularity of the pedigree may change in a year or two, and the value of such a pedigree may decrease considerably or may even be discriminated against. If popularity is actually based on merit, there is less danger of a diminution of value in a short period of time.

In using pedigrees for selection purposes, weight should be given to the most recent ancestors. This is because the percentage of genes contributed by an individual's ancestors is halved in each new generation. Some breeders place much emphasis on some outstanding ancestor for which three or four generations has been removed in the pedigree, but such an ancestor contributes a very small percentage of the genes the individual possesses and has very little influence on type and performance, unless linebreeding to that ancestor has been practised.

An individual's own performance is usually of more value in selection than its pedigree, but the pedigree may be used as an accessory to sway the balance when two animals are very similar in individuality but one has a more desirable pedigree than the other. Pedigree information is also quite useful when the animals are selected at a young age and their own type and conformation is not known. Pedigree is useful in identifying superior families if good records are kept and are available.

c) *Collateral relatives*

Collateral relatives are those that are not related directly to an individual, either as ancestors or as their progeny. Thus, they are the individuals brothers, sisters, cousins, uncles and aunts. The more closely they are related to the individual in question, the more valuable is the information for selection purposes.

Complete information on collateral relatives, gives an idea of the kinds of genes and combinations of genes that the individual is likely to possess. Information of this kind has been used in meat hog certification programs, where a barrow and a gilt from each litter may be slaughtered to obtain carcass data. This is done, because otherwise the animal himself has to be slaughtered if information on his own carcass quality is to be obtained, information on collateral relatives is also used in selecting dairy bulls, since milk production can be measured only in the cows even though the bull transmits genes to his offspring for this trait. Information on collaterals has been used in the All India coordinated Research on breeding sheep for mutton wherein information on slaughter traits has been used from full brothers slaughtered for selecting rams for future breeding.

d) *Progeny test*

Selection on this basis means that we estimate the breeding value of an individual through a study of the traits or characteristics of its offspring. In other words, the progeny of different individuals are studied to determine which group is superior, and on this basis the superior breedings are given preference for future breeding purposes. If information is complete, this is an excellent way of identifying superior breeding animals.

Progeny tests are very useful for determining characteristics that are expressed only in one sex, such as milk production in cows or egg production in hens. Even though the bull does not produce milk nor does the rooster lay eggs, they carry genes for these traits and supply one-half of the inheritance to each of their daughters for that particular trait.

Progeny tests are also useful in measuring traits which cannot be measured in the living individual. A good example of this is carcass quality in cattle, sheep, and hogs.

Progeny tests are also being used at the present time by experiment stations in studies of reciprocal recurrent selection. This type of selection is used to test for the "nicking ability" of individuals and lines and is based on the performance of the line cross progeny. Selection of this type is for traits that are lowly heritable and in which non-additive gene action seems to be important.

In comparing individuals on the basis of their progeny, certain precautions should be taken to make the comparisons fair and accurate. In conducting a progeny test, it is very important to test a random sample of the progeny. It would be more desirable if all progeny could be tested, but where this cannot be done, as in litters of swine, those nearer the average of the litter should be tested. It is also important that the females to which a male is mated should be from a non-selected group. One would expect the offspring of a sire to be superior if he is mated to the outstanding females in the herd. Such a practise would be misleading in comparing males by a progeny test, since much of the superiority of the offspring of one male could come from the dams and not from the sire. Some breeders prefer using a rotation of different dams when testing males, but this is practical only in swine, where two litters may be produced each year.

Using a large number of offspring in testing a sire increases the accuracy of the test rule. Where the number of females in a herd is limited, the number of males that may be progeny tested will be less as the number of matings per sire is increased. The point is, then that the breeder must make some decision as to how many sires to test and how many progeny must be produced to give a good test. The number of offspring required for an accurate progeny test will depend upon the heritability of a trait, with fewer offspring being required, when the trait is highly heritable, and more being required when it is lowly heritable.

To make accurate progeny tests, it is also important to keep the environment as nearly as possible the same for the offspring of the different sires. In progeny testing in swine, for instance, confusion would result when the progeny of one sire were fed in dry lot during the summer and the progeny of another were fed on pasture. This would be particularly true in progeny testing for rate of gain, where pigs fed with modern rations often grow considerably faster in dry lot than on pasture. When this environmental condition are not controlled, the inferior sire might actually be thought to be superior.

Progeny tests in most of our farm animals have certain definite limitations. In cattle, especially it takes so long to prove an animal on a progeny test that he may be dead before the test is completed and his merit actually known. Progeny tests may be now easily done in swine than in other farm animals, but even in this case the males are usually disposed of by the time they are thoroughly progeny tested.

The process of progeny testing may be speeded up by testing males at an earlier age than they would ordinarily be used for breeding purposes. By hand-mating them to a few females, or by using them on a larger number of females by artificial insemination, harmful effects that might occur from overuse at too early an age may be prevented.

Too often, farmers send their old sires to market just as soon as their daughters are old enough to breed, in order to prevent inbreeding. This practise has resulted in much loss of good genetic material for livestock improvement. Actually a sire is not proved until his daughters come into production. Rather than being slaughtered, a sire that has proved himself to be of high genetic merit should be used more extensively. It is true that his usefulness in a particular herd may be finished when his daughters are of breeding age, but he should be sent to another herd to be used for additional breeding purposes. To be proved, a sire must have completed a satisfactory progeny test record of some kind. He may be considered proved if he has offspring who have completed one year's record, but this varies with the traits involved. This may be a lactation record, or one of litter size, egg production, or birth and weaning weights, fleece yield and quality. A sire so tested may be said to be proved whether his offspring are good or poor. Before buying a proved sire to use in a herd, a breeder should not neglect to find out if he has been proved good or a poor producer. Newer methods of progeny testing may be developed that are superior to those already available. For instance, the semen of a bull that has been proved highly superior could be collected at regular intervals, frozen, and stored for later use, even after his death. In swine, it might be possible to get quicker progeny tests on females by weaning their pigs at two or three weeks of age and breeding them again as soon as possible to produce three litters per year. Super ovulation, by the injection of certain hormones, a female can be made to produce hundreds of eggs instead of the usual one or few. Embryo transfer technique has made possible using extra ova to other females, where the fertilized ova may develop to birth and possess the characteristics of the mother which ovulated the egg. The success of the embryo transfer transplantation of ova has been limited, but future studies may make it more practical. If this could be done, it would be possible for an outstanding female to have many offspring in one year, rather than one or just a few.

ii) *Methods of selection*

The amount of progress made, regardless of the method used, depends upon the size of the selection differential (selection intensity), the heritability of the trait, the length of the generation interval and some other factors. The net value of an animal is dependent upon several traits that may not be of equal economic value or that may be independent of each other. For this reason, it is usually necessary to select for more than one trait at a time. The desired traits will depend upon their economic value, but only those of real importance need to considered. When too many traits are selected for at one time, less improvement, in any particular one is expected. Assuming that the traits are independent and their economic value and heritability are about the same, the progress in selection for any one trait is only about 1/n times as effective as it would be if selection were applied for that trait alone. When four traits were selected for at one time in an index, the progress for one of these traits would be on the order of $^1/_2$ (not $^1/_4$) as effective as if it were selected for alone. For the selection of superior breeding stock, several methods can be used for determining which animal should be saved and which should be rejected for breeding purposes. Three of these methods which are generally used are given as below.

a) *Tandem method*

In this method, selection is practised for only one trait at a time until satisfactory improvement has been made in this trait. Selection efforts for this trait are then relaxed and efforts are directed toward the improvement of a second, then a third, and so on. This is the least efficient of the three methods practised in respect the amount of genetic progress made for the time and effort spent by the breeder.

The efficiency of this method depends a great deal on the genetic association between the traits selected for. When there is a desirable genetic association between the traits, improvement in one by selection results in improvement in the other trait not selected for, the method could be quite efficient. If there is little or no genetic association between the traits, the efficiency would be less. Since a very long period of time would be involved in the selection practised, the breeder might change his goals too often or become discouraged and not practise selection that was intensive and prolonged enough to improve any desirable trait effectively. A negative genetic association between two traits, in which selection for an increase in desirability in one trait results in a decrease in the desirability of another, would actually nullify or neutralize the progress made in selection for any one trait indicating a low efficiency of the method.

b) *Independent culling method*

In this method, selection may be practised for two or more traits at a time, but for each trait a minimum standard is set that an animal must meet in order to be selected for breeding purposes. The failure to meet the minimum standard for any one trait causes that animal to be rejected for breeding purposes. Let us assume that Pig A was from a litter of 9 pigs weaned, weighed 77 kg at 5 months, and had 1.3 inches of backfat. For Pig B, let us assume that it was from 7 litter of 5 weaned, weighed 94 kg at 5 months, and had 0.95 inches of backfat at 84 kg. If the independent culling method of selection were used, Pig B would be rejected, because it was from a litter of only five pigs. However, it was much superior to Pig A in its weight at five months and in backfat thickness, and much of this superiority could have been of a genetic nature. Thus in practise, there is likelihood to cull some genetically very superior individuals when this method is used.

The independent culling method of selection has been widely used in the past, especially in the selection of cattle and sheep for show purposes, where each animal must meet a standard of excellence for type and conformation regardless of its status for other economic traits. It is also used when a particular colour or colour pattern is required. It is still being used to a certain extent in the production of show cattle and sheep. It does have an advantage over the tandem method, when selection is practised for more than one trait at a time. Sometimes, it is also advantageous, because an animal may be culled at a young age for its failure to meet minimum standards for one particular trait, when sufficient time to complete the test might reveal superiority in other traits.

c) *The selection index*

This method is based on the separate determination of the value for each of the traits selected for and the addition of these values to give a total score for all the traits. The animals with the highest total scores are then kept for breeding purposes. The influence of each trait on the final index is determination by how much weight that trait is given in relation to the other traits. The amount of weight given to each trait depends upon its relative economic value, since all traits are not equally important in this respect, and upon the heritability of each trait and the genetic associations among the traits.

The selection indices is more efficient than the independent culling method, as it allows the individuals which are superior in some traits to be saved for breeding purposes even though they may be slightly deficient in one or more of the other traits. If an index is properly constructed, taking all factors into consideration, it is a more efficient method of selection than either of the other two described earlier, because it should result in more genetic improvement for the time and effort made its use.

Selection indices seem to be gaining in popularity in livestock breeding. The kind of index used and the weight given to each of the traits is determined to a certain extent by the circumstances under which the animals are produced. Some indices are used for selection between individuals, others for selection between the progeny of parents from different kinds of matings, such as line-crossing and crossbreeding, and still others for the selection between individuals based on the merit of their relatives, as in the case of dairy bulls, where the trait cannot be measured in that particular individual.

6.2.2 Breeding native sheep among them

Information available on some of the important carpet wool breeds in the country is given in Table 6.1. A major portion of good carpet wool comes from breeds in Rajasthan, Punjab, Haryana and Gujarat states. The common breeds are Magra, Jaisalmeri, Marwari, Pugal, Nali, Malpura and Sonadi in Rajasthan, Nali in Punjab and Haryana and Patanwadi in Gujarat. Of these, Chokla, Nali and Patanwadi produce finer wool. The available literature (Table 6.1) indicates that the average fibre diameter for 6 monthly clip of Magra, Nali and Pugal is 32.45, 34.92 and 35.13 μ, medullation percentage 48.29, 30.74 and 61.86 and staple length 5.81, 8.31 and 5.71 cm respectively. These values indicated much higher medullation and kemp percentage than recommended, as also the coefficients of variation in excess of 50% for both medullated and non-medullated fibres.

Thus a characteristic tendency of Indian carpets for shedding, when cleaned by a vacuum cleaner, could possibly be ascribed to this higher percentage of medullated fibres and kemp. It will thus be necessary to decrease the medullation percentage especially the kemp fibres and to reduce the average fibre diameter for improving the carpet quality.

Earlier work on selection against medullation percentage in Deccani and Bikaneri (Magra) breeds for producing non-hairy fine wool indicated possibilities of

Table 6.1. Weighted Means and the Corresponding Standard Errors for Fleece Characteristics in Native Breeds of Sheep, Greasy Fleece Weight (kg) and Fleece Quality Characteristics

Breed	Ist 6 months	1st annual	6 months adult	Adult annual	Medullation %	Staple length (cm)	Fibre diameter	Fleece density (cm^2)
Nali	0.95 ± 0.06 (208)	–	1.46 ± 0.01 (6702)	2.91 (374)	30.74 ± 0.40 (604)	8.31 ± 0.14 (935)	34.92 ± 0.69 (72)	1271.33 ± 37.8
Chokla	0.86 ± 0.34	–	1.39 ± 0.01	2.79 (5919)	24.01 ± 0.62 (656)	4.70 ± 0.07 (655)	28.22 ± 0.20 (720)	1008.7 ± 46.5 (24)
Bikaneri	–	–	1.09 ± 0.002 (1833)	2.17 ± 0.01 (649)	48.29 ± 0.39 (2968)	5.81 ± 0.02 (2851)	32.45 ± 0.35 (2851)	–
Pugal	–	–	0.80 ± 0.02	1.606 (72)	1.86 ± 0.62 (524)	5.71 ± 0.04 (513)	35.13 ± 1.00 (524)	–
Marwari	–	0.73 ± 0.02 (1441)	–	0.89 ± 0.02 (441)	72.69 ± 0.90 (390)	8.55 ± 0.07 (327)	39.87 ± 0.36 (390)	–
Patanwadi	–	0.91 ± 0.01 (1407)	–	0.91 ± 0.05 (3317)	33.55 ± 0.39 (2496)	6.69 ± 0.07 (3012)	30.76 ± 0.22 (2496)	1053 ± 13.88 (1093)
Jaisalmeri	–	–	0.77 ± 0.07 (28)	1.54 (35)	64.1 ± 3.1 (35)	6.46	39.1 ± 2.76	–
Malpura	0.48 ± 0.01	–	0.54 ± 0.004 (906)	1.08	71.84 ± 0.17 (6322)	5.60 ± 0.0001 (363)	41.95 ± 0.37 (453)	–
Sonadi	0.46 ± 0.10 (224)	–	0.45 ± 0.01 (5675)	0.91	83.3 ± 2.1 (60)	4.58 ± 0.27 (60)	52.65 ± 1.77 (143)	–
Deccani	–	–	–	0.74 ± 0.02 (225)	24.17 ± 1.02 (266)	6.95 ± 0.23 (47)	34.91 ± 0.47 (266)	1384 ± 26 (157)
Bellary	–	1.23 ± 0.03 (56)	–	1.00 ± 0.03 (88)	43.43 ± 1.42 (81)	–	59.03 ± 1.06 (36)	346 ± 9 (36)

Within paranthesis are the number of observations

reduction in medullation percentage as well as a simultaneous decrease in average fibre diameter. Such selection also resulted in a decrease in wool production, body weight and fertility possibly due to some inevitable inbreeding due to small flock/ sample size.

Very little work has been done for estimating the genetic and phenotypic parameters of wool quality traits in Indian carpet wool breeds. The available estimates based on small set of unadjusted data for most of the tangible environmental effects indicate that most of the quality traits have high heritability. Thus selection based on medullation percentage and 6 monthly greasy fleece weight; the former given a negative weightage, would not only improve the wool yield but also the wool quality. Apparently more research is parameters of greasy fleece weight, fleece quality, body weight, reproductive efficiency and survivability, so as to develop an index incorporating the six monthly body weight, first six month greasy fleece weight and medullation percentage is required. Research programmes on these lines are suggested.

The development programmes for improving carpet wool envisage selective breeding in breeds like Marwari, Jaisalmeri, Magra and Pugal in Rajasthan and Marwari sheep in Gujarat. The selection should be for greasy fleece weight but against medullation. As already indicated, the programme should incorporate six monthly body weight and medullation percentage which would involve proper recording of body weights, fleece weights and quality traits.

Upgrading of these Southern breeds with some of the North Indian carpet wool breeds like Magra and Nali has also shown some promise. It may be desirable to use them for upgrading for better carpet quality in Peninsular and Eastern regions. Usefulness of crossbreeding of Indian carpet wool breeds with those of middle Southern East countries having better fleece quality especially lustre, is also proposed. The introduction of exotic inheritance may improve the lustre and loftiness hitherto lacking in Indian wools.

Grading up with superior indigenous breeds (Bikaneri) was undertaken in Andhra Pradesh, Karnataka, Tamil Nadu and Uttar Pradesh. The results were not very encouraging in the southern states as the Bikaneri rams did not survive long because of their poor adaptation to hot and humid climate. There was, however, an improvement in wool production and quality as reflected by increase in average fleece weight and decrease in average fibre diameter. The improvement through such crossing has been quite substantial in Uttar Pradesh as neither the rams of Rajasthan breeds nor their crosses face serious problem of survival as in the more hot and humid southern States.

6.2.3 Crossbreeding with exotic breeds

The improvement in fine wool production can be brought about through selection for the fleece weight and fleece quality. Although the fleece weight and fleece quality characters are moderate to highly heritable (transmissible from parent to the progeny) and selection within the native breeds will improve the wool production and quality, the progress through selection will be a slow process because the level of performance of indigenous breeds is rather low. Earlier experiments conducted for improving wool quality through selection have shown considerable success. Such selective breeding against percent medullated (hairy) fibres have been undertaken on Deccani sheep at Sheep Breeding Research Station, Poona and on Bikaneri and Lohi at Government Livestock Farm, Hisar. There was a considerable decrease in hairiness and a correlated decrease in average fibre diameter (fibre fineness). Although evaluation of fleece quality could be made visually, for more accurate evaluation, recording of fibre diameter and medullation percentage would be necessary and this may not be possible for large scale selection programme under field conditions.

For faster improvement, it may be desirable to upgrade the inferior wool producing breeds with native superior wool producing breeds or by crossbreeding with exotic fine wool breeds.

Crossbreeding trials with exotic breeds both Merino and British, have been undertaken during pre-independence era and the results indicate marked (2 to 2.5 times) improvement in wool production and considerable improvement in wool quality. Merino crosses were observed to thrive better than those of British breed. Crossbreeding experiments carried out in a number of states involving a number of exotic fine wool breeds indicate that Rambouillet crosses performed better than Merino crosses and out of the native breeds the superior carpet wool breed like Magra produce better crossbreds.

Results of crossing indigenous carpet wool breeds with exotic fine wool, mutton and dual purpose breeds are presented in Table 6.2. Not only an improvement in greasy fleece production but also in quality traits like decrease in average fibre diameter and medullation percentage was relatively perceptible in crosses than the native breeds. The medullated fibres in the crossbred wools were not only finer but also mostly heterotypic. Kemp was also reduced. Some of these wools are liable to be graded as medium/superior carpet wool.

The results of crossbreeding experiments indicate that there is little difference among the exotic fine wool breeds in crosses with the native breeds and there is substantial improvement in greasy wool production especially in low fleece yielding

Table 6.2. Average Wool Production and Quality Traits in Crossbreds being Produced for Improving Carpet Wool Production

Breed	Annual greasy fleece weight (kg)	Fibre diameter (μ)	Staple length (cm)	Medullation (%)	Wool quality
Rambouillet × Malpura	1.92	28.87	4.32	26.88	54s
Suffolk × Malpura	1.74	32.48	5.18	59.05	48s
Dorset × Malpura	1.77	35.25	6.60	66.03	44s
Suffolk × Sonadi	1.75	32.08	5.50	61.88	48s
Corriedale × Muzaffarnagri	1.15	37.93	–	61.88	40s
Corriedale × Coimbatore	1.16	25.71	9.48	29.81	58s
Corriedale × Nellore	0.70	30.30	4.24FL	46.67	50s
Corriedale × Bellary	1.20	26.70	5.26FL	38.88	56s

Wool quality referred to (s) quality number based on average fibre diameter alone. FL – Fibre length

breeds and in fleece quality as reflected by a decline in average fibre diameter and medullation percentage. There is little increase in greasy wool production beyond 50% exotic inheritance although there is further improvement in fleece quality. In general the animals with 75% exotic inheritance are more disease susceptible and have higher mortality as is the experience of introducing the half-breds and $^3/_4$th rams in the field in North-western region. The experience in Jammu and Kashmir, however, indicates that under temperate and sub-temperate conditions it may be possible to sustain 75% exotic fine wool inheritance. It is recommended that in North and Southern hilly regions exotic fine wool inheritance from Merino or Rambouillet may be introduced between 50 to 75% keeping in view the climatic and nutritional conditions and stabilized between these levels through interbreeding. Further improvement should be brought in greasy fleece production through selection and for this purpose, the state Government should have large farms for producing stud breeding rams for distribution to the sheep breeders. In the North-western arid region, the level of exotic fine wool inheritance may be stabilized at 50% and further improvement in greasy wool production and quality especially the reduction in medullation percentage may be brought about through selection.

The ad-hoc committee on sheep breeding policy also recommended that in Northern temperate zone comprising of Jammu and Kashmir, Himachal Pradesh, and hilly regions of Uttar Pradesh, the crossbreeding with Rambouillet and/or Merino should be undertaken for improvement of fine wool production. The committee suggested that level of exotic inheritance should be maintained at 50% except in the case of state sheep breeding farms and selected areas.

6.2.3.1 *Types of synthetic breeds evolved*

A new superior carpet wool breed "Avikalin" has been evolved at the Central Sheep and Wool Research Institute, Avikanagar through interbreeding and selection of Rambouillet × Malpura half-breds. The results of this crossbreeding experiment (Tables 6.3 to 6.5) revealed that half-breds are significantly superior for six monthly greasy fleece weight and fleece quality in comparison to the first six monthly greasy fleece weight, staple length, average fibre diameter and medullation percentage. The comparison of F_2 with F_1 however gave different results in respect of the first six monthly greasy fleece weight where there was a large decline in F_2 over F_1. On the contrary there was seen, quite a large increase in the second six monthly greasy fleece weight in F_2 over the F_1. F_2s were, however, superior in wool quality. Rambouillent × Malpura half-breds revealed 120% increase in greasy fleece production and almost 50% improvement in wool quality. The survivability was also better in the half-breds as compared to the purebred natives. The Avikalin wool contains 58s count and can be used for preparation of medium quality apparel but because of its higher medullation percentage, it is better to use this wool for making superior carpets.

Half-breds produced from Corriedale, Dorset and Suffolk (Table 6.2) also offer promise for augmenting quality carpet wool production.

The new synthetic breeds produce more than 2.5 kg greasy wool annually. The Chokla synthetic produce 7.89% more wool than the purebred Chokla. The fibre diameter and medullation percentage is decreased by 16.39 and 52.94% respectively. The Nali synthetic produce more than 2.5 kg greasy fleece annually. The decrease in fibre diameter and medullation percentage is of higher magnitude, *viz.* 19.93 and 62.03% respectively. In reproduction and survivability the two synthetics perform almost similar to the natives. The results indicate that crossbreeding of Chokla and Nali with exotic fine wool breeds (Rambouillet or Merino as the crosses arising from these two exotic breeds did not significantly differ in any of the characters considered), stabilizing the exotic inheritance at 50% level and bringing further improvement in greasy fleece weight and reduction in medullation percentage through selection will allow improvement in fine wool production. The sampling of fleeces in this project for determining fleece quality was done at 3 months of age, which gave higher medullation percentage than at subsequent age, the difference being about 6–13% between 3 month and 1 year of age. The two shortcomings in the new synthetics are lower staple length and presence of medullated fibres. The medullated fibres are mostly heterotypes and

Table 6.3. Averages for Different Traits in Malpura and its Crosses with Rambouillet

Breed/breed crosses	1st Six monthly greasy fleece weight (kg)	2nd Six monthly greasy fleece weight (kg)	Staple length (cm)	Fibre diameter (μ)	Medullation (%)	Body weight at 1 year
Malpura	0.42 ± 0.01 (21,34.10)	0.532 ± 0.02 (97,36.10)	5.76 ± 0.10 (145,22.10)	39.32 ± 0.63 (145,19.15)	82.41 ± 0.04 (145,13.75)	23.90 ± 0.06 (62,11.05)
R × M(F_1)	1264 ± 0.056 (70,36.68)	1.215 ± 0.047 (65,31.34)	5.97 ± 0.12 (119,23.34)	27.17 ± 0.61 (117,24.35)	41.80 ± 0.06 (117,30.22)	27.62 ± 0.81 (53,21.18)
R × M(F_2)	0.838 ± 0.042 (82,45.45)	1.139 ± 0.088 (31,42.25)	5.06 ± 0.13 (87,24.33)	23.89 ± 0.73 (87,28.52)	33.80 ± 0.09 (86,44.17)	28.16 ± 0.99 (50,24.66)
R × M($^5/_8$)	0.866 ± 0.051 (59,49.84)	1.166 ± 0.093 (23,37.22)	4.51 ± 0.17 (52,26.61)	20.72 ± 0.86 (52,26.61)	17.70 ± 0.17 (49,66.76)	28.13 ± 0.89 (50,19.69)
R × M($^3/_4$)	1.053 ± 0.054 (45,40.45)	1.053 ± 0.012 (19,41.26)	4.37 ± 0.19 (58,42.63)	19.19 ± 0.79 (58,42.63)	13.92 ± 0.13 (58,31.05)	25.18 ± 1.07 (32,23.69)

Within parentheses are the number of observations and percentage coefficient of variation

Table 6.4. Superiority of F_1 and F_2 over Native in Terms of Percent Deviations of their Least Square Means

Traits	R × M(F_1)	R × M(F_2)
First 6 monthly greasy fleece weight (kg)	71.78	18.03
Second 6 monthly greasy fleece weight (kg)	100.97	119.77
Staple length (cm)	8.76	-5.78
Fibre diameter (μ)	-38.48	-42.62
Fibre diameter (μ)	-53.63	-58.61
Yearling body weight (kg)	7.79	3.94

R × M = Rambouillet × Malpura

Table 6.5. Percent Survivability of Malpura (M) and its Crosses with Rambouillet (R)

Breed/ Breed crosses	Pre-weaning (0–3 months)	Post weaning (3–12 months)	Total (0–1 year)
Malpura	82.12 (63)	62.18 (238)	42.16 (351)
R × M(F_1)	94.78 (134)	95.24 (84)	87.91 (91)
R × M(F_2)	96.15 (104)	78.46 (65)	73.91 (69)
R × M($^5/_8$)	86.36 (88)	83.93 (56)	69.11 (58)
R × M ($^3/_4$)	89.73	89.74 (39)	74.46 (47)

Figure within parentheses are the number of observations

will not create problems of harshness in feel and dyeing. These can be removed through selection in the new synthetics. These can also be taken care of by physico-chemical modification through deposition of layer of polymer. The staple length can also be increased through proper shearing regimen. Shearing at 8 monthly intervals compared to 6 monthly interval will allow one clip with longer staple length which could be processed on worsted system. The second clip utilized in semi-worsted woolen system for making knitting yarn.

Crossbreeding trials at Tal (H.P.), Dantiwada (Gujarat) and Sandynallah (T.N.) units where Gaddi, Patanwadi and Nilgiri respectively were used as indigenous breeds with Soviet Merino and Rambouillet as exotic breed, have also shown very encouraging results. The performance of Gaddi and its halfbreds indicated that through crossbreeding in Gaddi there was not only improvement in 6 monthly greasy fleece production but substantial improvement in wool quality also. Similarly the performance of Patanwadi and its crossbreds revealed that through crossbreeding the Patanwadi could be improved from 45.87 to 55.79% in wool production and 7.0 to 13.39% in average fibre diameter and 18.48 to 36.32 in medullation. Performance of Nilgiri and its crosses revealed that the $^{5}/_{8}$th and $^{3}/_{4}$th produced more than 2.5 kg of greasy wool annually which was of excellent apparel quality. There was also improvement in the body weight in crosses. The $^{5}/_{8}$th were observed better than the indigenous and the F_1s particularly beyond 6 months of age. These results will be of considerable importance since the improvement of fine wool production of local sheep will boost the profit accruing to the sheep farmers.

The heritability estimates and genetic correlations for important body weights and wool traits is given in Table 6.6.

A research project to evolve new wool breeds suitable for arid and semi-arid regions has been in operation at the CSWRI, Avikanagar since 1964. It involved crossing of Rambouillet with three diverse native type (Medium-fine carpet, medium apparel type–Chokla, medium carpet wool–Jaisalmeri and coarse hairy carpet wool-Malpura). Three levels of Rambouillet inheritance, *viz.* $^{1}/_{2,}$ $^{5}/_{8}$ and $^{3}/_{4}$ have been tested. Further halfbreds and $^{3}/_{4}$th have been interbred. The results available indicate that there is little gain in body weight and greasy wool production beyond 50% exotic fine wool inheritance, although further introduction does improve wool quality. There is no serious decline in performances of the progeny produced from interbreeding of crossbreds. The halfbreds pose no serious management and disease problems. Higher crosses do show lower adaptation to heat and prevalent disease and would not be suitable except in temperate regions with better pastoral conditions. Rambouillet × Chokla halfbreds produce about 2.5 kg of greasy wool per year of 58s to 64s quality which is suitable for medium quality apparel. Economic gains from improvement in wool alone are around 33% in Rambouillet × Chokla halfbreds over Chokla. One new strain 'Avivastra' has been evolved through interbreeding and selection of Rambouillet × Chokla (half-bred and $^{5}/_{8}$th) base. Annual greasy fleece produced is 2.26 kg with staple length (6 month growth) 3.82 cm, fibre diameter 23.60μ and medullation percentage 21.60. Average adult body weight for males is 36 kg and for females 29 kg. The surviability in younger age as well as in adults of this breed is better than either of its parents. The performance of these animals on reseeded *cenchrus* pasture with density of 5 sheep per hectare along with their lambs upto weaning age (90 days) with daily schedule of 12 hr grazing and without any supplementary grain feeding has been extremely satisfactory in terms of greasy fleece production, wool quality, lamb growth, survivability and breeding efficiency. Higher percentage of hairy hetero-fibres however give rough feel to the apparel. Elimination of this defect through selection is in progress. Rambouillet × Jaisalmeri crosses though superior to Jaisalmeri in wool production and wool quality pose serious problem due to black patches on the fleece which considerably reduce the return from the sale of wool. At $^{3}/_{4}$th level of

Table 6.6. Heritability Estimates and Genetic Correlations for Important Bodyweights and Wool Traits

		6 months weight	Yearling weight	Fleeee weight	Fibre diameter	Staple length	Medullation %
6 months weight	E	0.47 ± 0.072	0.897 ± 0.020	High	–	–	–
	R	–	–	–	–	–	–
Yearling weight	E	0.87 ± 0.800	0.450 ± 0.026	0.98 ± 0.009	–	–	–
	R	–	–	0.07 ± -0.90	–	–	–
Fleece weight	E	–	–	0.21 ± 0.055	0.51 ± 0.19	10.87 ± 0.084	0.08
	R	–	–	0.07 ± -0.60	-0.38 ± 0.67	-0.48 ± -0.91	–
Fibre diameter	E	–	–	–	0.453	0.37 ± 0.359	-0.68 ± 0.07
	R	–	–	–	–	0.03–0.47	–
Staple length	E	–	–	–	–	0.404 ± 0.05	0.08
	R	–	–	–	–	0.01 ± 0.86	–
Medullation %	E	–	–	–	–	–	0.661
	R	–	–	–	–	–	0 ± 0.80

E = Estimate; R = Range
Figures on the diagonal are the heritabilites and off-diagonal genetic correlations

exotic inheritance, the native types involved in crossing do not show any serious differences.

Based on the crossbreeding results, it is evident that the inferior carpet wool breeds like Malpura, Sonadi, Muzaffarnagri, Jaisalmeri of North-western region may be improved for better carpet quality by crossing them with exotic fine wool, dual or mutton breeds by stabilizing the exotic inheritance at 50% followed by selection. The experience so far gained after using dual purpose breeds like Corriedale indicates that it is difficult to handle this breed under field conditions. On similar pattern the inferior carpet breeds in Southern peninsular and Eastern regions, *viz.* Bellary, Deccani, Shahabadi and Chottanagpuri may also be improved for better carpet quality by crossing them with exotic fine wool breeds and stabilizing the exotic inheritance at 50%.

During the 4th Five Year Plan, the ICAR had sponsored an All India Coordinated Research Project for fine wool with the objectives to develop superior fine wool breeds suited to different agro-climatic regions of the country. The exotic breeds used for crossbreeding in this project were Russian Merino and Rambouillet. The different Centres of the project were being coordinated at the Central Sheep and Wool Research Institute (CSWRI), Avikanagar. The emphasis was to study breed combination, level of exotic inheritance, heterosis in production traits, and problems involved in production traits and in interbreeding of crossbreds for evolving new breeds from crossbred base. Some encouraging results have become available. Two new synthetics (Chokla and Nali synthetics) for fine wool have been generated at the CSWRI through the creation and evaluation of genetic groups, *viz.* F_1, F_2, $^5/_8{}^{th}$ and $^3/_4{}^{th}$ involving different combinations of exotic and native breeds in order to assess the prospective targets in wool in semi-arid conditions of the country. The performance of various grades of Gaddi, Patanwadi with Soviet Merino/ Rambouillet is given in Table 6.7. The performance of various grades of Chokla, Nali and their crosses with Rambouillet/Merino is given Tables 6.8 and 6.9. Performance of various grades of Nilgiri and its cross with Soviet Merino and Rambouillet is given in Table 6.10.

6.3 Breeding for Mutton Production

Mutton production, a polygenic trait, is mostly governed by additive gene action. The characters connected with mutton production, *viz.* body weight gains, efficiency of feed conversion and dressing percentage are moderate to highly heritable and thus the selection within the indigenous breeds can bring about considerable improvement in mutton production. Some of the Indian breeds like Mandya are quite comparable to most of the exotic mutton breeds with regard to

Table 6.7. Means with Standard Errors for Greasy Fleece Weight and Wool Quality Attributes in Gaddi, Patanwadi and their Crosses with Soviet Merino/ Rambouillet

Genetic groups	6 monthly greasy fleece weight (kg)	Average fibre diameter (μ)	Medullation (%)
Gaddi	0.499 ± 0.007 (584)	28.3 ± 0.11 (830)	28.0 ± 0.7 (436)
Merino × Gaddi ($^1/_2$)	0.679 ± 0.030 (76)	23.7 ± 0.25 (195)	11.2 ± 1.5 (75)
Rambouillet × Gaddi ($^1/_2$)	0.577 ± 0.028 (75)	21.0 ± 0.16 (241)	12.8 ± 1.4 (70)
Patanwadi	0.484 ± 0.024 (38)	24.87 ± 0.527 (55)	2.79 ± 2.163 (55)
Rambouillet × Patanwadi ($^1/_2$)	0.706 ± 0.015	23.3 ± 0.227	6.73 ± 1.284
Soviet Merino × Patanwadi (F_2)	0.518 ± 0.022 (26)	21.54 ± 0.580 (39)	8.93 ± 2.050 (39)
Soviet Merino × Patanwadi (F_2)	0.466 ± 0.122 (5)	25.84 ± 1.090 (7)	20.31 ± 3.790 (7)
Rambouillet × Patanwadi ($^3/_4$)	0.550 ± 0.000 (4)	19.01 ± 1.708 (8)	13.16 ± 3.430 (8)
Soviet Merino × Patanwadi ($^3/_4$)	–	19.03 ± 1.960 (2)	11.88 ± 5.990 (2)

Note: Figures within parentheses are number of animals

Table 6.8. Averages of Greasy Fleece Weight and Wool Quality Attributes of Chokla, Nali and their Crosses with Rambouillet/Merino

Genetic group	Average fibre diameter (μ)	Medullation (%)	Staple length (cm)	1st 6 monthly greasy fleece weight (kg)	6 monthly body weight (kg)
Chokla	23.12 ± 0.35 (79)	25.14 ± 0.05* (79)	4.10 ± 0.71 (78)	0.897 ± 0.02 (439)	13.311 ± 0.15 (315)
Chokla (F_1)	20.56 ± 0.19 (196)	10.70 ± 0.03* (193)	4.09 ± 0.05 (98)	0.980 ± 0.03 (795)	15.240 ± 0.11 (749)
Chokla (F_2)	21.20 ± 0.38 (33)	19.03 ± 0.04 (196)	3.95 ± 0.13 (82)	0.970 ± 0.03 (115)	16.193 ± 0.29 (124)
Chokla ($^5/_8$th)	20.80 ± 0.57 (33)	15.23 ± 0.06* (30)	4.74 ± 0.20 (19)	1.070 ± 0.04 (39)	14.791 ± 0.45 (41)
Chokla ($^3/_4$th)	20.03 ± 0.42 (63)	11.92 ± 0.04* (62)	3.15 ± 0.05 (56)	0.947 ± 0.07 (92)	15.784 ± 0.30 (104)
Nali	28.04 ± 0.54 (81)	62.85 ± 0.85* (81)	4.44 ± 0.11 (73)	0.953 ± 0.37 (341)	13.85 ± 0.15 (314)
Nali F_1	22.61 ± 0.18 (308)	28.09 ± 0.02* (308)	4.48 ± 0.05 (310)	0.998 ± 0.03 (1028)	14.820 ± 0.09 (954)
Nali F_2	21.79 ± 0.28 (159)	27.74 ± 0.03* (155)	3.87 ± 0.09 (153)	1.116 ± 0.02 (191)	16.22 ± 1.88 (206)
Nali $^5/_8$th	21.97 ± 0.51 (44)	23.76 ± 0.07* (43)	4.14 ± 0.15 (53)	1.135 ± 0.03 (66)	17.65 ± 0.342 (67)
Nali $^3/_4$th	19.29 ± 0.48 (61)	11.37 ± 0.05* (50)	3.10 ± 0.18 (64)	0.9532 ± 0.08 (125)	16.25 ± 0.26 (132)

The wool attributer are at months of age. The wool quality assessed at 2nd clip (1 year) have shown 6–13% reduction in medullation

Table 6.9. Means and Standard Errors (SE) of Different Characters of New Synthetics

Genetic group	Wool quality				Body weight (kg) at reproduction			Lambing % on basis of		Survivability		
	Annual greasy wool Production	Staple Length (cm)	Average fibre diameter	Medullation %	Birth	6 months	12 months	Ewes available	Tupped	Pre weaning	Post weaning	Adult
Nali Synthetic												
n	760.00	134.00	132.00	127.00	2218.00	1603.00	849.00	315.00	266.00	3502.00	1686.00	2755.00
Mean	2.86	5.20	22.90	16.26	2.92	16.35	20.49	74.92	88.72	91.61	83.04	90.97
S.E.	0.02	0.11	0.22	3.41	0.017	0.074	0.113	–	–	–	–	–
Chokla synthetic												
n	543.00	66.00	64.00	54.00	1976.00	1157.00	640.00	215.00	192.00	3021.00	1626.00	2061.00
Mean	2.75	5.13	21.06	8.75	2.83	16.34	20.43	82.32	92.19	86.93	84.14	91.13
S.E.	0.03	0.13	0.26	1.12	0.009	0.087	0.126	–	–	–	–	–

Table 6.10. Means with Standard Errors for Wool Production and Quality Attributes of Nilgiri and its Crosses with Soviet Merino and Rambouillet

Genetic group	Annual greasy fleece production(kg)	Average fibre diameter(μ)	Staple length (cm)	Medullation (%)
Nilgiri	1.285 ± 0.076 (57)	22.158 ± 0.050 (57)	6.166 ± 0.035 (57)	5.13
Merino × Nilgiri ($^1/_2$)	2.860 ± 0.043 (239)	21.527 ± 0.030 (112)	7.523 ± 0.030 (112)	Nil
Merino × Nilgiri ($^{II}/_2$)	2.689 ± 0.120 (30)	21.237 ± 0.080 (30)	7.364 ± 0.043 (30)	Nil
Merino × Nilgiri ($^3/_4$)	3.470 ± 0.106 (45)	20.254 ± 0.035 (45)	8.007 ± 0.035 (45)	Nil
Nilgiri ($^3/_4$ interse)	3.825 ± 0.168 (4)	20.193 ± 0.113 (4)	7.105 ± 0.140	Nil (4)
Merino × Nilgiri ($^5/_8$)	3.264 ± 0.069 (98)	20.784 ± 0.041 (41)	6.647 ± 0.034	Nil (41)
Merino × Nilgiri ($^5/_8$ interse)	3.341 ± 0.122 (53)	20.372 ± 0.036 (53)	7.564 ± 0.036	Nil (53)
Rambouillet × Nilgiri ($^1/_2$)	2.508 ± 0.041 (194)	21.946 ± 0.033 (92)	7.219 ± 0.028	2.14 (92)
Rambouillet × Nilgiri (F_2)	2.500 ± 0.188 (12)	21.751 ± 0.096 (12)	7.982 ± 0.006	Nil (12)
Rambouillet × Nilgiri ($^3/_4$)	2.900 ± 0.463 (06)	20.265 ± 0.133 (06)	7.616 ± 0.074 (05)	Nil

Note: Figures within parentheses are number of animals

their dressing percentage. However, selection within the natives is somewhat slow in bringing about desired improvement in the mutton production because of extremely low level of their present production. Improvement of inferior native breeds through upgrading with relatively superior native breeds or through crossing native breeds with exotic superior mutton breeds will prove to be a better alternative for bringing in more rapid improvement. Little efforts were made earlier to breed sheep exclusively for mutton. One such attempt was made by ICAR at Punjab Agricultural University, Punjab in the later half of 1960 and now under the AICRP on Sheep Breeding for the development of mutton breeds which proved to be economical under stall-fed conditions. Besides breeding sheep for improvement of wool production, there has also been a tendency to look into the improvement of live weight along with wool.

6.3.1 Studies on the purebred performance of native breeds

Weighted averages and the corresponding standard errors for body weight at different ages, efficiency of feed conversion and carcass yield in North-western carpet wool breeds and Southern mutton type breeds are given in Table 6.11. The estimates of heritability and genetic and phenotypic correlations of body weight at different ages in various breeds are presented in Tables 6.12 and 6.13. Perusal of the tables revealed, the North Indian carpet wool breeds to be heavier in body weight than South Indian mutton breeds. Efficiency of feed conversion of two month post-weaning individual feedlot trial ranged from 13.97 to 19.78%, the highest (19.78 ± 2.56) being in Nali and so also the dressing percentage (44.32 to 48.3%) on live weight basis (5 months), the highest (48.25 ± 3.29) for Nali. In Magra, 6 months age was the most profitable age for slaughter. Dressing percentage on empty weight basis varied from 50.07 to 53.95 and the total boneless meat percentage from 38.40 to 42.32 in Magra lambs slaughtered at 6 to 9 months of age. Results of the All India Co-ordinated Research Project on Sheep Breeding indicated that of the native breeds included in the crossbreeding programme at the Central Sheep and Wool Research Institute, Avikanagar (semi-arid), Malpura lambs were superior to Sonadi in terms of efficiency of feed conversion and dressing percentage. They were also better convertor of milk to body tissues than the Sonadi lambs. The meat bone ratio was somewhat higher in Malpura as compared to the high and low energy feeds. The heritability estimates of body weight from birth to one year of age varied from 0.11 ± 0.3 to 0.45 ± 0.66 but those for body weights other than of 6 and 12 months were small and they did not have so much genetic and phenotypic correlations (especially the weaning weight) between body weight of 9 and 12 months as at 6 months. The pooled heritability estimates of 6 month body weight were high and almost equal to that of 9 and 12 months.Furthermore the 6 month body weight bore high genetic and phenotypic correlations (0.664 and 0.785 respectively) with the yearling body weights.

These genetic parameters indicated that the body weight at 6 months of age should be a best selection criterion for improving market weight and adult body size and in turn the productivity.

6.3.2 Crossing among indigenous breeds

Some exploratory trials on crossing indigenous breeds for mutton production have also been undertaken. Grading up with indigenous breed (Bikaneri) was undertaken in Andhra Pradesh, Karnataka, Tamil Nadu and Uttar Pradesh. The results of this experiment were not very encouraging in the southern states, as the Bikaneri did not survive long because of their poor adaptation to hot and humid climate. Crossbreeding of North Indian carpet wool breeds (Nali and Lohi) with South Indian mutton type breeds (Nellore and Mandya) at the Haryana Agricultural University did not markedly improve the body weight gains, efficiency of feed

Table 6.11. Weighted Means with Standard Errors for Body Weights (kg) at Different Ages, Feed Efficiency and Dressing Percentage in Native Breeds of Sheep

Breed	Weaning at 3 months weight	6 months weight	12 months weight	Feed efficiency %	Dressing percentage		
					On live weight basis	On empty weight basis	Age at which slaughtered (months)
1	2	3	4	5	6	7	8
a) North-western Region							
Nali	10.19 ± 0.14 (263)	13.30 ± 0.20 (174)***	17.74 ± 0.41 (74)	19.78 ± 2.56 (12)	48.25 ± 3.29	–	5
Chokla	11.13 ± 0.13 (499)	14.05 ± 0.25 (174)	19.30 ± 0.46 (83)	–	–	–	–
Marwari	8.16 ± 0.84 (101)	9.40 ± 0.20 (101)	21.06 ± 0.04 (536)	–	–	–	–
Marwari	14.08 ± 0.18 (560)	–	–	–	–	–	–
Magra	11.74 ± 0.13 (415)	20.14 ± 0.10 (1812)*	27.90 ± 0.192 (3445)	–	44.75 ± 0.55 (24)	53.55 ± 0.15 (24)	12
Magra	17.83 ± 0.05 (4006)**	–	–	–	–	–	–
Malpura	9.86 ± 0.21 (711)	12.07 ± 0.40 (850)	21.44 ± 0.30 (105)	12.49 (54)	51.89 (50)	60.21 (50)	6
Sonadi	9.55 ± 0.19 (296)	14.65 ± 0.26 (125)	21.18 ± 0.48 (61)	12.06 (56)	51.21 (64)	59.34 (64)	6

Table 6.11. (*Contd.*)

Breed	Weaning at 3 months weight	6 months weight	12 months weight	Feed efficiency %	Dressing percentage		
					On live weight basis	On empty weight basis	Age at which slaughtered (months)
1	2	3	4	5	6	7	8
Lodi	10.88 (143)	14.34 (177)	19.57 (71)	18.28 ± 1.60 (11)	49.19 ± 2.69 (5)	–	5
Patanwadi	13.68 ± 0.11 (1727)	–	21.19 ± 0.25 (1225)	–	–	–	–
Muzaffarnagri	10.80 ± 0.18 (117)	16.54 ± 0.75 (24)	22.82 ± 0.14 (10)	13.64	45.60	51.65	6
b) Southern Region							
Mandya	9.71 ± 0.16 (1822)	12.76 ± 0.29 (1261)	21.02 ± 1.02 (106)	9.62 (8)	45.36 (24)	53.79 (24)	6
Nellore	9.26 ± 0.26 (243)	13.31 ± 0.26 (59)	18.92 ± 0.83 (6)	11.8 (8)	38.51	48.96	6
Deccani	14.00 ± 0.03 (153)	–	–	–	–	–	–
Madras Red	13.84 ± 0.17 (204)	16.13 ± 0.15 (200)	22.03 ± 0.45 (46)	–	41.15	–	–
Coimbatore	7.06 ± 0.22 (59)	9.87 ± 0.33 (44)	14.29 ± 0.42 (34)	–	36.80	–	5
Mecheri	9.81 ± 0.04	10.04 ± 0.05	18.96 ± 0.09 (2382)	–	47.76 (1697)	54.39	8 (199)

Table 6.11. (*Contd.*)

Breed	Weaning at 3 months weight	6 months weight	12 months weight	Feed efficiency %	Dressing percentage		
					On live weight basis	On empty weight basis	Age at which slaughtered (months)
1	2	3	4	5	6	7	8
Ramnad White	8.11 ± 0.25 (84)	9.37 ± 0.19 (56)	15.72 ± 0.07 (51)	–	–	–	–
Tiruchi Black	9.48 ± 0.28 (53)	10.73 ± 0.28 (48)	–	–	–	–	–
Kilakarsal	7.26 ± 0.07 (383)	8.53 ± 0.05 (356)	14.15 ± 0.55 (62)	–	–	–	–
Bellary	11.09 ± 0.15 (394)	16.28 ± 0.22 (349)	18.68 ± 0.41 (115)	–	–	–	–

Within parentheses are the number of observations
*16 weeks weight; ** 135 days weight; *** 150 days weight

Table 6.12. Heritability Estimates of Body Weights at Different Ages

Breed	Birth weight	Weaning at 3 months	6 months weight weight	9 months weight	12 months weight
Chokla	–	0.47 ± 0.21 (250)	0.80 ± 0.23 (250)	–	–
Nali	0.15 ± 0.13 (254)	0.29 ± 0.10 (226)	0.51 ± 0.26 (197)	0.33 ± 0.26 (138)	0.15 ± 0.24 (119)
Nali, Lohi and their crosses with Nellore and Mandya	0.08 ± 0.08 (523)	0.31 ± 0.12 (451)	0.48 ± 0.16 (376)	0.26 ± 0.16 (277)	0.23 ± 0.17 (248)
Bikaneri (Magra)	0.10 ± 0.03 (3524)	0.24 ± 0.09 (3468)	0.37 ± 0.091 (1369)	–	0.48 ± 0.07 (1021)
Nali, Lohi and their crosses with Nellore and Mandya	0.44 ± 0.13 (787)	0.26 ± 0.13 (759)	–	–	–
Bikaneri (Magra)	0.45 ± 0.20 (545)	0.68 ± 0.26 (534)	–	–	0.38 ± 0.23 (221)
Marwari	0.36 ± 0.13 (472)	0.07 ± 0.07 (464)	–	–	0.07 ± 0.09 (363)
Patanwadi	0.64 ± 0.13 (1528)	0.56 ± 0.12 (1438)	–	–	0.90 ± 0.18 (948)

conversion and dressing percentage in the crosses compared to the North Indian Carpet wool breeds. The efficiency of feed conversion ranged from 13.97 to 19.78% and the purebred Nalis. The results of crossbreeding of indigenous Bihar sheep with Bikaneri indicated that crossbred lambs grew significantly faster than the Bikaneri and Bihar lambs and that the average live weight of Bikaneri and Bihar sheep and their grades differed significantly at one, two and three months of age.

The dressing percentage on live and empty body weight basis was higher in F_1s arising from Dorset and Suffolk crosses with Malpura and Sonadi. There was comparatively a higher increase in Dorset crosses than the Suffolk ones. Mandya and Nellore crosses with Dorset, and Nellore crosses with Suffolk showed a decline in dressing percentage on live weight basis whereas only the Dorset × Mandya expressed a slight decline in dressing percentage on empty body weight basis.

The higher feedlot gains were obtained in F_1s over their contemporary purebreds Suffolk × Sonadi, showed the maximum gains.

Likewise there was improvement in the efficiency of feed conversion in F_1s over contemporary purebreds. In this respect the Suffolk crosses with Malpura and

Table 6.13. Genetic and Phenotypic Correlations Among Body Weight at Different Ages

Birth weight with weaning weight		Birth weight with 6 month weight		Birth weight with 9 month weight		Birth weight with 12 month weight		Weaning weight with 6 months weight	
r_p	r_g	r_p	r_g	r_p	r_g	r_p	r_g	r_p	r_g
–	–	–	–	–	–	–	–	0.888	0.907
								(248)	(248)
0.47 ± 0.04	0.78 ± 0.20	0.38 ± 0.04	0.75 ± 0.10	0.34 ± 0.06	0.90 ± 0.18	0.29 ± 0.07	0.71	0.72 ± 0.02	1
(226)	(226)	(197)	(197)	(138)	(119)	(119)	(119)	(195)	(195)
0.41 ± 0.03	0.53 ± 0.25	0.35 ± 0.03	0.86 ± 0.08	0.29 ± 0.04	0.31 ± 0.67	0.23 ± 0.05	0.56 ± 0.55	0.73 ± 0.01	0.90 ± 0.07
(451)	(451)	(376)	(376)	(277)	(277)	(248)	(248)	(263)	(263)
0.31	–	–	–	–	–	0.47	–	–	–
0.68	0.83 ± 0.09	–	–	–	–	–	–	–	–
(759)	(759)								
0.25 ± 0.02	0.120 ± 0.017	0.28 ± 0.03	0.07 ± 0.03	–	0.03	0.28 ± 0.22	-0.110 ± 0.220	0.62 ± 0.22	0.090 ± 0.030
*0.506	0.687 ± 0.069	0.363	-0.79 ± 0.071	0.309	0.86 ± 0.174	0.254	0.172 ± 0.204	0.939	0.900 ± 0.027

Within parentheses are the number of observations

* Pooled value r_p = Phenotypic correlation; r_g = Genetic correlation

Sonadi are superior than the Dorset crosses. Dorset × Sonadi showed the highest improvement in efficiency over their contemporary purebreds. There was however, a little difference in percent improvement between the Dorset and Suffolk crosses with Mandya and Nellore.

6.3.3 Crossbreeding of mutton type breeds with exotic mutton/dual purpose breeds

A substantial improvement in greasy wool production was attained in Suffolk and Dorset crosses with all the native breeds *viz.* Malpura, Sonadi and Muzaffarnagri. Maximun improvement was in Sonadi crosses especially with Suffolk. The F_2s were generally equal and sometimes superior to F_1s. There was seen 23.42, 27.15, 1.58% improvement in F_1 over the natives in respect of their final weight, feed efficiency and dressing percentage (Table 6.14). These figures for F_2 were 30.17, 43.14 and 9.09% respectively. F_2s showed 5.47, 12.58 and 7.39% improvement respectively over their F_1s. The $^3/_4$th did not show any advantage over the F_1s. advantage over the F_1s.

The crossbreeding trials under the AICRP on Sheep Breeding thus led to the generation of one new synthetic mutton strain at CSWRI, Avikanagar centre through crossing of Suffolk and Dorset with Malpura and Sonadi by stabilizing the exotic inheritance at 50% level. Another two synthetic strains have also been developed at Palamner centre through crossing of Suffolk and Dorset with Nelore and Mandya breeds by stabilizing the exotic inheritance at 50%. The performance of these new synthetics is given in Table 6.15. The new synthetics/strains were decidedly superior to the indigenous breeds involved, in respect of feedlot gain, feed conversion efficiency and to a small extent in dressing percentage on live weight basis. The results further indicated that the new synthetics/strains could attain the expected target of 30 kg body weight under intensive feeding conditions as well as produce mutton which is 60% cheaper than by the purebred natives perhaps due to better efficiency of feed conversion in the halfbreds as compared to the natives.

6.4 Breeding for Pelt Production

In India, sheep besides being traditionally kept for wool, mutton, skin, manure and to some extent for milk and transport, are also raised for lamb pelt production to which hardly any emphasis has been given. It is understood that the Khatik community in Rajasthan does produce lamb pelts from Malpura breed on a small scale. A research venture during mid seventies and eighties has been made to add this new dimension to sheep industry in India.

Table 6.14. Feedlot, Slaughter Traits and 1st Six Monthly Greasy Fleece Production

Genetic group	Final weight (kg)	Total gain (kg)	Feed efficiency (%)	Dressing percentage on the basis of live weight	1st six monthly greasy fleece production (kg)
Malpura	21.13 ± 0.44 (95)	8.94 ± 0.28 (95)	12.02 ± 0.29 (95)	48.42 ± 0.28 (72)	0.544 ± 0.031 (657)
Sonadi	19.73 ± 0.44 (107)	12.02 ± 0.31 (107)	48.07 ± 0.29 (98)	48.07 ± 0.29 (98)	0.526 ± 0.009 (439)
Crossbred (F_1)	25.16 ± 0.24 (420)	12.11 ± 0.13 (420)	15.28 ± 0.14 (420)	48.98 ± 0.15 (307)	0.579 ± .010 (1281)
Crossbred (F_2)	26.53 ± 1.11 (22)	13.59 ± 0.59 (22)	17.21 ± 0.51 (22)	52.60 ± 0.90 (16)	0.766 ± 0.023 (68)
Crossbred($^3/_4{}^{th}$)	23.98 ± 0.76 (33)	11.71 ± 0.43 (33)	16.56 ± 0.51 (33)	47.12 ± 1.85 (4)	0.697 ± 0.032 (55)

Table 6.15. Means and Standard Errors of Different Characters of Native and New Synthetics

Genetic group	Feedlot gain (kg)	Dressing % on the basis of		Efficiency of feed conversion	Lambing % on the basis of				Survivability % adult
		Live weight	Empty body weight		Ewes available	Ewes tupped	Pre weaning	Post weaning	
Native									
Mean	8.63	48.67	56.39	12.61	67.22	71.18	83.05	88.57	69.65
SE	–	0.18	0.20	0.19	–	–	–	–	–
N	240.00	209.00	209.00	240.00	360.00	340.00	2573.00	2056.00	3311.00
New Sythetic (CSWRI)									
Mean	11.79	49.63	57.25	15.81	40.00	69.42	82.68	84.79	59.04
SE	–	0.33	0.15	0.13	–	–	–	–	–
N	536.00	422.00	422.00	536.00	210.00	121.00	2835.00	2137.00	1106.00

Table 6.15. (*Contd.*)

Genetic group	Feedlot gain (kg)	Dressing % on the basis of		Efficiency of feed conversion	Lambing % on the basis of				Survivability % adult
		Live weight	Empty body weight		Ewes available	Ewes tupped	Pre weaning	Post weaning	
New Sythetic (CSWRI)									
Mean	9.24	49.15	56.71	17.38	75.89	94.08	96.89	96.37	89.15
N	37.00	63.00	63.00	37.00	419.00	338.00	806.00	1214.00	1349.00
Mandya Synthetic									
Mean	11.71	48.93	56.54	18.03	45.73	63.56	96.87	93.03	91.62
N	74.00	99.00	99.00	74.00	164.00	118.00	440.00	615.00	1288.00
Nellore									
Mean	7.47	47.76	54.59	15.06	90.00	98.18	98.69	96.05	90.27
N	35.00	57.00	57.00	35.00	180.00	165.00	461.00	788.00	3783.00
Nellore Synthetic									
Mean	12.07	50.28	58.77	19.31	51.85	73.68	98.24	89.56	89.54
N	65.00	73.00	73.00	65.00	135.00	95.00	467.00	523.00	1064.00

N = No. of observations

Advantages of lamb pelt production are: Karakul lamb pelts fetch 15–20 dollars in the international market depending upon their size and quality. In addition to this relatively high market value of pelts in comparison to other products from sheep, there are a number of other good features of maintaining sheep for pelt production.

The early slaughter of lamb eliminates mortality in weakling lambs. The mothers whose lambs have been slaughered can be rebred and thus can produce lambsmore often; they can be milked which can be used for human consumption or they can be used as foster mothers for orphan lambs or for the lambs of mothers with very poor milk production. Furthermore the elimination of a large number of lambs at birth will reduce the pressure on feed/fodder resources otherwise required for a longer period if reared for meat/wool. Other conventional products *e.g.* wool, mutton, skin, manure and milk also would become available as by-products from the flocks reared for pelt production. Thus pelts production would compensate only the disposal of extra male/female lambs which would otherwise have normally been marketed for meat around 9 months to 1 year of age.

Under an Indo-USSR protocol for co-ordination in the field of Agriculture and Animal Husbandry, the former USSR Government made available 250 Karakul sheep (200 ewes and 50 rams) in November, 1975 for evaluation of their purebred performance as well as that of their crosses with some indigenous coarse carpet wool breeds under hot and cold-arid conditions.

Out of the aforesaid 250 Karakul sheep, 27 ewes died during air transit, 50 ewes and 10 rams were transferred to Khumbathang in Kargil district of Jammu and Kashmir state to study their performance in cold arid conditions of Ladakh. Eight of the remaining 40 rams were transferred to the CSWRI main campus at Avikanagar for conducting cross-breeding trials with some native breeds of sheep.

The geographical location of Bikaner is at 28° 3' N latitude and 73° 5' E longitude at 244.84 meters MSL with an annual average rainfall of 281.15 mm. The farm area is undulating having ranges of sand dunes covering approximately 25% of the total land surface. The soil is sandy with about 90% sand. The growth, survivablity, and reproductive performance of Karakul sheep at the Division of Carpet Wool and Karakul Pelt Production at Bikaner over the years are presented in Tables 6.16 to 6.21. The studies indicated for very good adaptability of Karakuls to hot arid conditions. Their tupping and lambing percentages were quite satisfactory and so were pelts of the lambs born in India well comparable to those born in former USSR.

The geographical location of Khumbathang is at 34° 10' N latitude and 77° 40' E longitude. The total strength at the end of April,1977 at Sheep Breeding Farm,

Table 6.16. Averages alongwith Standard Errors for Bodyweight (kg) of Karakul Lambs

Age	Lambs born in				Pooled over the years
	1976 +	1977 ++	1978	1979 up to 19–03–79	
Birth	4.02 ± 0.04 (168)	3.46 ± 0.05 (94)	3.59 ± 0.06 (111)	3.27 ± 0.05 (150)	3.61 (523)
3 months	14.89 ± 0.42 (154)	9.54 ± 0.26 (65)	15.54 ± 0.32 (65)	12.53 ± 0.40 (66)	13.81 (284)
6 months	19.37 ± 0.95 (152)	16.12 ± 0.40 (64)	22.51 ± 0.45 (65)	20.40 ± 0.48 (65)	19.36 (281)
9 months	21.23 ± 0.47 (140)	23.20 ± 0.53 (63)	25.81 ± 0.56 (63)	25.77 ± 0.62 (65)	22.78 (266)
12 months	23.72 ± 0.82 (120)	28.30 ± 0.59 (61)	28.83 ± 0.61 (62)	–	26.17 (243)

* Within parentheses are given the number of observations
\+ Out of the ewes imported pregnant
++ Without any supplementation to pregnant and lactating ewes

Table 6.17. Survivability of Karakul Sheep in Different Age Groups

Age Group	No. of animals available	No. of animals died	Mortality%	Survival %
Year 1976				
0–3 months	160	9	5.36	94.64
3–12 months	151	28	16.66	83.34
Adults	158	28	17.70	82.30
Year 1977				
0–3 months	77	12	15.60	84.40
3–12 months	65	3	4.61	95.39
Adults	260	17	6.54	93.46
Year 1978				
0–3 months	66	1	1.50	98.05
3–12 months	68	2	3.08	96.92
Adults	303	5	1.65	98.35

Table 6.18. Reproductive Performance of Karakul Sheep

Year	No. of breedable ewes available	No. tupped	Tupping (%)	Conception (%)	Twinning (%)
1976	109	107	98.17	90.65	0.4
1977	159	132	83.02	84.09	1.5
1978	191	174	91.1	89.66	

Table 6.19. Averages for 6 Monthly Greasy Weight (kg) of Adult Karakul Sheep

	March 1976*	August 1976	March 1977	August 1977	March 1978	August 1978
Rams	2.39	1.84	1.227	0.971	0.968	1.141
Ewes	2.62	1.11	0.904	0.628	0.930	0.900

*Exact period of growth was not known

Table 6.20. Pelt Qualilty

Year	No. of lambs evaluated for pelt quality	Percentage of different types of pelts			
		Jacket	Ribbed	Caucasian	Flat
1976	168	44.05	22.62	33.33	–
1977	94	44.80	26.40	28.80	–
1978	111	20.72	18.02	51.35	9.91
1979	155	47.09	4.53	48.38	–

Table 6.21. Mortality in Karakul Sheep Disease-wise and Year-wise

Name of the disease	1976				1977				1978			
	A	W	L	Total	A	W	L	Total	A	W	L	Total
Pneumonia	–	2	–	2(6.8)	2	8	6	16(36.8)	1	–	–	1(2.5)
C.C.P.P	3	2	–	5(17.2)	5	3	2	10(23)	–	–	–	–
Intestinal	–	–	–	–	–	–	–	–	–	–	–	–
Obstruction	1	–	–	1(3.4)	3	1	1	5(11.5)	2	2	–	4(50)
Septicaemia	–	–	–	4	–	–	–	4(9.2)	–	–	–	–
Exterotoxaemia/ toxaemia	1	–	1	2(6.8)	–	1	2	3(3.9)	–	–	–	–
Retention of	1	–	–	1(3.4)	1	1	1	3(6.9)	1	–	–	1(12.5)
Urine/Rupture of bladder-uremia	–	–	–	–	–	–	–	–	–	–	–	–
Gastroenteritis	–	–	–	–	1	–	–	1(2.3)	–	–	–	–
Accidental trauma	–	–	6	6(20.6)	1	–	–	1(2.3)	–	–	–	–
Weakness and/or debility	–	–	3	3(10.3)	–	–	–	–	–	–	–	–
Tympany	5	–	–	5(17.2)	–	–	–	–	–	–	–	–
Pregnancy toxaemia	1	–	–	1(3.4)	–	–	–	–	–	–	–	–
Suppurative arthritis	–	–	1	1(3.4)	–	–	–	–	1	–	–	1(12.5)
Blindness/lumbar paralysis	1	–	–	1(3.4)	–	–	–	–	–	–	11	1(12.5)
Ataxia	1	–	–	1(3.4)	–	–	–	–	–	–	–	–
Grand Total	14	4	11	33	13	14	12	43	5	2	11	8
Flock Strength		304				350				369		
Overall mortality %		9.5				12.3				2.18		

Figures in parenthes indicate percentage mortality. A– Adults (above 12 months), W–Weaners (3–12 months), L– Suckling lambs (0–3 months)

Khumbathang was 10 rams, 50 ewes, 25 male and 29 female lambs. The growth, survivability, wool yield and reproductive performance of Karakul sheep at Khumbathang presented in Tables 6.22 to 6.25 revealed satisfactory performance of these sheep in the cold arid region.

Table 6.22. Average for Body weight (kg) with Range of Karakul Lambs

Year		Birth	6 weeks	6 months	1 year	1.5 years
1976–77	Male (24)	4.44 (3.20–6.30)	10.51 (8.00–18.00)	21.72 (14.00–27.00)	33.76 (23.00–46.00)	37.14 (30.00–7.00)
	Female (29)	4.07 (3.00–5.50)	10.51 (8.20–14.00)	21.71 (14.00–27.80)	33.75 (26.00–41.50)	30.00 (20.00–8.00)
1977–78	Male	4.11 (2.80–7.70)	8.53 (7.00–13.00)	14.24 (8.50–19.00)	–	–
	Female	4.20 (1.75–5.70)	9.00 (7.00–11.50)	13.98 (8.00–18.00)	–	–

Table 6.23. Bodyweight (kg) with Range of Adult Karakul at Intervals

	31.7.1976	1.4.1977	1.1.1978
Rams	45.80 (39.00–52.00)	59.20 (56.00–65.00)	65.31 (56.00–74.50)
Ewes	35.38 (30.00–47.00)	44.48 (39.00–57.00)	51.18 (43.00–64.00)

Table 6.24. Greasy Fleece Production (kg) with Range (1976)

Rams	3.08	(2.11–3.86)
Ewes	2.53	(1.62–3.40)

Table 6.25. Lambing Percentage

Year	Ewes available	Lambs born	Lambing %
1976–77	50	48	96
1977–78	50	62	124

Five Karakul rams were transferred in 1976 from the Division of Carpet Wool and Karakul Pelt Production, Bikaner to main campus at Avikanagar to conduct a small trial on crossing of Karakul with native carpet wool breeds, *viz.* Chokla, Nali, Malpura and Sonadi during February-March, 1976 breeding season. As per the results, the Karakuls combined well with the coarse carpet wool breeds (Malpura and Sonadi) as the crosses with Chokla and Nali were much inferior in pelt quality as compared to the crosses with Malpura and Sonadi. A research

Table 6.26. Distribution of Karakul Pelt types in Malpura, Sonadi and their Crosses with Karakul

Breed/ Breed crosses	Jacket								Ribbed							
	I		II		III		Overall		I		II		III		Overall	
	% age	Fibre Length (mm)	% age	Fibre length (mm)	% age	Fibre length (mm)	% age	Fibre length (mm)	% age	Fibre length	% age (mm)	Fibre length	% age (mm)	Fibre length	% age	Fibre length (mm)
Malpura	1.66 (2)	9.00 (2)	2.50 (3)	8.00 (3)	0.83 (1)	8.00 (1)	5.00 (6)	8.33 (6)	3.32 (41)	8.75 (41)	17.5 (21)	9.95 (21)	19.17 (23)	12.00 (23)	40.00 (48)	10.83 (48)
Sonadi	–	–	0.65 (1)	–	0.65 (1)	14.00 (1)	1.31 (2)	14.00 (1)	–	–	6.58 (10)	12.7 (10)	7.89 (12)	12.33 (12)	14.47 (22)	12.49 (22)
Karakul × Malpura $^1/_2$	17.02 (8)	6.46 (8)	10.63 (5)	10.40 (5)	2.12 (1)	9.00 (1)	29.78 (14)	8.05 (14)	21.28 (10)	7.56 (10)	14.89 (7)	8.74 (7)	–	–	36.17 (17)	7.76 `(17)
Karakul× Sonadi $^1/_2$	16.17 (11)	7.03 (11)	22.06 (15)	8.56 (15)	4.41 (3)	8.33 (3)	42.64 (29)	7.95 (29)	16.17 (11)	7.55 (11)	8.82 (6)	8.33 (6)	7.35 (5)	12.20 (5)	32.34 (22)	8.81 (22)

Breed/ Breed Crosses	Flat								Caucasian								
	I		II		III		Overall		I		II		III		Overall		
	% age	Fibre Length (mm)	% age	Fibre length (mm)	% age	Fibre length (mm)	% age	Fibre length (mm)	% age	Fibre length	% age	Fibre length (mm)	% age	Fibre length (mm)	% age	Fibre length (mm)	No class
Malpura	3.32 (4)	10.25 (4)	10.83 (13)	9.85 (13)	14.17 (17)	11.9 (17)	28.33 (34)	10.97 (34)	–	–	0.83 (1)	11.00 (1)	1.66 (2)	15.5 (2)	2.50 (3)	14.0 (3)	24.17 (29)
Sonadi	3.29 (5)	9.6 (5)	9.87 (15)	12.0 (15)	10.92 (16)	14.25 (16)	23.68 (36)	12.64 (36)	–	–	1.31 (2)	12.00 (2)	1.31 (2)	12.50 (2)	2.62 (4)	12.25 (4)	57.92 (88)
Karakul × Malpura $^1/_2$	10.64 (5)	8.40 (5)	12.76 (6)	9.33 (6)	4.25 (2)	9.00 (2)	27.66 (13)	8.92 (13)	4.25 (2)	10.00 (2)	2.12 (1)	9.00 (1)	–	–	6.38 (3)	9.67 `(3)	–
Karakul × Sonad $^1/_2$	19.11 (13)	7.99 (13)	7.35 (5)	9.20 (5)	1.47 (1)	10.00 (1)	27.94 (19)	8.42 (19)	5.88 (4)	8.25 (4)	5.88 (4)	10.25 (4)	–	–	11.76 (8)	9.37 (8)	–

component "Exploratory trials on crossbreeding Karakul with indigenous coarse carpet wool breeds" was included as a regular activity of CSWRI. Three such breeding trials were conducted from 1977–79 to confirm the earlier results. As such, the Malpura and Sonadi ewes available under mutton component of the AICRP on sheep breeding at CSWRI, Avikanagar were bred to Karakul rams during their last breeding season's oestrus cycle every year. The relative performance of resultant half-bred crosses was compared with their purebred natives produced under the mutton component in a routine manner. During the year 1978, 5 more Karakul rams were added, two out of the previously transferred had died due to specific diseases. For number of Malpura and Sonadi put to each trial being small, the results on pelt quality (using the same standards as for Karakul pelts) of the two coarse carpet wool breeds, *viz.* Malpura and Sonadi and their half breds with Karakul were pooled and are presented in Table 6.26.

It was seen that of the two native breeds, Malpura lambs produced better pelts as compared to Sonadi. While the ornament in the native breeds was restricted to only part of the skin, it was spread over whole of the skin in Karakuls. For Malpura, 5.00, 40.00, 28.33 and 2.50% pelts were of jacket, ribbed, flat and caucasian types as compared to 1.31, 14.47, 23.68 and 2.62% respectively in Sonadi. Although in general the halfbreds produced better types of pelts than their purebred native contemporaries, the Sonadi Karakul half breds produced even better types of pelts than the Malpura halfbreds. The jacket type pelts increased from 1.31% in Sonadi to 42.64% in Karakul × Sonadi halfbreds, whereas in the Malpura halfbreds, the jacket type pelts were 29.78% as against 5% in Malpura. There was also a decrease in average fibre length in half breds. The spread of ornament increased almost all over the skin and was quite comparable to the Karakul pelts. There was also improvement in lustre of half breds. The colour varied from patches of black or sur to complete black or sur, offering a future potential for number of colour lines.

CHAPTER 7

REPRODUCTION

Reproduction, a normal physiological process, is one, of the most remarkable and important phenomena for the sheep husbandry and breeding. Regular and repeated reproduction is the basis of the income from sheep. This is true not only of sheep but of other livestock; indeed, it is true of the whole realm of agriculture. Reproduction is not only the source of continuing income; it is upon this function that the whole science of genetics is based. As new animals are produced, it is possible to influence the fundamental characteristics of some of the members of the species. Influential as the reproductive function is, it is in turn influenced by other factors, particularly age, health, and nutrition. Some other features of environment may also be important in relation to reproduction.

There are many erroneous notions regarding the processes of reproduction. The breeder should have a sufficiently intimate knowledge of the anatomy and functions of the male and female reproductive systems to enable them to understand the reproductive processes in the management of a flock. Without such knowledge and understanding, there are no basic principles of guidance and no modern basis for the development of scientifically sound breeding programs.

Sheep in tropics generally breed throughout the year. Rams of indigenous breeds donate good quality semen under proper management conditions throughout the year. However, the rams of temperate breeds if not protected from high temperature, high humidity and high solar radiation will not produce good quality semen during hot dry and hot humid seasons. Females of tropical breeds cycle throughout the year unlike the temperate breeds which are affected by hour of day light and breed with declining day length. They come in heat in autumn from August to November, although some may breed upto February. Indigenous breeds usually breed during nutritionally better time of year, *i.e.* July to August, or immediately after the onset of the monsoon. A small percentage of them also breed during March–April when they are stubble grazed on harvested fields and supplemented with *Acacia* and *Prosopis* pods, and thus get flushed. Sheep breed in different seasons depending upon the availability of feed. To some extent breeding season is controlled by the sheep breeder on the basis of availability of feed resources and physical environmental conditions prevailing both at breeding and lambing, though the consideration would be more in case of lambing as extremes of weather will more seriously affect lamb survival. The breeding should, therefore, be arranged in a manner that the lambs are dropped when plenty of vegetation is available and weather conditions are mild.

7.1 Male Reproduction

Although there are individual exceptions, most males in the exotic breeds reach the stage of development at an age of approximately 4–6 or 7 months, when reproduction is possible. The sex organs of males are apparently fully developed by that time, and the so called secondary characteristics which are dependent upon the presence and functioning of these organs are very noticeable.

Because the reproductive cells of the male, the sperm or spermatozoa, are formed in the testicles, the testicles are generally considered the primary organs of reproduction of the male. In addition to the formation of sperm cells, a testicular secretion, testosterone when absorbed by the body causes the development of the secondary sexual characteristics. When these features are very marked in an individual, he is described as masculine. Masculinity is in no way associated with the transmission of any features such as colour, length of fleece, or other factors of the ram to his offspring. The well developed, masculine features merely indicate an abundance of the hormone which stimulates the development of the features. The removal of the testicles, and hence the removal of the source of the hormone, results in stopping the development of the masculine characters. This is the reason why wethers differ so much in appearance from rams.

The testicles are normally carried outside the body cavity in a sac called the scrotum. The normal temperature of the testicles is below that of the body. One of the main functions of the scrotum is to regulate the temperature of these organs. This explains the lack of fertility in rams if the testicles are not descended into the scrotal sac. It also explains why some rams that have been heavily fed and therefore may have had an increased body temperature accompanied by some increase in scrotal temperature, may be sterile. Breeding rams that have been shorn or heavily fed may hasten the recovery of the reproductive function of the testicles.

The production of the hormone by the testicles does not seem to be influenced greatly by temperature, and hence a cryptorchid or ridgling may appear thoroughly masculine. Rams with one testicle retained within the body cavity should not be used for breeding, as there is a tendency for the condition to occur in succeeding generations, although it will not appear in all cases.

The testicles are covered with a fibrous capsule, the tunica albuginea. Within this capsule are blood vessels, nerves and connective interstitial, and spermatogenic tissue. The connective tissue divides the interior of the testicles into very small sections or compartments. In each of these sections are the seminiferous tubules within which is the secreting tissue where the billions of sperm cells are developed. It is estimated that these tubules are so numerous that their aggregate length would be many thousand feet. It will be necessary to consider cell formation more in detail in studying the inheritance of characters, but, for the purpose of reproduction, it is sufficient to remember that the contribution of the male to the new offspring originates in the seminiferous tubules. There, spermatogonia or sperm mother cells enlarge and divide. Thus, two cells arise and both of these divide again. During these processes the cells move toward the end of the tubule. Nourishment is provided for the completion of the development. When this occurs, spermatozoa or sperm cells is known to be produced.

A normal sperm cell consists of a head, body, and tail. The formation of spermatozoa is a process which continues from the time it first starts until the end of life in the case of most sheep. The sperm as found in the testicles are not very active. They become active later when they come into contact with some secretions of other glands. Besides carrying the factors which the offspring will inherit from the sire, the spermatozoa initiate cell division on the part of the egg after its fertilization.

There are other important parts of the male reproductive system. Ducts, called efferent ducts or vasa efferentia, formed by the joining of many of the tubules, lead to the epididymis which is essentially a large tube that is found on the outside of the testicles and leads from the top of the testes to the base. The epididymis provides a passageway and temporary storage for the sperm. Ducts leading from the epididymis form the vasa deferentia. These tubes pass up through the inguinal canal the small opening into the abdominal cavity and connect with the urethra, which leading through the penis, completes the passage from the testes to the exterior. In rams there is a small filiform appendage at the end of the penis. This is not, however, a vital matter in reproduction. At various points along this passageway are accessory glands whose chief function seems to be the secretion of fluids which serve as media for the transference and perhaps nourishment or stimulation of the spermatozoa. Alongside of the ducts leading from the testes to the abdominal cavity are blood vessels, nerves, and supporting tissues. These are cut or broken when castration is performed. Rupture and separation of the main blood vessel so that blood does not reach the testicle is the basis of bloodless methods of castration. Under such conditions the testes shrink or become atrophied.

Not all of the sperm that are formed reach maturity. In many individual there are thousands of imperfectly formed sperm, and many of the sperm have no motility. Infections and above normal temperatures are two factors that are known to affect the sperm cells. Any males that have a large percentage of abnormal or non-motile sperm are likely to be unreliable as breeders, and in severe cases they are sterile. Microscopic examinations of the semen are needed to determine whether such conditions exist.

The semen of rams emitted at one time is from 0.5 to 2.0 cm^3 in volume. Very frequent service may reduce the volume as well as the number of sperm that the semen contains. The number of sperm in a mm^3 of semen ranges from 500,000 to 6,000,000 or more. The average number per cubic millimeter is probably about 1,000,000 sperm. Since only one sperm is required to fertilize each egg cell produced by the female, the lavish production of sperm is apparent.

7.2 Types of Breeding

Sheep are bred either by natural mating or through artificial breeding. Rams of the indigenous breeds donate good quality semen under proper management conditions throughout the year. However the rams of temperate breeds if not protected from high temperature, high humidity and high solar radiation, will not produce good quality semen during hot dry and hot humid seasons. Females of tropics breed throughout the year. Temperate breeds which are affected by hours of day light, breed with declining day length. They come in heat in autumn from August to November, although some may breed upto February.

7.2.1 Natural breeding

The natural breeding is done either by flock mating, pen mating and hand mating. In flock mating system 8 sheep can be bred naturally or through artificial insemination. The natural breeding is done by either flock mating, pen mating or hand mating. In flock-mating system, breeding rams are usually turned out in the flock during the mating season at the rate of 2 to 3% of the ewes all through day and night. It is most widely practised in the flocks of all farmers. Semi-flock breeding or pen breeding is done to conserve the energy of rams and give them rest. In this, the rams are turned out for service with the flock in the pen during night, and confined and stall-fed or grazed separately during the day time. Hand mating is practised when exotic purebred sires are used, or when it is considered desirable to extend the services of ram over much larger flocks. Sheep in heat do not manifest behavioural symptoms. Hence, the teaser rams are employed for detecting the ewes in heat. These ewes are then taken out of the flocks and bred to the designated sire of the flock. In pen and hand-mating systems and when teaser rams are used for heat detection, some dye mixed in grease or simple linseed oil is smeared on the brisket of the ram. This makes it possible to record the date when the ewe is bred and also to remove them from the breeding flock. The colour of the dye should be changed every 16 to 18 days so that the repeaters can be discovered, if the bred ewes have not been removed from the flock. This is termed as 'marking of ewes' by breeding ram and marked ewes are considered as bred. Breeding rams are usually turned out in the flock during the mating season at the rate of 2 to 3% of the ewes all through day and night. It is most widely practised in the flocks of all farmers.

7.2.2 Artificial Insemination

Four to eight ewes may be inseminated from one service of a ram although as many as 30–40 have been reported, 0.1–0.2 cc of semen is sufficient for the purpose when there is a heavy concentration of sperm in the semen. Great numbers of sperm are of course needed as they are microscopic in size, and the reproductive tract of the female through which they must migrate to come in contact with the egg (also microscopic), is very large indeed in relation to the size of the reproductive cells. Apparently, most sperm do not survive more than 18 to 24 hr in the reproductive tract of the ewe. Russian research workers have reported impregnating from 300–400 ewes in one season with the semen collected from a single ram. They also reported that 90% of the ewes so inseminated became pregnant. This is a far greater number than could be bred to one ram using natural methods. From 40–60 ewes, is generally considered reasonable for a ram in one season.

For the purpose of artificial insemination, artificial vaginas are used for the collection of the semen, although it may be recovered in some quantity from the vagina of an ewe. Semen can be preserved for several days with suitable media and low temperature and has been transported long distances and used successfully in experimental tests. The semen is diluted for use and is injected into the vagina or deposited at the base of the cervix through a glass or plastic tube. Two inseminations at about a 12 hr interval, beginning soon after the onset of heat, are often recommended.

Although little use has been made of artificial insemination in sheep except in an experimental way, some of the advantages and disadvantages are known. The chief advantages are: the usefulness of a superior sire may be extended; the number of rams needed for a large flock is reduced, and a flock of considerably greater uniformity should result. Disadvantages include the need for extra equipment and labour, some of which must be skilled in the work, the need for one or more "teasers" to detect which ewes are in oestrus, and the frequent handling of the ewes during the breeding season.

7.2.2.1 *Puberty*

Puberty in the males is marked by production of viable sperms and a desire for mating influenced by hormone testosterone. The age at puberty vary from 132 days for Tobasco (Pelibucy) × Dorset lambs on the high plane of nutrition in Mexico (Valecia *et al.*, 1977) to as old as 738 days for $^{3}/_{4}$Rambouillet $^{1}/_{4}$Malpura lambs in Rajasthan (Tiwari and Sahni, 1981). The quantity and quality of ejaculates collected from puberscent rams are poor with a high incidence of dead and deformed spermatozoa (El-Wishy, 1974). Ram lambs reach puberty about two months earlier than ewe lambs in each breed. Age at puberty depends upon the nutrition of lamb. Some breeders allow gimmers to mate immediately after they attain puberty. There is no evidence that early first lambing results in diminished reproductive capacity (Wilson and Durkin, 1983). The hormonal stimulation of lambs is of practical use in the tropics. The gonads of prepubertal sheep respond to exogenous hormones, but animals stimulated in this way do not commence spontaneous oestrus cycle (Hunter, 1980).

7.2.2.2 *Ram semen collection*

Ram Semen can be collected by making use of (i) An Anoestrus ewe (ii) Artificial vagina (iii) by Electrical stimulation.

i) *Anoestrus ewe*

In this method the rams are trained to serve anoestrus ewe and after the service the semen is aspirated from the vagina. It is unsafe to use the semen so collected for fear of contamination.

ii) *Artificial vagina (A. V.)*

The pattern of A.V. is similar to that in the cattle, except that the A.V. is short. The outer hard rubber casing is 15.0 × 3.75 cm. The A.V. is prepared in the usual way by pouring hot water (45 °C) through the inlet on the outer jacket. Necessary air is blown to provide cushioning effect and air valve tightened. Both the temperature and pressure are important. When the A.V. is ready, the inner temperature should be 40 °C.

Kaushish and Sahni (1976) considered it advisable to warm the A.V. first with hot water (47–50 °C) initially and then pour out and refill the A.V. with hot water at 45–47 °C, depending upon the season, in order to maintain the inner tube temperature between 40– 41 °C at the time of collection.

A rubber cone is attached to one end of the A.V. at the end of which a graduated 5 ml test tube is fitted to serve as a receptacle for the semen. The open end of the A.V. should be properly lubricated preferably with liquid paraffin. A glass collection cup may be directly fixed to the other end of the A.V. to obtain collection. Rams are very sensitive to temperature variations and therefore, adjusting, proper temperature in the A.V. is important. Once the rams are exposed either to an A.V. which is too hot or cold, they may refuse to serve on subsequent occasions. A freshly prepared artificial vagina should be used for each collection from a ram to avoid contamination. It should be properly sterilized after the use and kept in dry sterile chamber ready for use.

iii) *Electrical stimulation*

The equipment for collection of semen from the rams by electrical stimulation as modified by Dziuk *et al.* (1954) is said to be satisfactory. The power is obtained from 110 volt battery with 60 cycles which is reduced by transformer to 30 volt. This isolates the electrodes from the line of source. The probe for insertion into rectum is made up of rubber hose about $^3/_4$ inch diameter with shoulder rings one inch apart. Alternate rings or electrodes are connected to provide the difference in electrical potential between adjacent electrodes. With this equipment the collection can be made in standing position. However, at times it is necessary to restrain the animal in an extended position on a table. The rectal probe is lubricated, moistened and inserted so that last ring is located just inside the rectum. The penis may be gently pulled out before stimulation. But on many occasions it generally protrudes with stimulation. The end of the penis should be held in a glass vial or a test tube. During the process the voltage is gradually increased and then reduced to zero during a 5 second period followed by an equal period of rest. Stimulus is gradually increased until the semen is ejaculated in the order of 2, 5, and 8 volts. The number of stimuli necessary to produce semen is variable.

Collection of semen by the A.V. is considered as the best method. Collection by electrical stimulation has an advantage, that training of the rams is not necessary.

When a large number of rams are to be tested for semen quality, collection by electrical stimulation is advantageous. The major disadvantage in the electrical stimulation is likely the injury caused to the animal which can be minimized by the use of the bipolar electrode. Other disadvantages are: slightly lowered spermatozoal density; the possibility of contamination of semen by urine; difficulty in obtaining second collection within short time if the first one is not satisfactory and the refractions on the part of some rams.

Experience shows that by collection in artificial vagina the volume of one ejaculate is smaller but denser as compared to an collection by electrical method. High quality of ejaculates is the main advantage in collection of semen by artificial vagina.

7.2.2.3 *Semen volume*

The number of collections to be obtained in a day varies from ram to ram. From most of the rams semen can be collected three times a day producing average ejaculates of 1.0 ml, sperm concentration of 35×10^8 per ml and live counts 85–90% Semen could successively be collected up to 5 times daily for 17 days from mature rams without fall in semen quality (Miller, 1975). He further reported $15{,}000 \times 10^6$ sperm per day from 4 rams in three collections (Table 7.1).

Table 7.1. Results of Rams Semen Examined on day 6th and 15th day of Semen Collected

	Vol. (ml)			Sperm No. $\times 10^8$			Live (%)		
	No. 1	No. 2	No. 3	No. 1	No. 2	No. 3	No. 1	No. 2	No. 3
					Day-6				
Ram A	0.7	1.4	0.8	39	45	45	90	92	88
Ram B	1.0	1.0	1.1	41	31	31	90	96	88
Ram C	0.8	0.6	0.5	34	45	45	82	95	95
Ram D	0.7	1.0	0.4	54	43	36	95	95	95
				Day-15					
Ram A	0.6	1.0	0.8	40	40	35	90	90	90
Ram B	1.0	1.0	1.0	40	35	30	90	95	90
Ram C	0.0	0.6	0.4	32	40	45	80	90	95
Ram D	0.7	0.5	0.7	45	40	35	95	95	90

7.2.2.4 *Semen quality traits*

Semen quality varies with frequency of ejaculate quality. Ejaculates to the extent of even 20 to 30 a day followed by successive collections of equal number of ejaculates the next day have been obtained. Collections to the extent of 2 to 6 per day also have been obtained over a period of a week or 10 days (Chang, 1945). Salamon and Robinson (1962) observed that season and nutrition did not measurably affect frequency of service nor there was any apparent effect of season on volume, density or sperm count. However, marked effect was seen on these characters due to high protein ration. Satisfactory fertility can be maintained even with frequent ejaculations if the total number of spermatozoa is more than 80×10^6. Fall in lambing rate is observed when the sperm count is reduced.

i) *Colour*

The colour of ram semen is creamy. Variable grades are noticed according to frequency, season, age, rest and among individuals.

ii) *Consistency*

The normal consistency is thick which is however, variable. It may be classified as dense, medium, thin and watery. It appears that there is very little direct evidence about the value of density and motility determinations in rams' semen.

iii) *Volume*

Volume obtained on an average from 0.5 to 2.0 ml. Wiggins *et al.* (1953) observed that, volume of ejaculates gives a low but significant positive correlation with fertility.

iv) *Motility*

When a fresh drop of ram semen is examined under the microscope at body temperature, very brisk dark grey whirls are noticed.

The swirling motion is very rapid and dense in appearance. The grading is done as for bull semen. A drop of semen is diluted in Ringer's or Sodium citrate solution and examined under high power. Gradation of motility is done as for bull semen.

v) *pH*

The pH of ram semen varies between 5.9 and 7.3 with an average of 6.4 (Emmens, 1959).

vi) *Sperm concentration*

The average sperm concentration in a single ejaculate of ram semen contains on an average 3 to 5 billion spermatozoa per ml. In a good sample about 90% of the sperms are alive. Ram semen with high sperm concentration is slightly acidic in reaction, while with low sperm concentration it is slightly alkaline. Semen with alkaline reaction is often associated with poor quality and low fertility (Anderson, 1945).

vii) *Morphology*

A normal ejaculate may contain 5 to 15% of abnormal spermatozoa. A greater proportion of abnormal spermatozoa may lead to poor fertility. The most common form of the abnormal type are tailless and deformed. Other morphological abnormalities found are tapering heads, enlarged middle pieces, coiled or bent tails. It is usually observed that head abnormalities cause low fertility. The types of morphological abnormalities are of more importance than the total sperm count. High percentage of preforms or narrow heads are particularly associated with poor fertility.

Ram semen with sperm concentration of less than one billion per ml with a pH of 8.00 or higher, with slow sluggish motility and with more than 25% abnormal type and more than 1% abnormal sperm heads are of doubtful fertility and should not, therefore be used for insemination.

The percentage of live sperm so may be estimated by the vital stains (Lasley *et al.*, 1942) or by direct counting. A stain containing Congored and Nigrosin gives satisfactory results for ram semen and it is preferred to Eosin and Nigrosin stain (Blackshaw, 1958). The percentage of live sperms is not significantly correlated with fertility in a ram but the percentage of live normal spermatozoa has a significant correlation (Wiggins *et al.*, 1953).

viii) *Biochemical parameters of ram semen*

In the ram semen, $^1/_3$ portion is made up of spermatozoa and $^2/_3$ by the seminal plasma. There is marked difference in composition between the sperm and the seminal plasma. Nucleic acid is confined to spermatozoa. The spermatozoa contain more of iron, zinc, copper, hematin chiefly as cytochrome and plasmalogen than the seminal plasma (Table 7.2). Acid soluble phosphorus compounds are present in the sperm and also in the seminal plasma. Fructose and citric acid are present in large concentrations in ram semen and these are derived from the seminal vesicle secretions (Mann, 1964).

Table 7.2. Ram Semen Components (Mann, 1964)

Attributes	mg/100ml/g
Dry weight	14,820
Chloride	87
Sodium	103
Potassium	71
Calcium	9
Magnesium	3
Inorganic phosphorus	12
Total nitrogen	875
Non-protein nitrogen	57
Urea	44
Ammonia	2
Fructose	247
Laetic acid	36
Citric acid	137
Ascorbic acid	5

7.2.2.5 *Factors affecting semen quality*

i) *Breed*

Gunn *et al.* (1942) observed that British breeds in Australia are susceptible to the high environmental temperatures similar to Merino. However, the period of minimum spermatogesis did not correspond with the high environmental temperature of mid and late summers but were observed in spring. Under Indian conditions autumn appears to be the season characterized with poor quality of ejaculates and summer provides most superior quality of semen, as has been reported in the indigenous breeds like Bikaneri and Mandya (Shukla and Bhattacharya, 1952; Sahni and Roy, 1969; Deshpande, 1970). This is in contradiction to the general trend in temperate breeds reared under temperate or tropical conditions.

Singh and Roy (1963) in their studies on Romney Marsh rams (U.P) observed that similar to Bikaneri and Mandya breeds, the autumn seasons in characterize by poor quality semen but unlike these breeds which produce best semen in summer, it is in winter that the semen of Romney Marsh rams has the best quality. As against this, Emmens and Robinson (1962) observed that Romney Marsh rams give most satisfactory semen in autumn in Australia. It is due to such differences in response of different breeds to different seasons that Webster (1952), Mies Filho and De Almeida Ramos (1956b), suggested that extreme care is required in choosing rams of right breed for a particular environment. It is not clear whether such limited data is adequate to prove breeds and environment interactions *i.e.* differences in sensitivity of breeds to environmental conditions, but gives a sufficient caution before any large scale import of exotic sires is made with an aim to improve the local breeds.

Tiwari and Honmode (1968) have observed that good quality semen could be obtained from cross bred rams (Rambouillet × Malpura and Rambouillet × Chokla) at the age of 10th month under semi-arid conditions. Rambouillet rams at that age showed very poor sexual desire on oestrus ewes.

Honmode and Tiwari (1968) have reported that crossbred rams (Rambouillet × Malpura and Rambouillet × Chokla) produce greater quantity and higher concentration of spermatozoa as compared to Malpura and Chokla rams. A gross deterioration in semen quality of Rambouillet rams was obtained in the month of June.

Sahni and Roy (1969) have observed that the average semen volume of Bikaneri rams was higher when compared to Mandya rams in all the season. Excellent motility, high percent of live spermatozoa and higher sperm concentration was found in both the breeds through out the year although the statistical difference between the season were highly significant with respect to motility and spermatozoa concentration per ml.

Sharma *et al.* (1969) collected semen from adult Bikaneri and Mandya rams at various collection frequencies and observed that effect of continuous collection *i.e.* once in a day upto 20 or 40 days on semen production did not bring any change. Observation on 1, 2 and 3 collection frequencies per day in 3 successive weeks from adult Bikaneri rams revealed that highly significant difference between weeks and rams in all the seminal attributes examined (volume, motility, concentration/ml, concentration/ejaculate and live spermatozoa). Although with increased frequencies, sperm concentration in successive ejaculates declined, there was a net increase in either attributes when compared to only one collection per day. It was noted that there was no substantial advantage in increasing the frequency of collection from 2 and 3 per day. Based on the requirement of 100 M spermatozoa for obtaining optimum fertility in ewe, it has been concluded that an average 700–800 ewes can be inseminated when semen is collected once daily for a period of 20 days from one ram.

Tiwari and Sahni (1970) have examined the various seminal attributes of Rambouillet and cross-bred rams through out the year. It was concluded that Rambouillet rams did not donate good quality semen during hotter part of the year. Semen of excellent quality was obtained from the Rambouillet rams during the months from October to February when the indigenous sheep are not bred normally. Semen quality was moderate in April and it started deteriorating from May onward till it was watery with almost azoospermic condition in the month of June to August. The semen quality did not show any deterioration due to season in half-bred (50%) rams. No azoospermic was noted in 75% corss-bred rams; however, a slight deterioration in semen quality was observed during the hotter part of the year.

The studies on extension rates indicated that Rambouillet semen froze well to give optimum level of cryo-survival and motility, whereas $^3/_4$ exotic semen showed very poor post thaw survival in both the concentrations and replicates (Mathur *et al.,* 1989).

ii) *Season*

Marked seasonal variations in density, percentage of abnormal sperms and glycolysis in semen of rams was observed by Comstock *et al.* (1945). Poor quality of semen was recorded during summer and early autumn months *i.e.* July–September. Bell (1945) observed slight seasonal variations in Rambouillet rams in which semen quality was poor in spring (March–April). It improved in summer (May–July). Koger (1951) found that in autumn, ram semen has higher motility and longer viability on storage than in the semen collected in spring. Maqsood (1951) on his observations on young Suffolk rams reported that during the non-breeding season, there is a marked decrease in the libido, volume and spermatozoal density and increase in the number of abnormal sperms. Poor libido is believed to be due to interference with the production of androgens by the interstitial cells. Histological examination of testes of the young ram carried during spring and summer revealed arrested spermatogenesis in the majority of the seminiferous tubules. Waston (1952) observed that Merino rams show higher fertility in the autumn but there is uncertainty in the spring. Wiggins *et al.* (1953) found a significant correlation between libido and fertility in the range rams. Volcani (1953) observed seasonal variations in spermatogenesis in Awassi sheep in Israel. In July and August when days are long and hot, histological examination of testes in rams indicated active spermatogenesis. Whereas from September–December, degenerative changes were noticeable. He considered that the temperature was not the causative factor and implicated availability of grazing facility.

Shukla and Bhattacharya (1952) observed that the reaction time, differed significantly between rams, but there was no seasonal trend and had no relationship to quality of semen. Significant seasonal variations in the volume of semen, pH, sperm concentration, motility and percentage of abnormal spermatozoa were observed. Variations between months and within season were also significant. The quality of semen was poorest in autumn and best in spring.

Hafez *et al.* (1955) observed that the quality of semen of Ausimi and Rahmani rams in Egypt was very good during spring and autumn than in summer and winter. Hulet *et al.* (1956) found that ram fertility was 42.8% before September due to poor quality of semen and thereafter the fertility improved to 67.6%.

Sahni and Roy (1967b) observed in their studies on Romney Marsh rams brought in Uttar Pradesh (India) for improving the indigenous sheep, that in the month of January, the abnormal sperm count was 3.8% but during the months of February, as the climate begins to get warmer, the percentage of abnormal spermatozoa increased to 27.5%. The signs of seminal degeneration gradually developed, thereafter as the summer advanced the initial motility became poor, sperm concentration decreased, more than 90% of the sperms were found dead and the number of abnormal spermatozoa also increased. In the month of June, the wave motion was completely absent, during the months from July to September, there was complete cessation of spermatogenic activity. The semen was devoid of spermatozoa, ejaculates were watery and resembled those obtained from vasectomized rams. The restoration of normal spermatogenesis started from the end of October. There was remarkable improvement in November. The average initial motility was 3.2; the percentage of abnormal spermatozoa came down to 10.6% and the semen quality was excellent in December and January as in indigenous rams. The semen picture was normal and the abnormal sperm count was only 3–4%. These variations in semen quality are attributed to the variations in ambient temperature. It is evident that there is a periodic rhythm for spermatogenesis. This pattern is influenced mainly by high temperature leading to deteriorating changes in semen quality which usually start from the mid-summer. The semen quality gradually improves by mid-autumn. It is not clearly known to what extent the factors such as shade available, winds, amount of wool on rams, availability of greens, feed, short term heat waves present in some summers and absent in others and frequency of service and/or collections, influence the semen production.

Decline in semen quality should invariably be accompanied by a gross reduction in the initial motility, sperm count and increase in the proportion of abnormal spermatozoa. It is on the basis of such changes in semen picture of rams in areas away from the equator that a gross decline in semen quality has been reported by Comstock *et al.* (1945), Bogart and Mayer (1946) and Cupps *et al.* (1960). Despande (1970) reported that in Bannur × Somali rams, the best semen samples were obtained during spring and winter.

Sarkar *et al.* (1971) investigated the seasonal changes in semen characteristics of Magra rams. Semen samples were collected from each experimental animal on every third day using transistorised electro-ejaculators. The average volume of semen was found to be 0.71 ml in winter. This value was significantly higher than the values obtained in spring. Average values for motility, mass activity and sperm concentration was 70.27%, 3.66 and 1.783 M/ml respectively but no significant differences were observed.

Sahni and Roy (1972) reported that summer had an adverse effect on the production of Corrideale semen but the season had no effect on the semen production of half-bred rams under tropical conditions.

Sahni and Roy (1972) observed that the incidence of abnormal spermatozoa is negligible in the native and cross-breds during different seasons. A very high percentage of abnormal spermatozoa was observed in Corriedale and Romney Marsh breeds during summer season.

Studies show that shifting indigenous rams even from far distant areas to other tracts which fall within the broad range of tropic and subtropics do not impair the reproductive efficiency of Mandya rams which were brought to north India from South, an approximate distance of about 1500 km, did not alter their reproductive efficiency. The variations in temperature or duration of daylight had no effect on the production of quality semen throughout the year (Sahni and Roy, 1969).

iii) *Day light*

Ortavant (1958) studied interaction between temperature and photoperiodism on spermatozoal production. His P-32 tracer studies in spermatogenesis show that the rate of division of primary spermatocytes and time of subsequent maturation are relatively unaffected by the photoperiodic environment but the number of spermatids which survive the complete maturation process are affected. The failure rate was observed to be high under conditions of increasing daily illumination. He observed that the effect of high temperature in the spring and early summer months superimposed, on this basis photoperiodic effect is to destroy a high proportion of the relatively few spermatozoa which have survived the maturation process. Deshpande (1970) observed that during spring and winter, the day length varied between 11.02 and 12.56 hr, whereas, the actual day light hours varied from 7.7 to 9.8 hr and the semen quality was best during this season.

iv) *Altitude*

High altitude and poor nutrition are deleterious to fertility in rams (Easley, 1951).

v) *Shelter/housing*

Tiwari and Sahni (1970) have reported that there was no beneficial effect of underground housing on semen production of Rambouillet rams during summer months. It was indicated that unless the maximum environmental temperature was brought down to 70–80 °F artificially, there was no scope for obtaining good quality of semen from exotic rams with conventional sheds during hotter part of the year under semi-arid conditions.

Tiwari and Sahni (1972) have observed that the provision of shelter/housing has no beneficial effect on sex-libido and seminal attributes in cross-bred hogget males.

vi) *Disease*

Any infection such as foot rot or abscess which might cause general febrile condition, will result in seminal degeneration (Gunn *et al.,* 1942). Morley (1948) reported that scrotal fly strike may not necessarily impair fertility. Engela (1948) reported incidence of epididymitis, varicocele, scrotal tumors and other diseases of male genitalia which cause semen degeneration.

Andreevskii (1940) observed that the main cause of semen abnormality in 40,976 adult rams was chronic orchiepididymitis. Of these about 15% of rams showed morbid and degenerative changes in the testes, mostly in epididymis.

vii) *Shock*

Webster (1952) observed that in New Zealand, a cold southern wind during night rendered 18 rams infertile over night and the condition persisted for remaining period of the breeding season. He also observed from the past record that abnormally low lambing percentage was related to a sudden exposure to stormy weather during the mating season.

Dun (1956) recorded temporary infertility in rams, due to damp and marshy conditions which led to seminal degeneration associated with flabbiness of the testis. The condition was accompanied by warm humid weather and mosquito plague, shortly before mating.

Sahni and Roy (1972) reported that since the differences in semen characters of indigenous breeds of sheep in different seasons were small, the cold shock test at 5 °C or 15 °C could not be considered of any assistance as a method of testing the semen quality.

viii) *Feeding*

Honmode *et al.* (1971) have examined sexual desire and various seminal attributes of Chokla rams maintained on four different feeding schedule. It was found that sexual desire and all the seminal attributes studied was higher in group receiving *cenchrus* + Concentrate mixture and *cenchrus* + cowpea hay than those receiving only *cenchrus* + Pala leaves and *cenchrus* + guar follicles.

Feeding thyroprotein has not produced beneficial results in rectifying the season effect on rams semen.

It is reported that semen from rams fed with cotton seeds and phospholipids, freeze better. Similarly the level of certain minerals, vitamins and hormones also do play role on the freezing of the spermatozoa for successful revival. Certain workers are of the opinion that certain chemicals in the seminal plasma are harmful in the process of freezing and hence tried to freeze the washed spermatozoa in artificially prepared extenders but the results are equally disappointing. Feeding, housing and other managemental factors like vaccination, dipping, exercise, shearing, incidence of disease like bluetongue, sheep pox, affect the semen quality and the composition of seminal plasma and in turn the freezeability and fertility. Avoiding sudden change of temperature before equilibration, unnecessary exposure to oxygen and bright light, adhering to strict asepsis at every step using triple glass distilled deionized water, use of pure and nontoxic chemicals are some factors which deserve attention in the venture of deep freezing of ram semen.

ix) *Frequency of collection*

Honmode and Tiwari (1971) have observed a significant reduction in semen, volume in successive semen collection period of 50, 80 and 110 days in Malpura and Chokla breed. Sperm concentration decreased with increase in frequency of semen collection.

7.2.2.6 *Ram semen dilution*

The egg yolk diluent developed by Lardy and Philips (1939) is found satisfactory for ram semen. For this purpose equal volumes of fresh egg yolk and sterilized phosphate buffer (0.2 g dihydrogen potassium phosphate and 2.0 g disodium hydrogen phosphate per 100 ml of pyrogen free triple distilled water) are mixed thoroughly. The pH of the mixture is close to 6.75. If is not, it should be adjusted by the addition of sterile 15 M citrate buffer or bicarbonate can be used to replace the phosphate buffer without effect on the glycolysis (Moore, 1949). The replacement of the citrate buffer with glycine has improved the survival time of ram semen (Ahmed, 1955, Roy *et al.*, 1956 and Skolosvkaya *et al.*, 1956).

Milk appears to be the most useful dilutor for ram semen (Istvan, 1956; Foilmom *et al*., 1956; Macpherson, 1957 and Hill *et al.*, 1958). Miller (1975) reported the use of diluents like egg yolk citrate, heated milk from ewes, cow skim milk and accessory secretions from vasectomized rams. Of the above diluents, cow milk or skim milk heated at 92°C for 10 minutes was found to give good results with dilution rate of 1:4.

Sahni (1967) recommended that at least 50 M spermatozoa per dose are necessary for conception in case the diluted semen is deposited in the cervix.

7.2.2.7 *Preservation of semen*

Successful storage with optimum fertility of ram semen is a world wide problem. Researches have been carried out to evolve a suitable dilutor for ram semen in this country, but the results obtained are not very encouraging. The glycine egg yolk dilutor has been found to improve the spermatozoal survival of ram semen as compared to egg yolk citrate (EYC), later Roy *et al.* (1962) observed successful storage of ram semen for a longer period in glycine yolk diluent in the 1:10 to 1:20 semen dilutions.

Although ram semen can be stored for few days but it is desirable to use fresh semen because fertility is reduced rapidly with storage. Immediately after collection, the semen is diluted with extenders like fresh cow milk boiled and cooled for utilizing the semen within 4 hr of its collection. Similarly freshly prepared egg yolk citrate glucose (EYCG) can be used for diluting semen for immediate use or any time upto 24 hr. EYCG contains 0.8g glucose and 20% egg yolk in 2.879% solution of sodium citrate. A new diluter named Egg yolk McIllvaine glucose (EYMG) evolved at CSWRI gives better preservation and fertility results. It contains 20 ml egg yolk, 0.8 g glucose, 300 mg streptopenicillin and 100 ml McIllvaine buffer. An ejaculate of 1 ml containing 3–4 billion spermatozoa, diluted 1:1.5 and inseminated with 0.1 ml dose, permits 25 ewes to be inseminated in the cervix with approximately 120 M sperms in each dose. For fertilization 0.05 to 0.2 ml of fresh semen with a minimum of 120 to 125 M spermatozoa is effective. In stored chilled semen, the dose should be doubled. The results of insemination with semen stored for 24 hr are around 10% lower than with the use of freshly diluted semen. However, doubling the dose reduces the difference in conception rate. Large amount of semen is likely to overflow into the vagina and go waste. The ewes in heat should be detected both in the morning and evening. Those detected in the evening should be inseminated next morning and those detected in the morning should be inseminated the same evening.

With the realization of a limited scope to use diluted semen for large scale artificial insemination programmes in sheep, there is a need to develop a viable technology for cryopreservation of ram semen. There is enough potential for widespread use of frozen semen as it can be collected and stored.

Sahni (1967) in his detailed studies on the storage problem of indigenous ram semen has reported a quite satisfactory survival in heated cow milk and has recommended that cow milk appears to be a dilutor of considerable merit in the preservation of ram semen and also observed it to be significantly superior to buffalo, goat and skim milk. The milk and yolk containing diluents enriched with additives did not show results better than milk diluent alone and Cornell University Extender. Keeping quality of ram semen in these diluents showed significant seasonal variations but generally superior values of motility could still be observed in spring, rainy and autumn season. Effect of higher dilutions (1:10 to 1:100) in these diluents upto 72 hr of storage interval showed that there has been deterioration in keeping quality of ram semen with the increase of dilution rates beyond 1:10 and effect was more pronounced in egg yolk citrate as compared to glycine yolk and Cornell University Extender diluents on storage at 5–7 °C upto 72 hr. However, Tiwari *et al.*, 1968 observed better survival upto 30 hr in dilution rates varying from 1:3 to 1:10 depending upon the room temperature in different seasons it could be possible to utilize diluted semen within 34 hr after collection and dilution for artificial insemination and glucose saline without egg yolk may be used for dilution of semen to be utilized quickly within 2 hr of storage at 30 °C.

Ram semen preservation studies (Sahni and Tiwari, 1972), were further carried out, on efficacy of the dilution rates, storage temperature, additives, semen container storage media, cooling rates etc. Results revealed higher spermatozoal survival in egg yolk citrate glucose diluent (EYCG) as compared to EYC, heated cow milk, milk + egg yolk at 8, 24, 48,72 hr of storage interval. Irrespective of storage media, (refrigerator, semen shipper) and the diluents, higher spermatozoal survival have been observed in glass screw capped vials. Out of all the diluents tested, EYCG gave higher satisfactory results at different storage interval which was mainly due to better buffering capacity of the diluent. Further studies were then confined to the EYCG diluents only. Significant differences were observed in percent spermatozoal survival and pH duet to variation in number of live and normal spermatozoa (250–400 × 10^6) in 0.1 ml of diluted semen. However, such differences were non-significant upto 24hr of storage intervals and irrespective of number of live sperrnatozoa per inseminating dose (0.1 ml) about 70–80% sperm motility have been observed upto 24 hr of storage interval. This shows that concentrated ram semen could be stored successfully upto 24 hr and could be utilized for breeding operations. Indigenous rams did not differ among themselves in relation to spermatozoal survival. However, difference between breeds was found to be significant. Indigenous and half-bred rams showed relatively higher sperm survival than exotic rams at each storage interval. Differences between Rambouillet and Soviet Merino were nonsignificant. Addition of 1, 3 and 5% gelatine and polyvinyle alcohol in EYCG diluent did not provide significant beneficial effect on spermatozoal survival and pH during storage.

In other studies carried out by Singh and Saxena (1965) and Saxena and Singh (1967) it has been observed that motility and percent live spermatozoa was highest in skim milk yolk followed by egg yolk glucose bicarbonate diluent than EYC and CUE upto 168 hr of storage at 3 °C. However, during storage there was no automorphological changes in spermatozoa as reported earlier in EYC diluent upto 72 hr of storage, (Joshi and Singh, 1968 a, b ,c) in a comparative study on the efficacy of the diluents have reported higher sperm survival in yolk containing diluents enriched with fructose, tomato juice and glycine, skim milk diluent containing yolk, glucose or fructose. Beneficial effect of caproic acid on sperm motility during storage has been observed by Honmode (1972).

Sahni and Roy (1972c) studied the effect of deep freezing (79 °C) on post thawing revival of sheep and goat spermatozoa using various levels of glycerol, different equilibration time and rate of freezing in EYC and cow milk diluent. The post thawing survival varied from 10–14% in different trials and considerable variation between rams with respect to freezability was noted.

The efficiency of EYCG was further compared with diluents prepared from some known buffers by Tiwari *et al.* (1978). Out of many, the percent motility and live spermatozoa was relatively better in diluents prepared from the McIllvaine buffer containing yolk and glucose similar to that of EYCG. Along with better survival, the most common characteristics in these diluents were the absence of sperm clustering (sperm agglutination) which was about 20–30% in EYCG at 24 hr of storage and also relatively better buffering capacity. This shows that these diluents might be able to give better fertility as compared to EYCG diluent.

Factors affecting long term preservation are:

a) *Sire individuality*

The spermatozoa from all apparently healthy rams do not stand the cold shock, therefore rams have to be screened for this attribute and selected for donating semen for freezing.

b) *Extenders*

Neat semen cannot be frozen and revived, it requires certain additives like nutrients, cryoprotectants, cell wall fortifying agents, buffers and antibacterial agents etc. to be provided. Semen has to be diluted to ensure the required number of spermatozoa per insemination dose to avoid wastage and to be able to cover more females from one ejaculate. Different workers have developed a number of extenders using different combination of various chemicals having qualities of a good extender but with varying results of conception rate. A good extender should consist of following:

1. *Buffering agents*

Ram semen has high number of vigorously motile spermatozoa which consume nutrients very quickly, collect the metabolites specially of the anaerobic respiratory pathway like lactate, pyruvate and lower the pH. Low pH disturbs the electric potential of the sperm cell wall leading to seepage of the enzymes and other material from within the spermatozoa and preventing some from entering into their body, causing toxicity, bursting of the cells and autolysis. To check this, it is important to provide suitable buffering material which may not allow the pH to go down drastically and keep it within the tolerable range. Alternately certain chemicals like cortisol and catalase etc. which may avoid accumulation of lactate, pyruvate, H_2O_2 if provided may give good results. Milk has good buffering capacity but the results in the freezing experiments are not satisfactory. Egg yolk, blood serum albumin and different chemical buffers like glycine, citrate, phosphate etc. have given good results. Out of these, egg yolk and tris have not become popular. Although electrolytes are helpful at the time of thawing they are unwanted at the time of freezing.

2. *Cryoprotectants*

Spermatozoa contain some water. On freezing, the volume of frozen semen expands resulting in the bursting of spermatozoa cells and killing them. Cryoprotectants decrease the freezing point below 0 °C, allow the water to make smaller crystals and the chance to pass through the stage quickly. Glycerol, DMSO, ethylene glycol and propylene glycol have been used but varying percentage of glycerol has found favour with most of workers. Its higher percentage is harmful because it is reported to deplete the ATP, the energy reserves of the spermatozoa. Cyroprotectants need some time to enter the sperm cell. It is called "equilibration time". Equilibration time of 1 hr and 30 min have been tried but best results have been reported with 1, 4 and 5 hr at 4–5 °C temperature. Some workers have different opinion. It must however be taken care that the cabinet temperature does not go below 2 °C during equilibration.

3. *Cell wall fortification agents*

Egg yolk, egg yolk extract, lecithin, glutathione have been commonly used.

4. *Antioxidants*

Ram sperm cell wall lipids are oxidized quickly on coming in contact with oxygen in the medium or atmosphere and lower the fertilizing ability. α Tocopherol, BHT, etc. have been used for the purpose. Boiling water just immediately before preparing the buffers also helps in removing free oxygen dissolved in it.

5. *Osmoregulators*

Living cells perform well in the media isotonic to the biological fluid they are used to. They should perform well in extender isotonic to semen plasma although certain workers have claimed better results with slightly hypertonic diluents.

6. *Nutrients*

Although spermatozoa utilize fructose and glucose in order of priority but maltose, raffinose, lactose, trebalose have given better results in freezing experiments.

7. *Antibacterial agents*

Bacteria are practically omnipresent. Spermatozoa like the living cells, have to be kept away from bacteria and their growth be controlled by providing suitable bactericidal antibiotics.

8. *Dilution rate*

It includes the number of live normal spermatozoa to be included per ml of the diluted semen per insemination dose while freezing in straws. Normally ram semen contains 3000–5000 M sperms per ml of the neat semen and only 100–150 M sperms are enough to impregnate an ewe even though only one spermatozoa takes part in fertilization process. In certain extenders used for freezing, dilution rates of 1:2 and 1:3 have given better results than dilution rate of 1:1 which is general practice for short term preservation or for immediate use.

9. *Rate of heat exchange*

It is the most important factor which needs monitoring. It is important not only to control the rate of exchange of heat at the time of cooling but also at the time of thawing revival. It is clarified by some workers that it should be equal in both the stages and certain workers prefer high rate of cooling as well as thawing *i.e.* using high temperature water bath. There is a critical zone of temperature which is particularly harmful for the spermatozoa and the latter group in the absence of the knowledge of that zone advise the practise of fast rate of heat exchange. Now a days computer controlled programmable cooling and thawing cabinets are available and the day is not far off when the critical temperature zone will be identified and the measures to pass through it will be devised for successfully freezing the ram semen. Some workers have advised thawing the pelleted frozen semen in media like citrate solution others however, have found no difference between the results of thawing with or without the medium. Some prefer freezing pellets while the others prefer freezing in straws to give wider surface area for quicker heat exchange. Pellets are prepared by dropping 0.2 ml of diluted equilibrated semen in the smooth pits carved on the surface of plastic tray floating on liquid nitrogen allowing it to freeze for 3–5 minutes, collecting the pellets into the goblets and lowering down into the liquid nitrogen for storage. Allowing the liquid nitrogen in direct contact or not has made no difference. Straws have to be filled in the clean cabinet atmosphere of 4–5 °C temperature after equilibration, sealed, exposed to certain temperature below 0 °C for a particular length of time and then only lowered into liquid nitrogen (-196 °C) for storage. These days, small size liquid nitrogen containers are also available for transporting smaller quantities of frozen semen. During thawing certain workers have claimed good results with as low temperature of water bath as 37 °C, while some advise as high as 75 °C. Similarly the time for exposure may be 10–15 seconds for straws while 1–2 minutes for pellets.

10. *Additives*

Addition to the semen, certain chemicals like cortisol, amylase, glucuronidase which play part in longevity of the spermatozoa in the female reproductive tract and are reported to be deficit under stressful conditions in the tract and their varying concentration in the ram semen with season, individuality etc. Chelating agents like EDTA are also under study to see their effects vis-a-vis fertility rate.

The process of freezing adversely affects the survival of spermatozoa. It is dependent upon number of inter related factors such as diluent composition, concentration of cryoprotectant, dilution rate, size of pellet or straw and rate of freezing or thawing. There are some problems in achieving desirable lambing rate with the use of frozen semen. Unlike other species, the complex anatomy of the ewe's cervix does not permit smooth passage for frozen thawed spermatozoa in the cervical canal. Concerted efforts are going on to improve the fertility of frozen semen by improving the procedures of ram semen freezing and to develop better insemination techniques for laparoscope aided intrauterine deposition into the uterine lumen by-passing the cervix or transcervical insemination adapted to the anatomy of the ewe's gentalia.

A new diluent named egg yolk lactose raffinose citrate glycerol (EYLRCG) has been evolved for pellet freezing of ram semen giving more than 50% post-thaw motility. It contains 18 ml eggyolk, 5 g lactose, 8 g raffinose, 2.2 g trisodium citrate dihydrate, 3.3 ml glycerol, 300 mg streptopenicillin and 100 ml triple glass distilled water. Ejaculates of good quality semen are diluted with EYLRCG to bring the sperm concentration to 1 billion spermatozoa per ml. Diluted samples are equilibrated at 5 °C for 4 hr and then frozen into 0.2 ml size pellets. Pelleting is done in a pitted anodized aluminium tray floating over liquid nitrogen. Pelleted semen are placed in goblets and stored in canisters by lowering under liquid nitrogen in cryocans. Thawing of pellets is done in dry test tubes individually upto liquification at 50 °C.

In order to overcome the disadvantages of proper identification of stored pellets and lack of repetition in freezing conditions, a protocol has been developed, for deep freezing of semen in straws at CSWRI, Avikanagar. The freezing protocol, based on programmable freezing and computer assisted sperm analysis technique, gives 70% post-thaw motility consistently in all the breeds tested. Ejaculates having 90% initial motility and sperm density exceeding 3 billion spermatozoa per ml are extended in egg-yolk tris glycerol diluent to bring the final concentration to 1 billion spermatozoa per ml. The diluted samples are filled in 0.25–0.5 ml size straws, equilibrated for 2 hr at 5 °C and frozen @ -5–125 °C per minute in a programmable cell freezer. Thawing is done at 50 °C for 10 seconds in case of 0.25 ml size straws and at 60 °C for 10 seconds in case of 0.5 ml straws in a water bath. The application of computer assisted sperm analysis technique during various stages of cryopreservation enables precise and objective evaluation of sperm function. Apart from identifying motile and immotile spermatozoa, it helps also in computing the sperm velocity and track dimensions of the motile spermatozoa. Efforts are also being made to identify factors causing cryo-injury to ram spermatozoa during cryopreservation by cryo-microscopic technique in order to ameliorate the sperm plasma membrane from the deleterious effect of freeze thaw process.

7.3 Female Reproduction

Breeding ewes of indigenous breeds should be 1.5 to 2 years old depending upon their condition. Breeding too young ewes results in more weak lambing and thus higher lamb losses. The ewes should posses a long preferably low set body, roomy hind-quarters, well formed pliable udder, active foraging habit and good mothering instinct. The ewes having poor milking capacity, overshot or undershot jaw, broken mouth, blind teat and meaty udder should be disqualified from the breeding programme. Body weight of an ewe at breeding should not normally be less than the adult body weight of that breed. The ewes should be prepared for mating when they are gaining weight, by giving them access to luxuriant pasture or feeding them a little grain for about 2–3 weeks preceeding the beginning of the breeding season. After the ewes have been bred this extra feed can be stopped.

The normal heat period in ewes lasts from 6–48 hr with an average of 24 hr The normal oestrus cycle ranges from 16–19 days with an average of 17 days. The ewes come in heat usually about 2 months after lambing.

Sheep can be bred naturally or through artificial insemination. The natural breeding is done by either flock mating, pen mating or hand mating. In flock-mating system artificial insemination makes possible the extensive use of superior exotic and outstanding crossbred/indigenous rams. On an average, a ram ejaculate measures about 1 ml and contains 3–4 billion spermatozoa of which about 80–90% are alive, if the semen is from a healthy ram. Semen of higher concentration is usually slightly acidic in reaction, and that of lower concentration slightly alkaline. In fertile semen, spermatozoa have a whirling motion. Semen can be stored for artificial insemination in a chilled liquid form or deep-frozen in liquid nitrogen. Preserving semen in a liquid form allows short-term preservation while the deep-frozen form permits' long-term preservation for indenfinite period without any significant decrease in semen quality.

7.3.1 Puberty

Puberty in the female is exhibited by the first oestrus. The breed of the ewe affects the age and weight at puberty and sexual maturity. Ewe lambs in most of the mutton breeds reach puberty at the age of 10–12 months normally breeding during the first autumn. Larger, slower maturing breeds may mature at the age of 18–20 months.

Awassi lambs on a good diet first display oestrus at an average age of 274 days (Younis *et al.,* 1978), but Rambouillet crossbred lambs in Rajasthan are about 615 days old before they display oestrus (Kishore *et al.*, 1982). This difference is largely due to the different growth rates resulting from different nutrition. The average age at puberty in Ossimi and Barki ewe lambs reared on high plane of nutrition was 347 days in Egypt whereas, it was 366 days on a low plane of nutrition (Ei-Homosi and El-Hafiz, 1982). The variation in age at puberty occurs in subtropics where the ewes are reasonably anestrous so that if a lamb fails to cycle in her first season she will not cycle until the next, several months later.

Under Indian conditions, the ewes reach sexual maturity at 9–14 months of age, however, the full body growth is only attained between 18–24 months of age and the stage at which the mating is performed. Merino sheep mature more slowly than the Hampshire or Suffolk breeds, crossbred lambs generally mature earlier than pure bred lambs. Roy *et al.* (1962) recorded first oestrus in young Bikaneri ewes of less than one year of age. Majority of the Nellore ewes attained puberty at 630–810 days of age (Gupta *et al.,* 1987). The age at puberty has been reported to vary from 306–383 days in different exotic breeds (Martemucci *et al*., 1983).

7.3.2 The oestrus cycle

Rhythmic sexual behaviour patterns develop in females during puberty. This change in behaviour in called oestrus and it occurs with each oestrus cycle in non seasonal breeders unless pregnancy intervenes. The combination of physiological events which begin at one oestrus period and end at the next is termed as estrus cycle.

Oestrus occurs every 14–19 days with a mean of 17 days (Terrill, 1974). Similar duration has been observed for ewes in tropics by Narayanaswamy and Balaine (1976); Gaillard (1979) and Yenekoye *et al.* (1981). Cycle lengths of upto 70 days were recorded for Fulani ewes in Niger by Yenikoye *et al.* (1982).

Visual signs of approaching oestrus are, a swelling and redness of the vulva and restlessness or nervousness indicating a desire for company, but the most obvious sign is ridding and inturn being ridden. The breeding occurs only during oestrus although the ram is capable of breeding at any time (Dhanda, 1970).

Three types of oestrus cycles commonly observed are:

1) One type is seen in sheep, goat, cow and sow during breeding seasons. The infertile cycle in these species culminates in spontaneous ovulation of mature follicles; corpora lutea automatically formed, becomes functional and exists for a definite period of time.

2) Rat and mouse are example of another type of oestrus cycle in which ovulation is spontaneous, but the corpus luteum which is formed is not functional unless mating occurs.

3) In 3rd type, maturation and ovulation of follicles fail unless the male copulates with the female, rabbit is an induced ovulator.

The main events of the oestrus cycle of the ewe can be divided into those associated with the growth of follicle and those associated with the growth of the corpus luteum. The growth of the follicles takes place during pro-oestrus and oestrus. The period of the corpus luteum can be divided into two periods, metestrous and diestrous. The oestrus cycle is usually dated from the first day of oestrus.

7.3.2.1 *Oestrus*

Oestrus is defined as the period of sexual receptivity during which ovulation occurs in most species and the corpus luteum begins to form. The psychic oestrus is brought about by the action of estradiol on the central nervous system. Symptoms of heat are less pronounced in ewe than in cow or mare. The ewe in heat makes little effort to tease and mount. Because of weak symptoms, it is difficult to detect oestrus in ewe. The duration of oestrus in ewes varies from 12–72 hr (Taparia, 1972; Castilo Roias *et al.,* 1977; Gaillard, 1979; Yenikoy *et al*., 1981) with an average of about 45 hr Breed of ewe does not affect the duration of oestrus. Ewe lambs, usually have the shortest oestrus, older ewes have the longest and the yearlings are intermediate. The duration of the oestrus cycle is shorter at the beginning and at the end of the breeding season.

The 1st ovulation at the onset of puberty is frequently not associated with psychic oestrus. It appears that a small amount of progesterone is necessary to trigger psychological estrus. In the cycling ewes, this comes from the corpus luteum of previous cycle. Therefore, the first ovulation after the ovary is awakened from the quiescent state finds the ovary without an old corpus luteum to produce the small amount of progesterone necessary to bring about psychological oestrus.

The ovulation takes places about 12 hr before the end of oestrus Right ovary ovulates more frequently than the left (60–40% respectively).

During oestrus swelling and reddening of vulva occurs. Ewes normally come in heat after the weaning of lamb. Some nonseasonal breeding ewes may show a post partum oestrus a few days after delivery, but others wait for 4–6 weeks.

2–3 days before the onset of oestrus graafian follicles destined for ovulation begin to enlarge rapidly and become turgid. The theca interna hypertrophies and the ovum, with attached cumulus oophorus, moves from the embedded position in the granulosa layer toward the enlarged fluid filled antrum of graafian follicle. A transparent area appears in the follicle membrane near the apex when ovulation is impending.

7.3.2.2 *Metoestrus*

The ewe enters metestrous after oestrus. This phase lasts for 2 days. The corpus luteum organizes during this period and becomes functional. The progesterone level rises rapidly.

7.3.2.3 *Dioestrus*

Dioestrus is the period during which the influence of luteal progesterone on accessory sex structures predominates. This phase is referred to as the phase of the corpus luteum. Metoestrus and dioestrus are referred to as the luteal phase. Large amounts of progesterone enter general circulation resulting in the development of endometrial growth and also of mammary glands. The myometrium develops under the influence of progesterone and the uterine glands secrete a thick viscous material which will nourish the zygote. If a zygote reaches the uterus, the corpus luteum continues through out pregnancy, but if the ovum is not fertilized, the corpus luteum remains functional only until day 12 or 13. The decline of progesterone permits a surge of FSH and growth of another follicle, resulting in initiation of new oestrus cycle. In short the estrogen dominates for 34 days of cycle and progesterone dominates for about 13 days.

7.3.2.4 *Proestrus*

It is the period after which the corpus luteum fails, when the progesterone level drops, FSH release stimulates follicle growth and rising estrogen levels lead to oestrus. Proestrus and oestrus are often referred to as the follicular phase. Estradiol increases the blood supply to the tubular genital tract from vulva to the uterine tube and causes oedema. The vulva swells, the vestibule becomes hyperemic and the glands of the cervix and vagina secrete a serous secretion which appears as a vaginal discharge.

Anterior pituitary and the gonads

The primary target organs of the pituitary gonadotropins are the ovary and testes. The growth and development of ovarian follicles in mammals are dependent upon FSH but LH is essential for follicle maturation. Both FSH and LH are essential for the synthesis of estrogen. Rising estrogen blood levels suppress release of FSH and facilitate release of LH.

Apparently FSH and LH are secreted continuously by the anterior pituitary throughout oestrus cycle, but the proportions and levels of each, change during the different stage of the cycle.

Estrogen in large quantities inhibits FSH secretion. $PGF_2\alpha$ from non pregnant progesterone dominated endometrium reaches ovary by a local route between day 14 and 15 and kills the corpus luteum. The progesterone level falls and FSH facilitates follicular growth and quick estrogen production. The rising estrogen induces behavioral oestrus and trigers a release of an ovulation surge of LH on the days of oestrus. Estrogen levels then decline but the new corpus luteum starts progesterone production which rises during the next few days and is sufficient to hold gonadotropin release to low level. On day 16, the corpus luteum is killed again and the cycle is completed.

The corpus luteum is a temporary endocrine organ which functions for only a few days in cycling non pregnant ewes and throughout the duration of pregnancy once the ewe becomes pregnant. The corpora lutea of sheep reaches maximum size by days 3 and at this time it is reddish pink. Corpora lutea becomes paler as dioestrus approaches and after day 14, degeneration is rapid. Because the oestrus cycle of the ewe is shorter, the corpora lutea of the ewe are devoid of the lipochrome pigment, hence the lighter colour. The circulating levels of progesterone in plasma during follicular and luleal phases are 0.25 and 3.7 ng/ml in ewe.

Factors affecting corpus luteum

In polyestrous domestic animals, corpora lutea develop after ovulation and function for 14–15 days unless pregnancy occurs to signal the corpora lutea to continue functioning. There appears to be marked species differences in the control of corpora luteal function. The pituitary is needed in most species to produce luteotropin at ovulation to form the corpus luteum. Non-pregnant uterus produces a substance probably $PGF_2\alpha$ which has a luteolytic influence on the corpus luteum. The luteolytic effect of uterus is dominant to hypophyseal control of the life of the corpus luteum in sheep. Presence of an embryo in the uterus will prevent luteolysis. Researches have shown that an embryo must be present in the uterus by the 12th day of the cycle in sheep if luteolysis is to be avoided. Luteal regression can be prevented in the ewe and cow by uterine infections. Presence of intra-uterine devices (IUD) shorten the oestrus cycles in sheep and cow.

Oestrus during pregnancy

Oestrus in the pregnant ewes is associated with follicular growth, but ovulation, breeding and conception may occur during an existing pregnancy which would lead to surperfoetation. Oestrus has been observed in pregnant ewes (Matter, 1974; Younis and Afifi, 1979; Kandasamy and Pant, 1980; Sinha *et al.*, 1980). It has been observed in first month of pregnancy. It is not usually accompanied by ovulation

(Williama *et al.*,1956). There is a fear among sheep farmers that if mating takes place abortion may result (Sinha *et al.*, 1980).

7.3.3 Laproscopy and intra-uterine insemination

A modified laparoscope aided intra-uterine insemination technique has been developed for using frozen thawed semen. Ewes exhibiting heat are fasted overnight. They are restrained in a locally fabricated cradle suitable to the body size of native sheep. Sedation is induced with 1 molar injection of 0.1 ml of 2% solution of xylazine hydrochloride. The abdominal area adjacent to the udder is sprayed with 70% alcohol solution and infiltrated with 4 ml of 2% ligocaine hydrochloride. Ewes restrained in cradle are suspended in head-down position at an angle 45° and two stab incisions are made using 10 mm diatrocar approximately 5–7 cm anterior to udder and each side of midventral line. Through one trocar a laparoscope and through the other puncture site, a pair of atraumatic Allis forceps are introduced. Uterine horn is grasped gently with forceps and lifted upto puncture site to view from outside. Frozen and thawed semen straw is loaded in modified aspic and 0.1 ml semen is deposited in the anterior part of both the horns. The two puncture sites are then sutured and ewes are let free for grazing. An overall lambing rate of 44.4% has been achieved for two cycle insemination with this technique at CSWRI, Avikanagar. For wider acceptance under field conditions attempt is being made to develop a non-invasive and cost-effective transcerival insemination technique as an alternate to laparoscope aided intra-uterine insemination technique which is likely to improve the efficiency of artificial insemination of sheep with frozen semen.

7.3.4 Use of hormone, oestrus synchronization and multiple ovulation

Oestrus cycle of the ewes is synchronized so that a large number of them come in heat at one time. This would help in reducing the cost of artificial insemination or natural breeding and consequent care at lambing. Oestrus synchronization is also an integral part of embryo transfer. However, the cost of the hormones used for the oestrus synchronization is prohibitive under our conditions. It gives a uniform flock of lambs which will facilitate their disposal and fetch more price. It can also gainfully be employed in breeding sheep in spring. In India, although spring and autumn are technically peak breeding seasons, sheep can be bred throughout the year and synchronization can be done more cheaply by telescoping the breeding season. The telescoping is done by introducing ram in the flock after keeping it away for 2 to 3 months; 70 to 80% of ewes will come in heat in the very first oestrus cycle.

Synchronization of oestrus is initiated with the simultaneous administration, to the ewes, of progestogen hormones or their analogues through feed, as implant or as impregnated vaginal sponges. After the administration for 12 to 14 days the hormone is withdrawn. The animal comes into heat within 3 days. Since the reproductive females are still under the influence of the progestogen hormones the conception rate varies from 30 to 40% whereas it varies from 60 to 70 to 90% during the subsequent cycle. $PGF_2\ \alpha$ or its synthetic analogues, causing lysis of corpus luteum, are also being used for the purpose. Two intramuscular injections of 10 mg each at interval of 10 days bring all the animals in heat within 72 hr.

7.3.5 Gestation period

The duration of gestation in native and cross bred ewes has been reviewed extensively by Honmode (1970) and Kaushish (1971). The mean gestation length varies from 147–153 days. The average gestation period varied from 151.2 to 152.2 days in Nali, Lohi and crosses of former with Mandya and Nellore (Kaushish and Arora, 1974a). Ramamurti (1963), Singh *et al.* (1977) and Kaushish *et al.* (1986) reported duration of gestation in Native, Niligiri and Bikaneri and Soviet Merino, respectively.

The gestation period is shorter for exotic than indigenous ewes (Sahani and Pant, 1978), for ewes with exotic crossbred lambs, than those with indigenous lambs (Rao *et al.*, 1978), for lambs with lighter birth weights (Kishore *et al.*, 1980) and for female than male lambs (Narayanaswamy *et al.*, 1975). The gestation is positively related with weight at service (Kaushish, 1989).

The effect of breed, birth weight, sex of lamb, birth weight of dam at service, pH of amniotic fluid and post service gain in weight on gestation length has been studied by Ostgard (1957) and Kaushisha and Arora (1973, 1974a) . Gestation has been reported to have positive correlation with pH of amniotic fluid and negative with post service gain in weight and weight at service (Kaushish and Arora, 1974a).

A study on post parturn estrous interval has been carried out by Rawal *et al.* (1986b) to determine the effects of season and sex of lamb on this trait in Muzaffarnagri sheep. In the single normal lambing, twin lambing and abortions, gestation period averaged 137.4 ± 4.82; 168.1 ± 23.63 and 172.8 ± 13.22 days, respectively. The sex of lamb did not have significant effect on postpartum estrous interval in case of single births. Effect of season was also non significant. One percent ewes came in heat within 60 days. The average post partum estrous interval following twin lambings and after abortions was abnormally long, 28.2% ewes came in heat within 151–180 days after abortions.

7.3.6 Pregnancy diagnosis

There are several ways by which pregnancy can be diagnosed in sheep. Some methods are relatively simple. The failure of an ewe to return to oestrus gives an indication that she is pregnant.

Physical examination of ewes in late pregnancy is simple and quick. The lamb can be felt by gently palpating the ewes' abdomen.

The udder becomes firm and enlarged in pregnant ewes. A rectal probe can be used to palpate the uterus of ewe from about 70 days of pregnancy and was reliable when used by an experienced operator (Plant, 1980).

Richardson (1972) reviewed sophisticated methods of pregnancy diagnosis and found over 80% accuracy by vaginal biopsy, ultrasonic foetal pulse detection and radiography. The foetal pulse detection is more practicable and was reviewed by Thwaites (1981) and Wani (1981). This method can accurately predict whether or not an ewe is pregnant if used after 60 days of Pregnancy (Aswad *et al.*, 1976; Wani and Sahni, 1981) and the rate of lamb heart beat can be used to predict the date of lambing.

Trapp and Slyter (1983) reported that ultrasonic scanning instruments developed to measure carcass fatness, when used between days 69 and 112 of pregnancy, gave an accuracy of pregnancy diagnosis of about 90%.

By laparotomy, pregnancy can be detected by as early as 30 days by skilled workers (Pacho, 1973).

If the date of breeding is known, an analysis of plasma progesterone, 17 days later will determine whether ewe is pregnant or not (Robertson and Sarda, 1971). For lactating ewes, progesterone levels in the fore milk can be used to diagnose pregnancy (Ayalon and Shemesh, 1979).

7.3.7 Parturition

The act of parturition is termed as lambing in ewe. Normal parturition consists of expulsion of normal viable foetus from the uterus through maternal passage by natural forces alone at a stage when the lamb is capable of independent existence (Kaushish and Arora, 1974b). It is difficult to predict the exact day of parturition in Nali sheep by studying the antepartum blood cholesterol content. (Kaushish and Arora, 1974a).

Time of parturition

The time of parturition is one of the most critical periods in the life of ewe. During this process, the ewe could possibly be injured and future reproduction efficiency be impaired temporarily or permanently. The distribution of births appears to vary with the breed of the ewe and the management system (Kaushish *et al.*,1973; Younis and El-Gaboory, 1978; Tomar, 1979; Bhaik and Kohli, 1980). Highest number of lambings took place between 3 and 6 am and 3 and 6 pm (Kaushish *et al.*, 1973) in some native breeds and their crosses whereas, maximum (40%) lambings took place between mid night and 6 am in Russian Merino ewes (Kaushish and Sahni, 1975).

Signs of parturition

The symptoms of the onset of parturition consist of changes in the birth canal, progressive extensibility to permit safe passage of foetus, and hypertrophy of the mammary glands. The vulval lips become resilient and oedematous, two to six times their normal size. The ewes usually lay on the right side stretching the hind legs as far as possible, especially during contractions and frequently showed upward turning of lips. Frequent grinding of teeth were also observed. About 6–8 hr before the expulsion of foetus the vulval labiae were about 0.5 cm a part and this distance increased to 1.0 cm about one hr before parturition. The respiration rate increased about half an hr before the appearance of water bag (Kaushish and Arora, 1974b).

Process of parturition in sheep

The process of parturition has been studied in detail by Tiwari *et al.,* 1969a and Kaushish, 1971 (Table 7.3). The time taken for the completion of parturition was longest in Mandia, Nali and shortest in Lohi. The post service gain in weight of dam was negatively correlated with the time taken for the second and third stages of the parturition (Kaushish and Arora, 1974b). The duration of sequence of events in the second stage of parturition has also been studied in detail by them (Kaushish and Arora, 1974a).

Relationship among weight of placenta, number of cotyledons and birth weight has been studied by Kaushish and Arora (1983). The average number of cotyledons varied form 66.6 to 74.1 between groups. The weight of placenta ranged from 0.195 to 0.259 kg. The sex of the lamb did not affect the placental weight and the number of cotyledons significantly. There was a positive relationship between placental weight, number of cotyledons and birth weight (Kaushish, 1971). Weight

Table 7.3. Duration of Different Stages Noted in the Process of Parturition

Genetic groups	Sex of the lamb	First stage Mean ± SE	Second stage Mean ± SE	Third stage Mean ± SE
Nali	M	343.5 ± 19.50 (24)	33.6 ± 2.14 (27)	154.9± 8.40 (29)
	F	396.0 ± 14.37 (15)	33.8 ± 2.04 (30)	135.5± 14.69 (30)
	Both sexes	367.7 ± 17.51	33.7 ± 2.09	145.1± 11.60
Lohi	M	259.0 ± 38.07 (10)	33.1 ± 1.61 (14)	131.7± 16.16 (14)
	F	4213 ± 28.41 (9)	32.3 ± 1.81 (13)	131.6± 7.11 (12)
	Both sexes	288.5 ± 33.25 (19)	32.8 ± 1.73 (27)	131.6± 10.50 (26)
Nellore × Nali	M	327.8 ± 29.11 (7)	33.8 ± 0.80 (7)	161.4± 13.17 (7)
	F	330.6 ± 18.72 (8)	33.8 ± 0.88 (10)	135.5± 13.19 (10)
	Both sexes	329.3 ± 24.41 (15)	33.8 ± 0.88 (17)	146.1± 13.8 (17)
Mandia × Nali	M	459.4 ± 96.69 (9)	37.8 ± 2.89 (9)	124.0± 12.40 (8)
	F	468.6 ± 74.44 (5)	41.8 ± 5.81 (5)	113.7± 17.25 (4)
	Both sexes	462.7 ± 83.65 (14)	39.3 ± 4.32 (14)	120.6± 14.92 (12)

of placenta as percentage of lamb weight generally centered round about 8.5% in Nali, Lohi and the crosses of former with Mandya and Nellore (Kaushish and Arora, 1975). It was little lower in Bikaneri and Mandya (Tiwari *et al.*, 1969a). The effect of weight at service on time taken for expulsion of foetus and placenta has been studied by Kaushish and Arora (1973).

Post partum oestrus

Most ewes are seasonal breeders, consequently there are few attempts for the ovaries to function until the breeding season occurs for that particular breed. In the breeds which have been selected to produce two crops of lambs each year, there is need for breeding soon after lactation ceases. As soon as weaning occurs, these sheep breeds usually show a cyclic activity in the ovaries accompanied by estrous and ovulation (Table 7.4).

Table 7.4. Averages of Post Partum Oestrus Interval following Single Normal Lambings and Abortions in Muzaffarnagari Ewes

	Single births		Abortions	
Interval(days)	% of ewes	Mean ± S.E.	% of ewes	Mean ± S.E.
>60	19.1	35.5 ± 1.94	14.0	35.5 ± 6.20
61–90	16.1	77.6 ± 1.18	5.3	77.0 ± 1.0
91–120	12.3	103.9 ± 1.27	7.0	101.0 ± 3.10
121–150	12.3	137.1 ± 1.28	7.0	135.0 ± 4.77
151–180	12.9	163.8 ± 1.38	28.1	167.4 ± 1.90
181–210	9.1	196.0 ± 1.60	17.5	194.8 ± 2.95
211–240	7.1	229.5 ± 2.14	1.8	240.0 ± 0.00
241–270	2.3	254.1 ± 2.70	7.0	249.8 ± 3.11
271–300	3.9	285.2 ± 2.20	5.3	272.7 ± 1.53
301–and above	4.5	353.5 ± 7.03	7.0	426.7 ± 47.06

Induction of parturition

Hormones like dexamethasone or flumethasone (Bosc, 1972) can be used to induce lambing in sheep. Hormones have been used to combat prolonged pregnancy of Karakul ewes in South Africa (Roux and Wyk, 1977). The interval between injection and parturition depends on the stage of pregnancy (Aswad *et al*., 1974) and if the ewes are injected too early, the lambs die (Webster and Haresign, 1981). There is a great scope for this in tropics.

7.3.8 Inter-lambing interval

It is the interval between two successive lambings. Normally, it varies between 7 and 12 months, but may be as long as 20 months also. Values in the literature for lambing interval is given as 218 days for Morada Nova sheep (Teixeira *et al*., 1980) and 408 days for Mandya sheep (Purushotam, 1978).

Long interlambing intervals may be due to long postpartum anestrous period, failure to conceive, death of all embryos or abortion. Ewes which do not conceive for two successive breeding seasons are culled under improved managemental conditions. Lambing interval decreases with parity upto 4, suggesting that young ewes which are still growing take longer to regain condition after lambing. Interlambing period is also affected by season (Fall *et al.,* 1982).

Reducing inter-lambing interval

An 8 month of lambing interval is possible provided that nutrition and management are satisfactory (Naude and Grant, 1979). Hormonal treatment can be used to stimulate oestrus where seasonal anoestrus from photoperiod results (Fletcher *et al.*, 1980).

The period between lambing and conception (closed service period) of poorly fed ewes is about 180 days if no hormonal treatment is given. Intensively fed West African Dwarf ewes conceive at an average of 43 days after lambing (Berger and Ginisty, 1980). Estrous can be induced by using intravaginal sponges as early as 17 days after lambing (Nie Kerk, 1979). Honmode *et al.*, 1971a induced heat in local Malpura ewes by inserting sponges 30 days after lambing interval to 6 months (Brown *et al.,* 1972).

Sheep loose weight if bred frequently (Sahni and Tiwari, 1974b). A seasonal production on lambs results in high lamb mortality (Labban and Ghali, 1969) and poor growth rate (Ganesakale, 1975).

7.3.9 Embryo transfer technology

Embryo transfer technology in sheep has been developing since last 3–4 decades and a great deal of research is involved to simplify the technology and to refine the protocol. This technology embraces a sequence of procedural events like oestrus synchronisation, superovulation, embryo collection, evaluation, transfer, and freezing etc.

There has been little commercial embryo transfer activity in sheep as compared to cattle mainly due to high cost and surgical procedures involved in the technology. However, during the last decade, with the advent of laparoscopic procedures for intrauterine insemination, embryo recovery and embryo transfer, it may be possible to invite attention of sheep industry for substantial improvement of genetic make up through the use of embryo transfer technology to increase the meat, milk and wool production.

This technology can be applied

1. To increase number of offspring from superior ewes.
2. To ease transportation of superior genes across the national and international boundaries.
3. To conserve the endangered breeds.
4. To produce exotic lambs from well adapted native surrogate mother.
5. To avoid risk of diseases while transporting live animal from outside and within the country for breed improvement programme.

6. To produce lambs of desired sex.
7. Production of identical twins, Chimeras etc.

Donor ewes are selected on the basis of their genetic merit. Donor and recipients ewes should have high fertility and free from any disease. A close synchrony of oestrus in donor and recipient is an important factor influencing the success in embryo transfer. Oestrus in ewes can be synchronised either by lengthening of progesterone phase of cycle (luteal phase) or by shortening of progesterone phase. The former is achieved by administering progesterone in body of ewe by means of daily progesterone feeding or intra-muscular injection, vaginal sponges, subcutaneous implants and controlled internal drug releasing device (CIDR). Sponges impregnated with 0.35 g of progesterone kept for 12 days *in-situ* vagina can bring the ewes in oestrus during breeding seasons. An additional treatment of pregnant mare serum gonadotrophin (200 IU) on day of sponge removal is required in acyclic ewes.

Corpus luteum, a source of progesterone, is sensitive to prostaglandin $PGF_2\,\alpha$. Two intramuscular injections of $PGF_2\,\alpha$, 8–11 days apart can bring most of the treated ewes in heat within 48 hr of last treatment. A protocol for oestrus synchronisation in donor and recipient ewes with two I/M injection of 10 mg prostaglandin $PGF_2\,\alpha$ is in practice. Superovulation among donor ewes is induced with the use of gonadotrophin. Pregnant mare serum gonadotrophin (PMSG) and follicle stimulating hormone (FSH) from porcine, equine and ovine origin, are most frequently used for superovulation. In general, gonadotrophins are administered in multiple doses to female whose cycle is controlled by using progesterone or $PGF_2\,\alpha$. Recent attempts to increase the superovulation in donors have included the use of gonadotrophin releasing hormone (GnRH), a combination of PMSG and FSH and progesterone priming.

Ewes treated for superovulation are either mated with a proven fertile ram 2–4 times at 12 hr interval or subjected to A I with fresh diluted semen. Superovulation is known to impair the sperm transport through cervix and decreases the sperm viability after natural mating. In order to obviate problem of sperm transport, semen is directly deposited into uterus. Laparoscope aided intra-uterine insemination with fresh diluted semen generally result in high fertilization rate in superovulated ewes.

Different methods of embryo collection and transfer have been attempted. These include (a) surgical and (b) laparoscope aided nonsurgical procedures. In surgical procedures, reproductive tract is exteriorized for embryo collection and transfer. This leads to surgical trauma and formation of post-operative adhesion and therefore, limit the repeated embryo collection and/or transfer. Laparoscope aided procedures have opened the possibilities of repeated embryo collection and transfer in sheep.

Quality of embryos are adjudged under stereozoom microscope according to uniformity of zona cell shape and size, perivitelline space, vacuoles and cytoplasmic granules.

Depending upon age and number of cells, the good quality embryos are transferred in oviduct (<3 days old), near uterotubal junction (days 3–4) or in uterine body (days 5–6).

Embryos can be stored in phosphate buffer saline for few hr at 20 °C or in refrigerator at 5 °C and liquid nitrogen for many days. Cryopreservation of embryos has become an integral part of embryo transfer and programme related to conservation of germplasm. The transportation of germplasm resources as embryos rather than live animals is cheaper, possess less risk of disease transmission and has added advantage of allowing exotic stocks to develop in recipients well adapted to local conditions. Procedures that are currently used for cryopreservation of sheep embryos either rely on slow rates of freezing or direct transfer of embryos from room temperature to liquid nitrogen *i.e.* vitrification. Results of survivability of frozen thaw embryos in general varies from 35–73% depending upon freezing protocol, cryoprotectant, method of thawing, age and quality of embryos. *In-vitro* production of embryos is a multistep process and include the techniques of *in-vitro* maturation and fertilization of oocytes and *in-vitro* culture of embryos. In recent past technology of *in-vitro* production of embryos have developed and birth of lambs are reported. Oocytes are generally recovered from slaughter house ovaries or from live animals using laparoscope aided follicular aspiration/ folliculocentesis.

CHAPTER 8

NUTRITION

8.1 Components of Sheep Nutrition

It is well known that all animal live upon the feeds that are taken by them. After the feeds are ingested, they are subjected to the various processes of digestion by which portions become available for great variety of functions of the body. The undigested residues and waste products formed within the body are excreted. The excreta of animals, often more or less mixed with materials used for bedding constitute the animal manures which are one of the great by-products of livestock industry and which when properly used, serve to return to the soil, a large percentage of the materials originally contained in the feeds.

8.1.1 Grazing

Sheep prefer ground vegetation, grasses, legumes and a wide variety of forages. Practically little or no supplementary feeding is provided to sheep for efficient sheep production. The pastures and natural range lands should be optimally utilised and some supplementation of concentrate containing grains, cakes and agro-industrial by-products is necessary for maximizing production from sheep.

Like any other ruminant, sheep also has a compound stomach with 4 compartments, *viz.* rumen, reticulum, omasum and abomasum.

Since the well being of sheep depends upon feeding and management, the feeds provided for them must contain the nutrients which they need. The practical feeding of sheep is often a relatively simple matter, but the physiological process of the body are very complicated and thus a thorough understanding of nutrition in sheep requires a knowlege regarding soil, plants, and animal physiology as there is a important relationship between these.

8.1.2 Essential nutrients

The essential nutrients in sheep nutrition are classified into six groups based on chemical, physical and biological properties. These are water, carbohydrates (Energy), fats, proteins, minerals and vitamins.

8.1.2.1 *Water*

Water, is essential for proper functioning of the body. We now recognise that water and its ionization products are important determinants of characteristic structure and biological properties of proteins and nucleic acids as well as membranes, ribosomes and many other cell components.

It performs several very important functions in the body, as it aids in holding other nutrients in solution or suspension and hence helps in the digestion, utilization, and elimination of them or their products . The body water plays an important role in the animal's thermo regulatory mechanism. Water constitutes over 50% of the body composition of a lamb, and abundant supply of water is essential for the thriftiness of sheep. There is a widespread notion that sheep can do well without water, but careful sheepmen always keep a supply for them or drive them to a good source. When grazing succulent forage in cool weather, the amount drunk will be small. Since sheep sweat to only a very slight degree compared with some other animals, the principle means of elimination is through the kidneys and by respiration.

Water acts as a solvent for crystalloids, medium for digestion, absorption, metabolism, secretion and exertion and transportation of nutrients, and hormones in animal body. Water also plays dominant role in equalizing the temperature throughout the body due to its physical properties like greater thermal conductivity, higher specific heat and higher latent heat of vaporization than any other ordinary liquid. Water also serves as a lubricant for moving surfaces in the body. It is the main constituent of all body tissues and helps in the digestion, metabolism of nutrients and excretion of waste products. The body water plays an important role in the animal's thermoregulatory mechanism. The water requirement is usually satisfied by the water present in the feed and also by drinking. Water requirement is influenced by atmospheric temperature and humidity, stage of growth, gestation and lactation and other stresses. An adult sheep requires about 2 liters of water per day during winter and 3.5 to 4.0 liters during summer. Sheep in desert areas can withstand water deprivation upto 3 days. Offering water on alternate days has no deleterious effect. Sheep in desert areas with scanty and brackish water can tolerate salt content upto 1% in the drinking water.

8.1.2.2 *Energy*

Energy is required for all vital functions like maintenance of blood pressure and muscle tone, transmission of nerve impulses, ion transport across membranes, reabsorption in kidney, protein and fat synthesis, secretion of milk and production of wool. Several variables *viz.* body size, age, sex, level of growth and production activity and environmental conditions influence energy requirement of animals. Energy deficiency for extended periods results in lowered production of meat, milk and fibre. Deficiency of energy is manifested primarily as a lack of growth, body tissue losses or reduced production. Energy is usually defined as capacity to do work. Some of the terms that are used to express energy are ergs, joules, calories (cal), Kilo calories (Kcal), Mega calories (Mcal) and therms. The gross energy, digestible energy, metabolisable energy and net energy are usually expressed as cal, Kcal and Mcal per unit feed. Other expressions of energy value of feeds include total digestible nutrients (TDN) and starch equivalent (SE). In India TDN and ME (Mcal/kg) are generally used to express energy value of feeds. An adult non-pregnant sheep requires 93 Kcal ME/KgW $^{0.75}$ for maintenance. An additional allowance of 50% over maintenance is generally provided to pregnant ewes. The lactating ewes require still higher amount of energy: the maintenance requirement of lactating ewes has been estimated as 102 Kcal ME/KgW $^{0.75}$.

Energy source in the body is met by feeding foods containing carbohydrates and fats. Common carbohydrates, sugar, starch, and cellulose are the fibrous or woody parts of plants. They are characterized by the fact that in composition they contain hydrogen and oxygen in the same proportion as these exist in water; they are free of nitrogen and therefore are often referred to as the nitrogen-free extract of feeds. Grains are high in nitrogen free extract while roughages are much lower, and that in roughages is usually much less digestible, for in the latter much of the material is in the form of fibre. Pentosans are similar in some respects to starches and sugars but are found mainly in the fibrous portion of plants. The value of these materials to sheep is of two distinct kinds: the fibrous portion helps to form bulk, a matter of extreme importance in the diet of all herbivorous animals, of which sheep are a part; the sugars and starches, are very important sources of energy. A surplus taken into the body may be transformed into fat and stored as a reserve supply for later use in the production of heat and energy. If the supply of carbohydrates is inadequate, it is possible for sheep and other animals to make up for the deficiency by using the protein and fat contained in the feed. Carbohydrates are needed in abundance in the rations of fattening sheep. Apparently, these materials cannot be used for body functions other than for the deposition of fat and the production of heat and energy. One of the most common practical consequences of lack of carbohydrates in the ration is an increase in expense.

8.1.2.3 *Fat*

Fat or similar materials are found to some extent in almost all feeds. Seeds of certain plants are high in oils or fats, while the stems and leaves contain very little. From the standpoint of the nutrition of sheep, it is likely that some fat in the feed is desirable, although it seems unlikely that these animals suffer from a deficiency of dietary fat. Fat may be used for energy, or it may be converted into body fat and stored with fat formed from other nutrients. Body fat of sheep has a very high melting point, and it is apparently not affected in any significant way by the kind or amount of fat in the usual feeds.

8.1.2.4 *Proteins*

Proteins are very complex compounds which serve considerably different uses from those served by the other nutrients. For sheep the proteins are almost exclusively of plant origin. The leaves and seeds of plants are rich sources of proteins. The protein content of plants differs greatly, both with respect to quantity and quality. Differences in quality seem to be closely allied to the various amino acids of which the protein is composed. Some amino acids may be synthesized in the rumen. All proteins contain nitrogen, carbon, hydrogen, and oxygen and some also contain phosphorus, iron, and sulphur.

The percentage of protein contained by the dry matter of feeds used for sheep varies from about 5 to 20 in the case of roughages up to around 45 for some of the concentrates. Among the best sources of proteins are the leguminous plants which form such an important element in the pasture and agricultural industry. These plants are able to make use of the nitrogen of the atmosphere which constitutes such an important element of soil fertility.

The protein is needed for maintenance of existing body structure, growth of tissues and organs, reproduction and production of milk and fibre. Protein is the basic structural material of all the body tissues and is required for regeneration of living tissues which are subjected to constant destruction and repairs. The breeding animals need protein for prenatal growth, development of the foetus and to produce milk for post-natal growth of young ones. Clean scoured wool comprises of keratin which is almost a pure protein. If the ration does not contain enough energy, the protein will be used as energy source. But protein cannot be replaced by any other nutrient in the ration. Protein deficiency causes reduced feed intake and poor feed efficiency. This would result in poor growth and development of muscle, reduced reproductive efficiency and wool production. The protein deficient animals have lower disease resistance due to smaller amount of immune protein.

Although proteins vary in their nitrogen content, in the usual method of analysis, nitrogen content of sample is multiplied by 6.25 and is termed crude protein (CP). Many other nitrogenous substances which are not true proteins are also present in feed and these are often called non-protein nitrogen (NPN). Some of NPN compounds present in feed are amines, amino acids, ammonia, nitrates, nitrite, pigments, urea and vitamins (nicotinic acid).

In several countries requirement of protein is expressed in terms of CP. Under Indian conditions, it may not, however, be applicable because of many unconventional feeds being used which have very low digestibility of protein although the crude protein content is reasonably high. Hence the digestible crude protein (DCP) content of the diet is more important than the crude protein content. The approximate daily requirement of DCP for maintenance is $^{1}/_{10}{}^{th}$ of the TDN or 1g for every 1 kg of live weight. This requirement will increase by about 50% during preqnancy and by 100% during lactation and growth. A 30 kg sheep will thus require 30 g DCP during dry non-pregnant stage, 45 g during last 6 weeks of gestation and 60 g during 1st 10 weeks of lactation. Protein quantity is more important than quality in the sheep ration because the rumen microbes can convert a low quality protein to high quality amino acids, although the availability of some sulphur containing amino acids, *e.g.* methionine and cysteine, could become a limiting factor in woolly sheep.

Sheep require liberal amounts of proteins for growth, for the replacement of body tissues and fluids, for reproduction, and for the growth of wool. Sheep can subsist on feeds partially deficient in protein, but they cannot give maximum production. The protein requirements of pregnant ewes is very high, much higher than in beef cattle, for sheep reproduce in a shorter time and often have two or more lambs, which represent a large percentage of the body weight of the ewes. It has been shown that the daily protein requirement for wool growth alone is from 0.04–0.06 kg per 450 kg of live weight. Since the wool fibre is chiefly protein, and the demands for protein for other purposes is high, the significance of a diet adequately supplied with this nutrient is evident. In general, sheep are unable to produce protein from other materials. However, it has been established that a non-protein nitrogen compound, urea, may serve as a source of protein for ruminants because of the utilization of the urea nitrogen by some of the bacteria and protozoa in the digestive tract of such animals. The bacterial and protozoan proteins are then utilized by the sheep.

The ration of sheep should contain at least 9–10% protein, preferably 13–14%, and it has been observed in many cases that a much higher intake increases the rate at which lambs grow and fatten. If the protein is insufficient, the most obvious symptom is a decline in the appetite for feed and a slowing up or a cessation of growth, together with a very poor development of the muscles in growing lambs. In mature ewes there is a lack of thrift and poor wool growth; there may also be some impairment of reproductive functions shown in the dropping of weak lambs and the scant milk flow for support of the offspring. Experiments have shown that sheep utilize the protein of the ration better if it is close to the minimum needs, but there is no apparent harm to any of the body organs if it is fed to a much greater extent than needed to meet requirements. Although it may not be so efficiently utilized, the resulting production will be satisfactory. Perhaps the lowered efficiency in utilization when protein is fed in large quantities is due to the lack of sufficient non-nitrogenous nutrients to enable sheep to use the protein with maximum efficiency. This may be an explanation of the usually satisfactory results secured in lamb fattening with a ration of alfa-alfa and corn. These two feeds often produce as good results as can be obtained with more complicated qualities. It is possible that some of the beneficial effects of high protein feeds are due to other essential nutrients, such as vitamins and minerals often found in such feeds.

8.1.2.5 *Minerals*

Minerals are required for the building and maintenance of the skeleton and teeth. They play an important role in digestion, maintenance of osmotic pressure in different body fluids and wool growth. They form essential parts of certain organic compounds which occur in muscles blood and various secretions. Deficiency of any mineral will exhibit clinical symptoms. The role of the minerals in sheep nutrition is complicated. Excess of some of them may result in poor feed intake, digestion and utilization of other minerals, and can even cause toxicity. Sheep have certain definite mineral needs. Without these minerals proper nutrition is not accomplished, but feeding them to excess is in no wise supported by investigations of the problem. Eleven mineral elements have been found essential for sheep and other herbivorous animals. These are calcium, phosphorus, sodium, potassium, chlorine, magnesium, iron, iodine, sulphur, copper and cobalt. In addition to these there has been some evidence to indicate that manganese and zinc are necessary for various body functions.

The common mineral deficiency symptoms are anorexia (reduced appetite), reduced gain or loss in body weight, unthriftiness, abnormal hair or wool coat and skin dullness, bone deformation, staggered gait and organ damage. Calcium and phosphorus are necessary for bone formation and its maintenance. Deficiencies or imbalance of these minerals are indicated by rickets in young ones and osteoporosis in adults. Deficiency of copper and cobalt may result in tetany and foggy wool. Selenium deficiency may lead to white muscle' disease in lambs. Sulphur is present in wool and hair and its deficiency will lead to poor wool production and quality. Inadequate supply of iron, copper and cobalt results in anaemia, and lack of iodine in goitre.

(i) *Salt*

Salt is a combination of sodium and chlorine, two of the essential minerals. These minerals perform an important work in maintaining the osmotic pressure in the cells of the body and thus aid in the transfer of nutrients to the cells and in the removal of waste materials from them. Both of these minerals are also found in the blood. Chlorine is required for the production of the hydrochloric acid of the gastric juice, which is one of the digestive juices of the stomach. Sheep have a very low stomach acidity, but they have a great need for salt and become so hungry for it if deprived of it for a considerable period that they may overeat, and death may result. The daily salt consumption ranges from one-quarter to three-quarters ounce daily per head, depending upon the size of the animal and upon the character of the feed. Rangemen usually figure the requirement as a 400 gm per month for each sheep. There are some coastal areas and some alkali regions where the salt content of the herbage and soil is so high that sheep require no salt in addition. However, in all other areas salt in some form should be made available to them all times. This is a better practice than giving salt at weekly intervals. Although there is often much varied views regarding the relative merits of salt in various forms, there are no fundamental differences except some forms can be more readily eaten than others. There is no doubt that a lack of salt is a far more common cause of mineral deficiency among sheep flocks than is the lack of any other mineral. This arises because of the failure to provide it regularly and in ample quantities and because it is excreted daily, chiefly in the urine. Some may be lost in perspiration, but this is not as important an avenue of loss as in the case of other animals. A deficiency of salt is shown first by a great craving for it. A decline in thrift follows slowly, for under such conditions the excretion of salt almost ceases, and the small supply is retained as long as possible. There is listlessness, a loss of appetite, and a decline in weight. These symptoms are rapidly overcome with the restoration of salt to the diet.

(ii) *Calcium and Phosphorus*

These two minerals are closely associated in nutrition. They are the most important mineral constituents of the body so far as quantity is concerned. Together they make up about 70–75% of the total mineral matter of the animal body. About 99% of the calcium and 80% of the phosphorus is found in the bones and the teeth. They are very important in milk as they represent more than half of its mineral content. Hence, there is a strong need for calcium and phosphorus in the rations of pregnant ewes, of ewes giving milk, and of growing lambs. Mature animals that are not producing also need these minerals, for there are daily losses from the body which must be replaced.

The supply of these minerals is adequate in various regions in the pastures and roughages, used by sheep. But there are some regions where the supply of calcium and phosphorus in soils has been depleted by long crop production without replenishment. Under these conditions there is a lowered efficiency, reduced production, and cases of serious injury or even death. The requirement of calcium is greater than that of phosphorus, but sheep are more likely to experience a lack of phosphorus than of calcium. This is because roughages make up a large part of the ration of sheep, and roughages contain more calcium than phosphorus. It is only through the use of roughages of very low grade grown on poor soils that a serious deficiency of calcium is apt to result. Inadequate amount of calcium results in weakened bones and lameness. Rickets or "bent leg," is a condition usually known in sheep. This because of lack of calcium being deposited. This may be more likely due to an accompanying lack of vitamin D necessary for the deposition of calcium.

A lack of phosphorus is evident by symptoms similar in some respects to those due to calcium deficiency. There is stiffness and soreness of the joints, listlessness, lack of appetite or a deprived appetite accompanied by attempts to eat bones or chew wood. Other symptoms are reduced growth, loss of flesh, failure to breed regularly and poor milk production. In certain cases the lack of milk at lambing may be due to lack of phosphorus.

As phosphorus is a part of some of the proteins, deficiency will not develop if, protein-rich feeds are used.

Feeding extra amounts of calcium and phosphorus normally results in bones of greater strength and of denser structure, but there is very little effect on the outside diameter. The assimilation of calcium and phosphorus depends to some extent upon the relation between the two. A ratio of 2:1 is satisfactory.

(iii) ***Iodine***

This is one of the minerals that may be deficient in rations in certain regions of the country. About half of the idodine of the body is found in the thyroid glands of the neck. This origin is very important in relation to health, growth and general development, and reproduction. Deficiency symptoms are seen by lambs having very little or no wool at birth and by slight to extreme enlargement of the thyroid gland. This condition is described by practical farmers as 'big neck' or goiter. The mortality rate in such lambs is high. This can be controlled by the use of iodine supplement, such as iodized salt.

(iv) *Magnesium*

Magnesium is present in the bones, fluids and tissues of the body. It is rare that sheep experience a lack of this mineral, although, a few cases of 'grass tetany' have been ascribed to it.

(v) *Potassium*

This occurs largely in muscles. Apparently no deficiencies have been found, although some soils do not have an ample supply for large plant production.

(vi) *Sulphur*

Sulphur is contained in amino acids methionine and cysteine which are present in some proteins. Wool contains about 4% sulphur present in these amino acids; Apparently elemental sulphur can be utilized in small amounts but organic sources are of most significance in the usual rations of sheep.

(vii) *Iron, Copper, and Cobalt*

These minerals are associated in some of the body functions. Iron is an important part of the blood. Copper and cobalt are related in some way to the formation of blood. A lack of any one or more of these minerals results in a decline in the red blood cells and in the hemoglobin and the development of a nutritional anaemia. Cobalt is a constituent of vitamin B_{12} and for the synthesis of this vitamin by rumen bacteria. Cobalt deficiencies have been reported in several areas. The symptoms of cobalt deficiency are rather non-specific being similar to those of general malnutrition. The condition is often called "salt-sick", "bush sickness", or "pining". It has been suggested that a cobalt deficiency depresses the appetite and thus causes the deficiency syndrome. Diagnosis depends on the response of the animal to cobalt feeding. A lack of cobalt or copper can be corrected by using cobalt chloride followed by per sulphate.

(viii) *Manganese and Zinc*

The requirement are normally met in ample quantities in pastures or other feeds.

Mineral supplements

There are circumstances when the use of mineral supplements may be advisable. It is assumed that only 35–50% of the minerals of the ration are used by the animals and hence two or three times the theoretical requirement is therefore recommended. The producer can afford to be liberal, as the cost is usually rather low; but minerals are of no value except to correct deficiencies, since they do not stimulate production above a normal level and their value depends solely upon the extent to which the lack is made good. The practice of including in commercial mixtures many more components than are necessary, only increases the costs without improving the effectiveness. Commercial mixtures may be eaten in large amounts because of the inclusion of certain condiments, but the amount eaten is not necessarily a criterion of need or value. The real test depends if one secures better health, more rapid growth, increased efficieny and greater production of the sheep.

Mineral poisoning

Some minerals are poisonous to sheep. In some areas in the western region selenium occurs to such an extent in some feeds that a disease known as "alkali disease" is caused by it. Affected animals may lose their coats, become lame, lose their appetites, and finally die. However, if affected animals are fed on feeds free of selenium, recovery may be expected.

Fluorine

Fluorine is apparently not contained in plants to a harmful extent; but, if minerals, such as raw rock phosphate, that contain significant amounts of fluorine are fed, the animals may show the effects by a discolouration and softening of the teeth and bones.

Other minerals such as the lead of paints or of spray materials are poisonous, and suitable precautions should be taken to keep animals away from them to avoid losses.

8.1.2.6 *Vitamins*

Vitamins are metabolically essential. A general knowledge of the vitamins and their importance in nutrition, including the conditions under which deficiencies may occur and how to prevent them by the use of suitable feeds, is necessary for continued success in sheep feeding. In sheep some vitamins are synthesized in their tissues and some by micro-organisms in their gastro-intestinal tract. Fortunately, the development of deficiencies to the extent where they cause serious losses in sheep are apparently rare. At present it has not been fully established whether all of the known vitamins are essential in the diet of sheep. The two which seem to have most significance in sheep feeding are vitamin A and D. The symptoms of vitamin deficiencies are anorexia, reduced growth, dermatitis, weakness and staggering gait. In sheep vitamin A is more important and its deficiency can cause various kinds of blindness. Vitamin deficiency also leads to abnormal bone development, weak and still-born lambs and respiratory problems.

(i) *Vitamin A*

This vitamin is synthesized by animals from the carotene of plants. Carotene is abundant in the green parts of plants and is also plentiful in the yellow colouring matter of corn carrots and sweet potatoes. Its importance explains why sheepmen had long insisted and even insist today that there were certain things of great nutritive value in such things as carrots, clorets in Alpine pastures that had not been revealed by the usual analysis pertaining only to nitrogen-free extract, protein, minerals, and dry matter. Vitamin A promotes growth and aids in protection against respiratory infections and impaired vision. A deficiency of this vitamin in sheep is characterized by night blindness, sore eyes, poor appetite, poor condition, and weakness. Affected ewes may be, or their lambs may be very weak and die soon after birth. Although the carotene of feeds may decrease rapidly during curing and storage, high grade legume hay of good green color, good silage, and roots, such as carrots, are fair to good sources of vitamin A. Green pastures, however, are the best natural sources for sheep. It is not unreasonable to suggest that some of the difficulties that develop during the late winter months, especially after periods of summer drought, may be associated with a deficiency of this vitamin. It is stored in the liver of sheep, but there may be cases where storage is not ample to meet all the needs of breeding ewes and their lambs.

(ii) *Vitamins B*

Sheep seldom suffer from any lack of B vitamins, for it has been shown that they have the ability to synthesize practically all of the known vitamins (the B-complex) in the rumen. It is generally thought that all of the vitamins thus formed becomes available to the sheep as the feed is later digested. The B vitamins furnish a stimulation to the appetite, protect against nervous disorders, and are essential for reproduction and lactation. There may be cases of deficiencies in newborn lambs, but none has been thoroughly demonstrated and it is unlikely that they occur in older lambs and sheep, since, in addition to the synthesis in the rumen. Yeast is very high in some of the B vitamins.

(iii) *Vitamin C*

Except in the case of newborn lambs, it is not likely that sheep require a dietary source of this vitamin due to internal synthesis by microorganisms.

(iv) *Vitamin D*

For sheep the only important sources of vitamin D in feeds are in field cured hays, as growing plants and grains have very little or none of this vitamin. When sheep are exposed to sunshine, vitamin D is formed in the body by the action of the ultraviolet rays. Hence, any deficiency is most likely to be evident late in the winter after periods of cloudiness and when the sunshine is not so effective. Vitamin D is essential for the metabolism of calcium and phosphorus and for the formation of bone. Without adequate vitamin D, young animals develop rickets or a "bent-leg" condition. Vitamin D cannot make up for a lack of calcium and phosphorus, but these minerals cannot be effectively used in the absence of this vitamin. Some spasms and sudden death in winter lambs of few weeks of age have been associated with inadequate vitamin D.

(v) *Vitamin E*

In some animals normal reproduction is impossible without vitamin E, but there are few reports regarding its need by sheep. "Stiff lamb" or white muscle disease in newborn lambs results from vitamin E deficiency. Sheep can eat many types of fodders, grains and cakes. It can also eat some types of shrubs and leaves of variety of tree species. Sheep is also able to meet all its requirements of nutrients from grass lands except during winter season. Since most feeds, especially legume hays and green pastures, are good sources of the vitamin, the natural supply is seldom insufficient. Although high in vitamin E, wheat germ oil may not increase productivity.

8.1.3 Silvipasture grazing

Sheep throughout our country depend on meagre range vegetation, natural pasture grasses and crop residues on cropped lands. These natural range and common grazing land have generally been over exploited and are dominated by extremely poor and generally annual grasses and shrubby vegetation. The nutritive value of some important grasses is given in Tables 8.1 and 8.2. There is great prospect of improving these pastures through reseeding with more productive and nutritious perennial grasses and legumes and introducing fodder trees and shrubs.

Table 8.1. Digestible Nutrients of Grasses

Botanical name	Common name	DCP	TDN	SE	References
Tribullus terristriss	Gokhru	3.2	12.9	8.9	Shukla and Ranjan (1969)
Brachiaria mutica	Para grass	1.2	18.0	13.5	Joshi and Ludri (1967)
Panicum kabulabula		1.3	26.2	17.8	Joshi and Ludri (1967)
Brigival		2.1	12.5	11.5	Joshi (1966)
Garthamus tinctorius	Kerr	2.0	12.7	10.4	Ranjhan *et al.* (1959)
Ghiabate	Ipomoea	1.5	10.3	9.0	Talpatra and Bansal (1959)

The tree leaves, classed as emergency fodders for livestock, form integral part of feed for sheep in India. They serve as a potential source of feed during December to June when the grazing material becomes scarce in arid and semi-arid regions, and grazing lands are covered with snow in hill regions. The nutritive value of important tree leaves is given in Table 8.3.

The tree leaves contain comparatively lower percentage of crude protein than grasses and hays. Their fibres become complex, and with age highly lignified and crude protein content decreases and the crude fibre content increases. Their calcium content is 2–3 times more than in the cultivated fodders and grasses. The phosphorus content is low and so phosporus balance is negative. The dry-matter consumption by sheep from different top feeds is quite satisfactory as they consume on an average 3.3 kg from most of the annual and perennial grasses. Crude protein is digestible from about 30–80%. Tannic acid in tree leaves is a limiting factor for the proper utilization of protein and carbohydrates by ruminants. In spite of high intakes of nitrogen and calcium, the balances of these nutrients are poor when the tree leaves are fed as sole feed. The wide ratio of calcium to phosphorus in take results in poor utilization of these inorganic nutrients. The higher calcium in tree leaves is of no use unless phosporus is supplemented with them.

8.1.4 Traditional grazing practises

Under the existing system of feeding management sheep in India are mainly raised under exclusive rangeland grazing with supplementation of top feed resources during critical lean season. Organized husbandry practises, nutritional inputs commensurating with various physiological stages and production traits and prophylactic and curative health coverage are almost negligible. The native sheep have poor production potential and efforts for their genetic upgradation by introduction of high producing breeds has not been largely successful under field condition due to lack of supporting feed input. The sheep under field condition subsist by grazing/browsing on variety of grasses, bushes and top feeds. *In vogue*, grazing practises for sheep are indiscriminate

Table 8.2. Chemical Composition of Grasses on Dry Matter Basis

S.No.	Botanical name	Common	DM	CP	CF	EE	NFE	Total ash	Calcium	Phosphorus	References
1.	*Amaranthus spinosa*	Goja	–	16.13	20.29	2.51	42.12	18.89	2.28	2.28	Singh and Joshi, 1957
2.	*Andropogon annulatus*		Young	9.70	33.06	1.81	42.51	13.58	–	–	Mathur, 1960
3.	*Andropogon contortus*	Speargrass	50.0	1.87	11.12	0.26	–	2.00	0.22	0.12	Chucko, 1954
4.	*Andropogon muricatus*		Ripe	4.84	36.60	1.18	44.18	12.60	–	–	Mathur, 1960
5.	*Briachiaria ramosa*	Kura	–	9.43	28.53	0.78	43.92	–	–	–	–
6.	*Cenchrus ciliaris*	Dhaman	Ripe	11.42	27.20	1.87	41.14	18.10	–	–	Mathur, 1960
7.	*Cynodon dactylon*	Doob	Ripe	4.84	36.60	1.18	44.18	12.6	–	–	Mathur, 1960
8.	*Cyperus iria*	–	Ripe	6.04	32.77	6.89	51.69	8.61	–	–	Mathur, 1960
9.	*Dicanthium annulatum*	Karad	–	2.40	35.77	2.01	49.84	–	–	–	–
10.	*Carthamus tinctoris*	Kerr	–	12.40	30.89	1.94	44.37	–	–	–	Ranjan *et al.*, 1959
11.	*Heteropogon contotus*	–	Young	5.00	34.24	1.36	47.70	11.64	–	–	Mathur, 1960
12.	*Ischeum rengachunianum*	–	60.0	1.90	13.27	0.87	–	2.22	0.20	0.05	Chucko, 1954
13.	*Lasiurus sindicus*	Seven	–	5.98	30.08	–	44.73	–	–	–	–
14.	*Panicum sa uilatis*		–	7.47	31.64	2.01	59.19	–	–	–	Mathur, 1960
15.	*Panicum antidotale*	Blue panic	Ripe	7.21	40.07	1.91	43.11	7.07	–	–	Mathur, 1960
16.	*Panicum colonum*		Ripe	7.26	40.47	1.19	43.11	7.79	–	–	–
17.	*Panicum maximum*		Ripe	7.69	37.33	1.67	39.44	13.89	–	–	Mathur, 1960; Saxena *et al.*, 1972
18.	*Pennisetum flacidum*		21.15	11.36	30.40	1.97	45.07	11.21	0.34	0.30	Venkatachalam, 1960
19.	*Tribulus terestris*	Gokhru	–	12.10	27.80	2.60	40.80	–	–	–	–

Table 8.3. Nutritive Value of Fodder Tree Leaves of India on DM Basis

Trees	DCP (kg)	TDN (kg)	DE (Mcal per kg)	ME (Mcal per kg)
Ardu (*A. grandis*)	13.0	63.0	–	–
Banj (*Quercus incania*)	5.7	43.7	1.92	1.58
Beri (*Ziziphus jujuba*)	3.1	30.7	1.35	1.10
Bankli (*A. latifolia*)	0.6	47.7	2.15	1.76
Bamboo (*Dendrocalamus strictus*)	9.3	48.8	2.40	1.97
Bahera (*Beleria*)	16.7	54.6	2.40	1.97
Bhimall Biul (*Grewia opposositifolia*)	16.7	54.6	2.40	1.67
Bel (*Aegle marmelos*)	10.8	56.7	2.40	2.04
Bagad (*Ficus bengalensis*)	0.81	17.8	0.78	0.64
Bahera (*T. belerica*)	0.8	54.5	2.40	1.96
Gauj (*Millettia auriculata*)	15.5	44.9	1.98	1.62
Jaman (*Syzygium cumini*)	0.02	17.5	0.77	0.63
Kanju (*Molopteoea integrifolia*)	5.2	58.1	2.56	1.20
Khoda (*Elevis*)	8.4	54.8	1.41	1.98
Pakar (*Ficus infectora*)	1.61	18.3	0.81	0.66
Phaldu (*Mitrazyna parviflora*)	1.58	49.9	2.22	1.80
Jheengham (*L. grandis*)	4.9	55.2	2.43	1.95
Jharberi (*Ziziphus nummularia*)	5.5	51.1	2.26	1.84
Kusum (*Scheleichera trijuga*)	3.4	47.5	2.09	1.71
Katchnar (*Bauhinia variegata*)	9.2	55.5	2.45	2.00
Kumbhi (*Careya arborea*)	0.2	46.3	2.04	1.67
Lasora (*Cordia obliquoa*)	5.4	26.9	1.18	0.97
Mahua (*Madhuca indica*)	0.0	37.0	2.16	1.38
Marorfali (*Helicteres isora*)	9.17	58.4	2.57	2.11
Neem (*Azadirachta indica*)	8.4	53.3	2.34	1.92
Pipal (*Ficus religiosa*)	7.0	38.3	1.68	1.38
Pula (*Kydia calycina*)	7.9	45.2	1.99	1.63
Rohina (*Mallotus philippensis*)	1.2	34.0	1.50	1.23
Ratendu (*Saurauia napaulensis*)	1.2	34.0	1.50	1.23
Sainjana(*Oleifera*)	11.0	61.5	2.71	2.32
Sal(*Shorea robusta*)	1.1	42.7	1.88	1.57
Siras (*Albizia labbeck*)	10.5	49.3	2.17	1.61
Sandal (*Qugeinia dalbergoides*)	3.7	45.6	2.00	1.64
Sain(*Termenatia tomentosa*)	0.0	34.9	1.54	1.26
Tut (*Madhuca indica*)	10.7	59.6	2.76	2.15
Tilonj (*Quercus dilatata*)	4.2	43.2	1.92	1.58

Source: Rajhan, S.K. 1989. Agro-industrial Byproducts and Nonconventional Feeds for Livestock feeding. Indian Council of Agricultural Research

and governed by compulsions of feed resource availability, taboos and tradition specific for the regions and socio-economic conditions of the farmers. Under the existing management practises of exclusive rangeland grazing the lambs hardly achieve 15 to 16 kg at six months of age whereas they are marketed for slaughter around 9 to 12 months weighing about 20 to 22 kg with an average carcass weight of 10 kg. The emerging awareness of consumers for quality meat, export avenue to Middle Eastern countries and changing land utilization pattern is likely to alter the small ruminant production system in the country and in future majority of the stock will be raised on stall feeding or grazing with supplementation to meet the market requirement. Appropriate technologies have been generated to meet the requirements of changing scenario.

(a) Extensive natural range management

Long term experiments with sheep and goats maintained under free grazing on semi-arid forest rangeland at stocking density of 3 sheep and goats per hectare indicated that the lambs weighed 11, 16 and 21 kg at 3, 6 and 12 months of age and in similar experiments on arid range land spread over three years the lambs weighed 11 and 16 kg and kids 10 and 13 kg respectively. The innovative experiment involving a local sheep farmer to graze and manage sheep as his own stock on exclusive rangeland grazing and stocked as per carrying capacity, it was evident that overall productivity of native sheep was better than the crossbred. Continuous grazing provided desirable production traits in the first year which decreased in second year followed by termination of experiment in the first quarter of third year as the pasture could not maintain the animals leading to heavy mortality. The results indicated that the pasture even stocked as per carrying capacity has to be provided rest for its proper rejuvenation and health. Survey spread over five years, conducted on farmers own flock maintained on public grazing land in villages indicated that the native and crossbred sheep weighed 9.2, 13.1, 1 6.1 and 19.7 kg respectively at 2, 4, 6 and 9 months of age and crossbred were heavier than native. Similar studies on lab to land and Operational Research Project adopted sheep farmers of Rajasthan indicated that birth, 3, 6, and 9 months weights of native lambs were 2.9, 10.9, 13.5 and 20.2 kg while the corresponding weights in crossbreeds were 3.0, 11.6, 14.2 and 19.5 kg respectively. The results indicated that crossbreeds were heavier than natives at birth, 3 and 6 months of age while at marketable age of 9 months natives were heavier than crossbred.

(b) Cultivated pasture

The perennial monoculture grass (*Cenchrus ciliaris*) pasture in semi-arid region has biomass yield of 2.0 t/ha which is sufficient to maintain 5 small ruminants or 1 Adult Cattle Unit (ACU) with followers round the year. The lambs born to the ewes maintained exclusively on *Cenchrus* pasture had birth, weaning, six and 9 months weight of 3.2, 13.9, 20.6 and 23.9 kg respectively with empty live weight dressing yield of 54.3%. Similar studies conducted on cultivated sewan (*Lasirus sindicus*) pasture in arid location at stocking density of 5 sheep/ha indicated that the pasture intake was insufficient to maintain them from January to June requiring supplementation during lean period. Nutrient intake and utilization in sheep for supporting desired growth rate in lambs and to sustain adults stock in critical physiological stages, addition of pasture legumes *viz.* Dolichos and Serato or intercropping with Cowpea has definite advantage.

(c) Silvipasture

Establishment of silvipasture improves quality of nutrient available ensuring its supply round the year. Experiment conducted at CSWRI has indicated that the hoggets are able to achieve 30 kg at yearling age whereas under grazing on natural pastures with some supplementation with grass and cultivated fodder hay it was 25.0 kg. Similar studies conducted on ewes in advance stage of gestation and lactation indicated that pasture intake was not adequate for them requiring supplementation to harvest desirable production traits. Another study with weaner lambs and kids on three and two tier silvipasture indicated that lambs and kids were able to achieve 22.0 kg body weight at 6 months of age without concentrate supplementation.

8.2 Supplementary Feeding

The slaughter weight in Indian sheep is far lower and the age at which target weight is achieved is far higher than expected. The system requires input of concentrate which although cost effective, yet has not been adopted due to socio-economic conditions of the farmer. The work done in the country on grazing with supplementation have indicated that in addition to free grazing supplementation of limited quantity of concentrate amounting to 1.5–2.0% of body weight will provide marketable finishing weight of 25 and 30 kg at 6 and 9 months of age respectively. The lambs under sole grazing on *Cenchrus* pasture with concentrate supplementation @ 1.5% of their body weight achieved 26.7 kg at 6 months of age with about 100 g average daily gain.

8.2.1 Maintenance requirements

An animal has to maintain itself before it is able to produce in the form of tissue growth, development of foetus and milk or wool production. Maintenance requirements include the requirements for basal metabolism (BM) plus the normal activity without any production. Energy is the basic need to keep the animal's physiological processes going. The next is nitrogen (protein) which is required to meet the normal wear and tear of tissues. Mineral and vitamins are also needed to aid metabolic processes and normal tissue functioning. Several experiments in India have shown that for maintenance of adult sheep weighing 30–40 kg, 350–450 g of TDN per day are required. In other words, 10 g of TDN per day per kg body weight should be ensured to maintain the weight of the animal.

8.2.1.1 *Energy requirement*

Energy requirement for maintenance could be met from grazing alone, if a well developed pasture is available. Generally, during the period from January to June in arid and semi-arid areas, the energy availability to the animal is even less than the maintenance requirement thereby necessitating additional supplementation. The TDN content of the pasture, forage is about 50–55% during lush green season and 35–40% during dry months. Assessment has to be made at the local level whether the pasture available could maintain the animals. If the animals are fed in stalls on legume hay or a high quality grass it is sufficient to maintain the animals. The vegetation resources on the grazing land should be developed in 3 tier system (grass-shrub-trees). No additional supplementation is required just to maintain the animal, if sufficient vegetation is available in the grazing area. The best system of management should be to leave the animals to themselves in a paddock although the initial investment is very high to make a paddock.

8.2.1.2 *Protein requirement*

The protein requirement for maintaining the body weight of animal is based on the metabolic faecal nitrogen (MFN) and the endogenous urinary nitrogen (EUN) excretion plus the nitrogen requirements for normal growth of their hair and hoof and nitrogen loss from the skin. About 2 mg of nitrogen or 12.5 g of protein are required per Kcal of basal heat production. This value refers to protein actually needed in the replacement of nitrogen. In actual feeding practise extra amount should be provided to allow for wastage in digestion and metabolism. Provision of twice the amount of nitrogen/protein required per kcal of basal heat would meet the maintenance requirement of the digestible protein. It is reported that 0.32 g of DP kg body weight could maintain the weight of the rams (Singh *et al.*, 1979). This value is quite low compared to that reported by Gill and Negi (1971) who found 23 g of DCP was required for a ram of 30 kg weight. A diet containing 3.4% DCP with a nutritive ratio of 1:13 was found to be adequate for maintenance of a ram (Sharma and Negi, 1977). Grazing studies at Central Sheep and Wool Research Institute, Bikaner have shown that even with low level of digestible protein intake (7 to 9 g per day) during the period from January-June, Marwari ewes maintained weight. It appears that at low level of intake of protein, the animals recycled nitrogen and met their protein requirements to maintain weight. On the basis of available data, it can be concluded that in practical feeding conditions it would be desirable to provide 1g of DCP per kg of live weight for maintenance of sheep. This requirement could easily be met if legume fodder is fed or if the animals graze on green pastures. With dry pastures, during summer or with feeding of straws or low quality grass hay, additional supplementation of protein would be necessary.

8.2.1.3 *Mineral and vitamin requirement*

If the energy and protein requirements are met through conventional feeds, the mineral and vitamin requirement are also met for maintaining the weight of the animals except in specific deficient pasture or feed. During summer months, there is a possibility of vitamin A deficiency but this has not been demonstrated in clinical form, although there is evidence of a low level of vitamin A in blood during summer. (Hazra and Patnayak, 1975). The requirements of different nutrients for maintenance of adult sheep weighing about 35 kg are given in Table 8.4. These requirements are based on the figures given in the NRC tables.

Table 8.4. Requirements for Different Nutrients for Maintenance of an Adult Sheep Weighing 35 kg

Nutrient	Required/day*
Total feed	900 g
TDN (Total Digestible Nutrients)	450 g
DCP (Digestible Crude Protein)	45 g
Calcium	2.4 g
Phosphorus	2.0 g
Magnesium	1.2 g
Common salt	5.0 g
Potassium	4.5 g
Sulpher	1.5 g
Copper	8.0 mg
Cobalt	0.2 mg
Zinc	75 mg
Iron	60 mg
Manganese	30 mg
Iodine	0.1 mg
Molybdenum	1.5 mg
Selenium	0.1 mg
Carotene	1.5 mg
Vitamin A	750 IU
Vitamin D	200 IU

* Air dry basis

8.2.2 Feeding for production

Growth

Growth is a very complex physiological process in the sheep production system. It can be defined as a correlated increase in body mass in definite intervals of time in a way characteristic of a species.

True growth is characterised primarily by an increase in protein, mineral matter and water. Fat deposition is not considered as growth. It is however difficult to partition the growth and fat deposition in the normal process. Although in a strict sense the starting point of growth is conception from the feeding point of view. It is better to consider the growth requirement after birth. The requirement for growth from conception to birth can be covered under feeding of pregnant ewes. The type of feeding for growth would depend upon the age, physical and physiological state of the stomach compartments.

Pre-weaning growth

Growth during this period play an important role in the subsequent development of the lamb. From birth to 3 weeks of age, the lamb has non-ruminant status and feeding during this period should be mostly on liquid milk and milk substitutes. From 3 weeks to 2 months of age, it is a transition phase during which rumen development takes place and at 2 months of age the development is completed. The creep feed should be fed to lamb as early as 2 weeks of age so that the lamb would consume more energy and also the rumen development could occur faster. The muscular development of rumen depends upon intake of dry feed. The animals offered creep feed show greater ability to digest starch and cellulose than those reared on milk alone or those on grass pasture. Addition of high quality roughage further enhances the development process in the rumen.

The feed offered during pre-weaning period should contain higher proportion of grains, *viz.* maize barley and sorgham etc. It should also contain fish meal or skim milk powder say about 5% alongwith mineral mixture and vitamins to ensure wool growth. The protein percentage should be 14–16%. The feed consumption level during this stage should be to the extent of 5% of the body weight to obtain a growth rate of 200 g per day. Under intensive feeding system, in other words, the growth rate of 250–300 g per day has been obtained with a feed conversion ratio of 2.7:3.5. Under Indian conditions such a growth rate has not been possible so far. However, a feed conversion ratio of 5 has been obtained in a feed lot experiment. To obtain good growth, the consumption level of energy and protein should be 3 times the maintenance requirements and the protein to energy ratio should be 1.8:8.5.

Post-weaning growth

Post-weaning growth is generally the fattening and finishing period in the lamb feeding practise in advanced countries, where fat lamb production is a commercial enterprise. This period is very important to prepare the male lambs for the market or to get a good replacement ewe. Fattening is successful if pre-weaning growth is satisfactory. During this period 12% protein in the diet is adequate. If the pre-weaning growth is poor, a diet with higher percentage of protein(14%) is necessary. A consumption level of 4–4.5% of the body weight has to be ensured during this stage. To ensure this rate of intake, the energy density of feed should be very high and feed having concentrate to roughage ratio of 60:40 should be the minimum requirement. Higher percentage of concentrate upto 80% would increase the rate of growth in lambs. In feed lot experiments under the All India Co-ordinated Research Project, a growth rate of 150–200 g/day over 90 days feeding has been obtained in lambs. A diet of 70% concentrate and 30% roughage has been found quite suitable. If a good legume fodder is fed, a ration of 50% roughage and 50% concentrate has produced equally good results. Complete feed containing ground fodder, *viz.* lucerne, cowpea etc. grains, oil cakes and wheat bran have been developed. A lot more work need to be done to economise the ration by using the unconventional feed resources in the complete feed, without affecting growth.

A feeding level of 2.5 times of maintenance requirement of energy and protein has produced weight gain of 100 g/day over a feeding period of about 6 months (Maheshwari, 1977). For higher rate of growth, feed at about 3 times the maintenance level would be necessary. The protein to energy ratio in the ration should be 1:9. It has been observed that in our conditions the post-weaning growth continues up to one year of age, although after 9 months it is quite slow. If there is adequate nutrition during pre-weaning and during the 1st 3 months post-weaning, there may not be much growth between 6–12 months of age. However, experience has shown that supplementation of concentrate or good legume hay upto one year of age gives beneficial results. A check in growth has been noticed in the 1 superscript to 2–3 weeks following weaning causing an adverse effect on overall growth. Prolonged growth check after weaning need to be avoided and this may be possible by ensuring full development of rumen by appropriate feeding and management during pre-weaning. The energy and protein requirements for growth calculated on the basis of some recommended formulae are given in Tables 8.5–8.6 respectively. Post weaning growth of the lamb is also very important to prepare the male lamb for the market or to get a good replacement ewe. The feeding rate during this period should be about 4–5% of the body weight.

Table 8.5. Total Digestible Nutrient (TDN) Requirement (g/day) for Maintenance and Growth of Lambs

Body weight (kg)	Maintenance TDN	Weight gain			
		50 g/day	100 g/day	150 g/day	200 g/day
5	120	150	210	250	300
10	168	211	254	296	339
15	222	279	335	391	447
20	271	340	408	477	546
25	330	414	497	581	664
30	374	469	564	658	753
35	425	476	640	747	855
40	474	594	714	834	954

Table 8.6. Digestible Crude Protein (DCP) Requirement (g/day) for Maintenance and Growth of Lambs

Body weight (kg)	Maintenance DCP	Weight gain			
		50 g/day	100 g/day	150 g/day	200 g/day
5	10	15	20	25	30
10	16	20	25	30	35
15	21	28	34	40	47
20	26	34	42	50	58
25	31	41	50	59	69
30	36	47	57	68	79
35	40	52	64	76	89
40	44	58	71	84	98

The male hoggets should be disposed off for slaughter betwen 6 to 9 months of age except the ewes to be retained for breeding. The female hoggets have to be maintained separately until they join the breeding flock at about 12 to 18 months age. If the post-weaning growth is adequate in normally good pasture. In the absence of good grazing, a good legume hay should be supplemented. If the early growth is not satisfactory, supplementation of about 200–250 g concentrate would be necessary upto yearling age in addition to the fodder supplementation/good grazing.

Yearling stage

This stage is only relevant to the replacement females, and males that are retained for breeding and pertains to 9 to 18 months of age. The female hoggets have to be maintained separately until they join the breeding flock at about 12 to 18 months of age. During this stage, provision of grazing on a good pasture would be adequate to meet the requirements.

Feeding of rams

When the rams are to be mantained they should be only for breeding purpose. Otherwise they should be sold for slaughter. In some places in the field, farmers keep large number of rams in their flocks only for getting the wool clip. Some farmers have also religious sentiments not to slaughter their rams. This practise should be discouraged. Generally if good grazing is available, rams do well on grazing alone and no additional supplementation is required. During breeding season, concentrate supplement should be provided. In the absence of concentrate, supplement with good legume hay should be provided in the quantity twice the recommended concentrate allowance. During non-breeding season if the rams loose weight, fodder supplement should be provided. About 5 to 6% DCP in the fodder/pasture is adequate to maintain the animal during non-breeding season.

Feeding of ewes

For proper feeding purpose, the ewe flock should be divided into dry, early pregnancy, late pregnancy and lactating ewes and fed separately after grazing.

8.2.3 Feeding for reproduction

Reproduction stages include tupping and pregnancy. If the animals are under optimum nutritional state, they come into 1st oestrus at 10 to 12 months of age. Otherwise it starts at about 15 months of age. Under Indian conditions one and half year age is considered as the breedable age.

Flushing

Flusing is conditioning of ewes for breeding. If the ewes are in low plane of nutrition prior to breeding, additional supplementation for about one month has beneficial effect in bringing the ewes into oestrus. Even without additional supplementtion when there is flush green pasture, there is flushing effect. Supplementation of about 250 g of concentrates could bring about flushing of ewes quite well. If the plane of nutrition of the animal is good prior to breeding, flushing is not at all needed.

Feeding during pregnancy

If the ewes are flushed and good grazing is available, no additional supplementation is necessary during early pregnancy upto about 14 weeks of gestation period. During advanced pregnancy (last 6 weeks) however, extra feeding is essential. The consumption of energy, protein and other nutrients should be 50%. During this stage, there is also depression in the intake capacity and feed digestibility. Hence highly digestible feed need to be fed during this stage.

8.2.4 Feeding during lactation

As already indicated, during lactation the requirements are twice the requirement for maintenance of the animals. It is not possible to provide the nutrients required by the mother for optimum growth of the nursing lamb from all forage rations. Hence supplementation with high energy feed is essential. A diet with 40% concentrate and 60% roughage would be appropriate for the lactating ewes. A reduction in the concentrate allowance would adversely affect the growth of the lamb. Adequate quantities of minerals and salt are to be provided both to the pregnant and lactating ewe. An ideal system would be to graze them in separate paddocks and provide them additional supplement as per the need after grazing. However, this has not been possible in most farms in our country and every effort should be made to adopt the system. The later part of pregnancy (last 4–6 weeks) and early lactation period, 4 to 6 weeks post partum is most critical for the ewe as well as for the lamb. Higher amounts of energy and protein should be provided to the ewe during this period. The nutrient requirement increase by about 50% during late pregnancy and by about 100% during lactation over the maintenance requirement of the ewe for the growth in lambs. A feeding schedule for different categories of sheep is given in Table 8.7.

Table 8.7. Suggested Feeding Schedule for Different Categories of Sheep

I.	**Exotic sheep**	
1.	Rams (70–80 kg)	
(a)	Breeding	(i) Grazing (ii) Dry fodder 1.5 kg Breeding (iii) Concentrate 500 g
b)	Non-breeding	(i) Grazing (ii) Dry fodder 1.5 kg (iii) Concentrate 300 g
2.	Ewes (40–50 kg)	
a)	Dry and early pregnancy	(i) Grazing (ii) Dry fodder 1.5 kg (iii) Concentrate 300 g

Table 8.7. (*Contd.*)

b)	Last month of pregnancy	(i)	Grazing
		(ii)	Dry fodder 1.5 kg
		(iii)	Concentrate 400 g
c)	Lactation (1st month)	(i)	Grazing
		(ii)	Dry fodder 1.5 kg
		(iii)	Concentrate 500 g
d)	Lactation (2nd and 3rd month)	(i)	Grazing
		(ii)	Dry fodder 1.5 kg
		(iii)	Concentrate 400 g
3.	Hoggets and yearlings (30 and 40 kg)	(i)	Grazing
		(ii)	Dry fodder 0.5 kg
		(iii)	Concentrate 250 g
4.	Weaners (3 to 6 months)	(i)	Grazing
		(ii)	Dry fodder 500 g
		(iii)	Concentrate 350 g
5.	Pre-weaning lambs (1–3 months)	(i)	Grazing
		(ii)	Succulent hay or green fodder equivalent to 300 g dry fodder
		(iii)	Lamb creep for nibbling *ad lib*
6.	Young lambs (1 week to 1 month)	(i)	Green fodder for nibbling
		(ii)	lamb creep *ad lib* for nibbling
II. Native and crossbred sheep			
1.	Rams		
a)	Breeding	(i)	Grazing/top feeding
		(ii)	Dry fodder 500 g
		(iii)	Concentrate 300 g
b)	Non-breeding	(i)	Grazing/top feeding
		(ii)	Dry fodder 1 kg
2.	Ewes		
a)	Dry and early pregnancy	(i)	Grazing/top feeding
		(ii)	Dry fodder 1 kg
b)	Last month pregnancy	(i)	Grazing/top feeding
		(ii)	Dry fodder 700 g
		(iii)	Concentrate 300 g
c)	1st month lactation	(i)	Grazing/top feeding
		(ii)	Dry fodder 700 g
		(iii)	Concentrate 450 g
d)	2nd and 3rd month lactation	(i)	Grazing/top feeding
		(ii)	Dry fooder 700 g
		(iii)	Concentrate 300 g
3.	Hoggets and yearlings	(i)	Grazing/top feeding
		(ii)	Dry fodder 700 g

Table 8.7. (*Contd.*)

4.	Weaners (3–6 months)	(i)	Grazing/top feeding
		(ii)	Dry fodder 500 g
		(iii)	Concentrate 350 g
		(iv)	Green fodder if available
5.	Pre-weaning lambs	(i)	Grazing/top feeding (1–3 months)
		(ii)	Concentrate ad lib
		(iii)	Green if available
6.	Young lambs (1 week to 1 month)	(i)	Green fodder *ad lib*
		(ii)	Concentrate *ad lib*

8.2.5 Concentrate and hay feeding, feed lot experiments and their impact on growth, carcass yield and quality

Post-weaning growth is primarily affected by hereditary factors, plane of nutrition, prevailing meteorological conditions, animals ability to adapt to the environment and managemental stresses. In agriculturally advanced countries post-weaning phase of growth is mainly used for fattening and finishing purpose, whereas in our intensive mutton production strategies, the active growth is completed by 5–6 months of age depending on the weaning age of lambs. In view of lower growth rate of native sheep, under All India coordinated Research Project on Mutton, improvement in their rate of gain and productivity was attempted through crossbreeding local sheep with superior mutton breeds. The crossbred weaner (90 days) lambs in 90 days intensive feeding exhibited superior growth rate [180 g average daily gain (ADG)] compared to their native (127 g ADG) counterparts. In view of similar growth performance of different genetic lines of evolved crossbreds lambs, the defined exotic crosses were merged together and renamed as Mutton synthetic. These Mutton synthetic (MS), Malpura selected (IM) and Sonadi (S) weaner (90 days) lambs could achieve 170, 150 and 118 g with 1:5 to 7 feed efficiency ratio under intensive feeding on 50:50 roughage (Pala leaf) and concentrate based ration. In the experiment out of 99 MS lambs used, 90% lambs achieved finishing weight of 30 kg live weight in 56 days of intensive feeding and out of 7 IM lambs 71% achieved target weight in 66 days. Similarly fed native Nali, Chokla and their crosses with Merino/Rambouillet weaner lambs had average daily gain of 111–135 g with 1:12–15 feed efficiency ratio. Faster gain in feed lot lambs is associated with higher fat content of carcass some time exceeding 20% hence, the weaning age of lambs can be reduced from 90–60 days to improve feed conversion efficiency, save higher feed input preventing undesirable fat deposition and to provide early economic return. Accordingly lambs after weaning at 60 and 90 days were put to feedlot experiment on a 50:50 roughage concentrate ration for further 90 days. The Mutton synthetic lambs had 160 and 151g ADG, 16 and 12% feed efficiency, respectively, in groups weaned at 60 and 90 days of age and respective figures for Malpura lambs were 139 and 146 g ADG and 16 and 14% feed efficiency. In slaughter studies it was observed that irrespective of age at slaughter and feeding regimen carcass of the finisher lambs weighing around 25 kg and provided desirable carcass characteristics. In accordance with the findings the lambs can be weaned at 60 days of age and fed intensively till they achieved 25 kg live weight and slaughtered thereafter. In such feeding regimen also with 50:50 concentrate and roughage based feed lot ration, the MS; IM and MC lambs achieved finishing weight of 25 kg in 73, 91 and 136 days of intensive feeding with 162, 135 and 112 g ADG and 18, 16 and 14% feed efficiency, respectively. Incorporation of different sources of energy (maize, barley, jowar, bajra and damaged wheat) in feedlot rations of weaner lambs indicated that all these sources could be effectively utilized in growing animals as their ADG ranged between 92–114 g in different groups. Similarly while exploring use of different protein supplements in feed lot ration (groundnut cake, cotton seed cake, mustard cake and guar meal) and its complete replacement with urea, it was observed that growth response was better with groundnut and cotton seed cake (137 g ADG) compared to guar meal, mustard cake and urea replacement (112 g ADG) groups.

8.2.6 Feed compounding

The sheep production practises in our country are different from that in most of the advanced sheep producing countries. The land left after grazing by large animals is generally available for sheep. Thus sheep have to depend largely upon meagre grazing on natural vegetation, crop stubbles and straw and other agricultural and in dustrial by-products that cannot otherwise be used by the human beings and large ruminants. For obtaining moderate production from indigenous as well crossbred animals only grazing may serve the purpose but in recent time under the situation where sheep are blamed for surface vegetation removal, deforestation and soil erosion because of their close grazing habits, then alternative feeding practises are required to be adopted to obtain higher production under intensive feeding management specially for fat lamb production. For this, available raw material be suitably compounded to arrive at economical concentrate mixture. Research conducted in this area indicated that feed mixture (crude protein content 16–17%) prepared by using equal quantity of tree leaves, seasonal grasses, legume straws and *kadbies* along with coarse grain, bran and oil cakes etc is ideal growth ration for fat lamb production. Feed compounding should be an important aspect of mutton production in future under the situation where grazing land is shrinking due to reclaimation of land for grain production. Utilization of newer unconventional raw materials which are otherwise going waste can effectively be used by compounding them with concentrate mixture at various levels and can serve as maintenance as well as production ration for sheep. Feed compounding plants should be established to cater the need for compounded feed for different catergories of sheep.

8.3 Pasture Improvement and Management

Pasture is a valuable fodder for sheep as it is cheapest source of proteins, energy, minerals and vitamins necessary for maintenance and production. Sheep in India are low producers as they are generally under-nourished. The pasture lands are also poor and meagre and need to be improved by protecting them from biotic factors, conserving good natural grasses, choosing the best fodder trees and shrubs, removing non-edible grasses, weeds and shrubs, and reseeding with nutritious and perrenial grasses and legumes. Natural legumes like *Rhynochosia minima*, *Indigofera endecaphylla* and *Gribulus terrestris* are very useful and should be preserved. The range and common grazing lands and forest lands should be reseeded with nutritious perennial grasses like *Cenchrus ciliaris Lasiurus sindicus* and *Dichanthium annulatum* in arid and semi-arid plain regions depending upon rainfall and soil conditions. It is also essential that some perennial legumes like *Dolichos lablab* (carpet legume), *Clitoria ternatea* (titli matar), *Macroptilium atropurpureum* (siratro), *Atylosia scarbacoides* and *Stylosanthes* species are incorporated in the regenerated or reseeded pastures. In sub-humid plains, *Sehima nervosum* is a good perennial grass. For temperate and sub-temperate regions, perennial grasses like fescue, rye grass, kikyu and legumes, like white and red clovers, can be utilized for developing pastures both in the forest area and orchards as well as mountain areas. Through associated growth of grass and legumes in the pasture, production of nutritious forages can be increased by 20 to 30% over that of grass alone. The legumes, besides being rich in protein content are more palatable and more digestible. They also enrich the soil by nitrogen fixation and help in checking soil erosion. During the first year of pasture establishment, grazing should not be allowed but grass and legumes (used as nurse or cover crop or as a pasture vegetation) be harvested, conserved as hay and fed to be sheep during lean period. During subsequent years neither additional sowing nor harvesting will be required. On an average such a pasture may last for 6 years and will provide 25–30 quintals of dry matter/ha a year. Life span of established 68 pasture can be further extended by 2 to 3 years with adoption of appropriate renovation techniques of tiller cultivation with 60 kg N and 30 kg P_2O_5 per hectare at the start of monsoon with sufficient moisture. The management of such improved pastures is equally important. Protection of pasture, removal of undersirable bushes and weeds, soil and water conservation, application of fertilizers, proper stocking rate and grazing system (rotational or deferred rotational) are very important for getting best out of these pastures. A good *Cenchrus* pasture can provide sufficient grazing for 5 sheep per hectare, while a protected range land can allow only stocking of 2 sheep per hectare. The unprotected range land can hardly support 27 g one sheep per hectare throughout the year.

Rotational grazing by dividing the pasture into 4 equal compartments and allowing grazing sequentially, should be practised. This will help the grasses to regenerate and check the soil erosion caused by overgrazing and allowing various operations to the resting pasture. The growing lambs may be allowed to graze first followed by pregnant and lactating ewes and dry stock in the last. This will make the lambs grow faster and avoid picking infections especially of gastro-intestinal parasites. It will also help the ewes to produce healthy lambs and more milk to allow good suckling by the lambs in addition to producing more wool and meat. At the time of field preparation sheep manure @ 10 tonnes per hectare, reseeding of pasture and top dressing of inorganic fertilizer @ 30 kg N/ha at regular intervals should be applied. Controlling the grass pests is also important for maintaining a good pasture. Spraying and dusting with DDT, BHC, Gemmaxene etc should be done as and when required. The animals should not be allowed to graze for 2 to 3 weeks after spraying. Timely hoeing and weeding operations will not only improve the forage yield but will also help in checking the growth of undesired bushes, grasses and weeds, and prevent worm infestations. If cattle, sheep and goats are to graze on the same pasture, it will be desirable to allow goats first followed by cattle and sheep.

8.3.1 Pasture improvement programmes

The fodder trees serve as a potential source of feed for sheep during December to June when the grazing resource becomes scarce. The common fodder trees of arid and semi-arid regions are ardu (*Ailanthus excelsa*), arjun (*Terminalia arjuna*), babool (*Acacia arabica*), dhamani (*Grewia elastica*) kankera (*Gymnosporia*), kachnar (*Bauhinia variegata*), khark (*Celtis australis*), khejri (*Prosopis cineraria*), kheri (*Acacia senegal*), subabul (*Leucaena latisiliqua*), neem (*Azadirachata indica*), semal (*Bombex malabaricum*) siras (*Albizia lebbek*), timla (*Ficus roxburghi*) zinja (*Baukinia recemosa*). At present their number in the range and common grazing lands is very small. Since they are being lopped very severely every year, they are not able to provide sufficient fodder. Hence a scientific method of lopping should be practised for getting maximum fodder from a tree. Through lopping studies it has been observed that Ardu trees should be lopped at six month interval whereas in *Khejri* the young trees should be lopped once a year and the full grown trees twice a year. In neem the reverse trend of khejri can be adopted. The trees not only provide green fodder during scarcity period but also check the soil erosion and improve soil texture. They also provide shade to the grazing sheep during summer. The fodder trees should be planted in the grazing lands in large numbers. In a well established and properly managed pasture 100 fodder trees and 100 bushes per hectare maintaining row to row and plant to plant distance of each species 10 m and 10 m, respectively should be planted after the first monsoon rains. This system has the advantage of providing three-tier feed resources. During wet months, the sheep can graze the green grass and during dry season they can look upto the green foliage of shrubs and trees. An additional yield of 18–20 q of good quality dry fodder per hectare per year can be obtained from these 100 full grown trees by lopping them twice a year. The lopping can be done in November to December and again in May–June and in such a manner that trees have no lopping injury. As good grazing on the surface is available during October–November, the leaves may be lopped during scarcity periods of January–March. The lopping available in May–June may be fed fresh. Ardu (*Ailanthus excelsa*) leaves are best among all the fodder trees from palatability and nutritive value point of view. It is a very fast growing tree and the leaves can serve as a maintenance and production ration for sheep as a sole feed. The pods of many trees specially babool (*Acacia arabica*) and *khejri* (*Prosopis cineraria*) are very nutritious and palatable and serve as a good source of feed for flushing ewes.

8.3.2 Agroforestry

It is a sustainable land management system in which forestry trees are grown alongwith arable crops. The trees may be of timber, fuel wood or fodder use. In order to get the quality fodder leaves in multi-purpose compatable fodder producing should be grown with arable crops. Some prominient fodder trees like *khejri* (*Prosopis cineraria*), Ardu (*Ailanthus excelsa*) and Babul (*Acacia nilotica*) do not cause significant yield reduction of companion crops like cowpea, guar and bajra. Hence, by growing these fodder trees in agroforestry system, additional quality dry fodder yields of 20 q/ha can be obtained from 100 trees.

8.3.3 Introduction of fodder crops in crop rotation and mixed farming

The fodder crops should be included in the grain or commercial crop rotation programme. The practice of keeping the lands fallow for wheat, paddy, gram etc. should be discontinued and fodder crops like sorghum, pearlmillet, cowpea and clusterbean should be introduced in *kharif* in the rotation. The fodder crops should also be sown with grain or commercial crops in such a way that they do not affect the production of grains. Alongwith sorghum and pearlmillet, the legumes like cowpea, dolichos, clusterbean, clitoria, black-gram and green gram, may very easily be grown as companion crops. The legumes will not only provide nutritious fodder for sheep without adversely affecting the grain yield but will also improve the soil fertility through nitrogen fixation which in turn will be available to the grain crop. It is essential that the programme of feeds and fodder development should provide for the production and multiplication of seeds and planting materials of recommended fodder crops.

As the pasture lands are shrinking day by day and an average farmer does not have more than 5 acres of land the only method left is growing nutritious fodder for livestock. By adopting any of the following crop rotations sheep farmers can get green nutritious fodder throughout the year.

Napier-lucerne

Napier × Bajra cross No. 21 and Lucerne can be sown in strips. This will give fodder throughout the year.

Sorghum-berseem-maize

Improved varieties of sorghum such as J.L–263 and J.S–29/1 can be sown in the last week of June. It can be harvested after 80–90 days. Then Berseem can be sown in mid October. After taking 5–6 cuts, maize can be sown in April, which can be harvested by June.

Sorghum-oats-bajra

After harvesting sorghum in mid October, H.F.0–114 variety of oats can be sown. Two cuttings of oats can be taken. Then fast growing varieties of bajra (S 530 or S $^{1}/_{3}$) can be sown.

Sudan grass-berseem-maize

Sudan grass is a high yielding crop which can give 5–6 cuttings. The berseem can be sown in October. Maize can be sown in April (Sudan grass-berseem-maize).

In addition to these there are more fodder crop rotations. By following these sheep farmers can get green fodder for their animals throughout the year. Package of practises for increasing fodder production are given in Table 8.8.

Table 8.8. Package of Practises for Increased Fodder Production

Name of crop	Varieties	Time of sowing	Seed rate (kg/ha)	Manuring (kg/ha)	No. of irrigation	No. of cuts (q/ha)	Fodder yield
Cowpea	H.F.G. 41–1 No. 10, F.O.S.–1	March to July	40–50	N 25, P_2O_5 62.5 K_2O 100	3–4	2	300–500
Guar	Guar–2 Guar–227 H.F.G. 119	April to mid July	40–50	N 20 P_2O_5 50	3–4	1	250–300
Bajra	2nd genera	March to August	8–10	N 75	2–3	2–3	400
Berseem	mescabi	October	20–25	N 25 P_2O_5 70 K_2O_5 60	5–6	5–6	700–800
Jowar	JS–20, JS–263 JS–29/1 HC 136	March to July	50–60	N 50 P_2O_5 15	2	1 3–4 (Multicut)	400–500
Lucerne	Lucerne–9	October to November	12–15	N20, K_2O 30 P_2O_5 60	5–6	8–10	700–750
Maize		April to August		N 50, K_2O 20 P_2O_5 30	5–6	1	350–400
Mustard		September	6–8	N 30	2	1	300–350
Napier		March to	33000 root	N 35, K_2O 15	Fortnightly	6–8	1100–1300
Hybrid		August	Slips	P_2O_5 25			
Oats	HFO–114 Weston–110	October to November	75-100	N 60, K_2O 15 P_2O_5 30	3–4	2–3	300–400

8.4 Fodder Conservation

The pastures will turn dry during summer months in arid and semi-arid areas and will be covered with snow during winters in temperate regions. Thus, total dependence on pasture for maintaining the sheep throughout the year can involve a great risk for proper health and production. Alternative arrangements should be made to maintain the sheep through lean periods. Fodder conservation is the only alternative that can solve this problem to a great extent. The other alternative is to have a large number of fodder trees and shrubs which could be lopped during pasture scarcity period.

The most economic, simple and perhaps also the most satisfactory method for preserving forage crops or grasses is in the form of hay. The grasses, legumes and tree leaves should be conserved during their luxuriant growth. These should be harvested at the pre-flowering stage to maintain their nutritive quality. The hay should not be sun-bleached as a good hay should retain its green colour.

The fodder could also be converted as silage. This could be prepared in pits or trenches, if it is not possible to conserve as hay due to adverse weather conditions. Cowpea, pearlmillet, clusterbean, sorghum, maize, seasonal grasses, etc. should be chaffed before putting in the silos. These fodders should also be harvested and ensiled at pre-flowering stage after bringing the dry matter content to 30–35% by wilting. Thorough pressing and packing of the fodder and proper sealing of the silos is equally important. Similarly, the tree leaves lopped during the period of abundance of grazing resources should be conserved as hay or silage. Fodders, grasses and tree leaves which have poor nutritive value may be improved through added supplements like urea, molasses and mineral mixture. This will improve their feeding value and also improve intake. The grasses from the forest lands are not fully utilized for the animals and hence these grow very tall and become unfit for sheep grazing. If these grasses are harvested during August and September before allowing the sheep to graze, a substantial amount of hay silage can be made.

CHAPTER 9

HOUSING AND MANAGEMENT

9.1 Sheep Management Practises

Depending upon the type of sheep different systems of Sheep management are required to be provided so that genetic potential of the animal is fully expressed for maximization of the production in respect of wool, milk and mutton. This has to be done keeping in view the economic limitations and resources available with farmer cooperatives or the farm. Different systems of management *in vogue* are extensive, semi-intensive and intensive. Further, for the proper management of a sheep farm, various management practises can be subdivided into: 1. Breeding management; 2. Feeding, watering and grazing management; 3. Housing and shelter management; 4. Shearing management and control of canary colouration; 5. Health management.

9.1.1 Breeding management

i) The heat detection should be carried out by using an aproned entire ram by stock Assistant/Laboratory Technician both morning and evening outside the grazing hr. The animals in heat should be separated and detained for weighment and breeding.

ii) All animals to be bred should be weighed prior to breeding and such weight recorded in the daily and mating registers.

iii) The animals should be bred through artificial insemination or natural service as per the breeding plan and this should be done by qualified Breeder/Farm Manager.

iv) The records of the breeding should be maintained by Breeder Farm Manager.

v) The animals bred should be separated into a new flock and should be detected for heat only after 2 weeks after 1st breeding.

vi) All ewes above 12 months of age and ewes which have lambed 60 days prior to the starting of breeding should be detected for heat and bred in case they exhibit oestrus.

vii) The rams should be put to service at about one and half year of age.

viii) No rams should be used for more than two breeding seasons and at least 10 females of a particular genetic group will be assigned to each of them. The programmes as decided should only be followed. In assigning ewes at random to rams, care must be taken that the same ewe is not bred to the same ram in the consecutive breeding seasons and is not related to it through sire or dam (is not half/full sib).

ix) As far as possible, rams should be kept away from the ewes and the two should be brought together only for breeding.

x) Breeding males must be provided individual housing so as to avoid fighting and consequent injuries.

xi) All effort should be made to reduce unproductive phase in the life of ewe, as it is directly related with its life time economics production.

9.1.2 Feeding, watering and grazing management

The feeding and grazing conditions vary from place to place. The most favorable grazing time is soon after onset of monsoon till the onset of winter. Grazing resources become extremely poor during summer months. During this period supplementary feeding should be done. Sheep generally thrive well on pasture. Attention should be paid on pasture improvement and management. Rotational grazing should be followed to avoid worm infection and unthriftiness, and to ensure availability of good pasture all the time. The fodder should be consumed in the form of hay and silage for lean period. Fodder should be planted in the pasture to provide shade and tree fodder during the lean period to the grazing flocks. Supplementary feeding of concentrate should be done depending upon the physiological status and availability of grazing resource in the pasture.

Water requirement of sheep vary depending upon its physiological status and ambient temperature in different seasons. The sheep should be watered at least once a day at the rate of 2–3 liters per head per day. The requirement of water for crossbreeds during summer month will be slightly more and may range between 5–6 liters. The younger ones may require 1–2 liters of water every day. The watering should be done in metallic troughs or cement channels.

9.1.3 Watering

Water requirements of sheep vary depending upon its physiological status in growth and production. It also varies according to ambient temperature during different seasons.

i) A native sheep may consume about 2–3 liters of water per day.
ii) The requirements of exotic and crossbred sheep range from 5–8 liters/day particularly during hot summer season.
iii) Water sheep at least once a day in watering trough or channels.
iv) Water sheep on alternative days during winter and rainy seasons when water requirements are reduced considerably.

9.1.4 Housing and shelter management

Normally sheep do not require elaborate housing facilities but the provision of the minimum housing requirements will definitely increase productivity by providing shelter to young lambs, pregnant ewes and breeding rams against inclement weather conditions and predation.

In the north temperate region after the flocks return from the Alpine pastures, 80% of the flockmen pen their sheep in the open near the houses, 18% along with other animals and 2% on the ground floor of their houses. In Rajasthan, 59% of the flocks are penned in open fields away from the houses, and the rest in kutcha courtyards near the houses. In the peninsular region specially in Andhra Pradesh, 41% of the flocks are penned in enclosures near the houses and the rest in the open fields. Sheep normally do not require elaborate housing facilities but provision of suitable shelter particularly for young lambs, pregnant ewes and breeding rams will definitely increase productivity and reduce losses due to mortality. It is therefore necessary to provide cheap houses with thatched roofs made of locally available material and thorny fences. In areas normally experiencing extreme cold or hot winds, some protection against the winds, should also be provided.

1) Provide sufficient protection against solar radiation and hot blowing winds to the exotic animals.
2) Also arrange shelter against rains and cold chilly weather conditions.
3) Native and crossbred animals may face these situation in a better manner but protection against predation during night time will save any loss of livestock.
4) Provide 10–12 sq. ft. of space for exotic sheep and 9 sq. ft. for native ones.

5) Sheds measuring 60" × 20" having thatched or with asbestos roofing can accommodate about 120 sheep.

6) A fenced enclosure of 30" × 20" can be provided for night padlocking of sheep.

7) Provide 2" thick thatched roofing during extreme summer.

8) Save exotic rams from summer sterility by providing comfortable room temperature not exceeding 30 °C during hot summer months.

9.1.5 Shearing management and control of canary colouration

a) Canary colouration of wool

Unscourable yellow stain on the wool, during its growth occurring during hot humid months, is referred to as canary colouration. This wool fetches 8–20% lower price than the white wool depending upon the intensity of colouration. For the control and to minimize its occurrence, following steps may be taken:

i) Shear wool before the onset of rains so that the quantity of wool grown during July–August is minimum and saved from heavy yellowing.

ii) Wash sheep before shearing.

iii) Save shorn sheep from direct solar radiation by resorting to morning and evening grazing.

iv) Keep sheep under shade during hot noon hr.

v) Crossbreeding may also help in minimizing the yellowing of wool.

b) Shearing, skirting and primary classing

Shearing is one of the important operation at the sheep farm and considerable preparation is necessary before this event.

1) Shear sheep at a separate shearing shed or room. The floor should be clean and free from dirt, dust or vegetable matters.

2) Use of hand-shears may be avoided as it often results in double cuts thus reducing staple length.

3) Utilize machine shearing service through the concerned state department. Machine shearing not only facilitates the quicker completion of the operation but also helps in getting uniform fleeces.

4) Resort to shearing in the flocks twice a year *viz.*, spring and autumn seasons.

5) Shearing in the months of February and early July will help in reducing occurrence of canary colouration of wool.
6) Graze sheep in the cooler hr and provide shelter during hot noon hr after shearing.
7) Greasy fleece weight of individual sheep may be recorded after each shearing.
8) Do skirting and primary classing of wool at the shearing floor itself.
9) Grade wool by visual appraisal, feel and handle.
10) Graded wools fetch 10–20% more price.
11) Sell wools to State Wool Marketing Department Wool Board.

9.1.6 Health management

Treatment-prophylaxis and curative

Sheep enterprise can pay, provided suitable preventive measures have been taken against the occurrence of diseases. The agro-climatic conditions of the area play an important role in risk to many diseases. In semi arid regions of the country, the incidence of parasitism increases during rainy season. Therefore, it is necessary to drench sheep before the onset, during and at cessation of monsoon. Dipping is done in the BHC/ DDT/Garathion/Malathion/Cimathion fortnight after shearing to get rid of ectoparasites. It should also be ensured to spray the sheds, simultaneously. The flock should be vaccinated against most common diseases like enterotoxemia, sheep pox, hemorrhagic septicaemia, PPR and rinderpest depending upon the occurrence of the disease in the area. To avoid foot and mouth, foot rot and other feet diseases, antiseptic foot baths are given. The sick and diseased animals should be isolated from the flock and checked thoroughly for any contagious/diseases.

For efficient disease control following recommendations are made

a) The Veterinarian/Farm Manager should provide a schedule of prophylactic health cover, *viz.* vaccination, drenching and dipping. In addition strategic drenching whenever necessary will be instructed. The Veterinarian/Farm Manager should further ensure procurement and supply of necessary medicines and vaccines to the unit for the purpose and maintain proper record of the procurement and use. The prophylactic health covser should be provided by the unit staff under the supervision of Health Staff and the Stock Assistants/ Laboratory Technicians with the Veterinarian/Farm Manager should assist the unit staff in this work. The Veterinarian/Farm Manager should also issue written instructions for the disinfection of the sheds and corrals whenever felt necessary and also at regular intervals. This should involve changing the earth in the sheds and corrals and replacing it by new earth mixed with lime and gammexene, painting of all structures with lime containing gammexene and also spraying with gammexene or BHC in all area of the sheds and corrals and the surrounding areas.

b) The treatment of sick animals should be done by the stock Assistant and Assistant Farm Manager except in such cases where specialized treatment is required which will be done by the Veterinarian . The instructions for treatment should be given in writing by the Veterinarian visiting the unit every morning. In case of serious illness the Stock Assistant/Farm Manager should inform the Veterinarian who should provide necessary treatment.

c) Sick and weak animals will be retained at the corrals and fed and watered by the labourer under the supervision of Stock Assistant and Farm Manager.

d) Tail docking will be done by the Stock Assistant within first month of the life of the lambs. Instructions and medicines for the purpose should be issued by the Veterinary Officer.

e) Regular testing against *i.e.* johnes', tuberculosis, contagious diseases and elimination of positive reactors be practised.

In sheep enterprise the prophylactic measures pay rich dividends. In semi arid land areas, the risk of parasitic infestation is less due to dry nature of soil and high wind velocity. Curative treatment should only be done in case of costly exotic and crossbred animals.

The following programme is advisable against ecto and endo-parasites and also for the prevention of main sheep diseases.

i) Drenching

1) Drench sheep with Phenovis/Thiabendazole/Banminth or Nilverm anthelmintic before onset of rains, during rainy season after cessation of rains and before spring showers.

2) Frequency of drenching may be increased in exotic and crossbred animals under unfavourable weather conditions.

3) Frequency of drenching may also be increased in young lambs which are more prone to parasitic infestation.

ii) Dipping

1) Dip sheep in plung type sheep dip against ecto-parisitic infestation.

2) Resort to dipping about a week after each shearing operation.

3) Use BHC, DDT, Garathion, Malathion and Cimathion as the dipping agent under the proper supervision of a veterinarian.

iii) Vaccination

1) Vaccinate flocks against most common diseases like enterotoxaemia, sheep pox and haemorrhagic septicaemia at appropriate time of the year.
2) Deal foot and mouth, foot rot and foot diseases properly.
3) Give anti-septic foot baths to check these diseases.

9.1.7 Grazing management

Sheep in India are mostly maintained on natural vegetation on common grazing lands, wastelands and uncultivated (fallow) lands, stubbles of cultivated crops and top feeds (tree loppings). Rarely they are kept on grain, cultivated fodder or crop residues.

There are no pastures in this country, as one understands the term in pastoral countries like Australia and New Zealand. There are, however, some natural pastures, which are mainly in the forest areas. The type of grasses in these pastures varies from area to area on account of varying climatic conditions in different parts of the country. There are three types of forests, *i.e.* government, civil and panchayat forests. The quality and the type of pastures in these forests vary, depending upon their maintenance and management.

It is the topography of these pastures and their carrying capacity, which determine the grazing routine of sheep flocks as well as their grazing behaviour. As said earlier, most of the sheep flocks in this country are migratory and their area of migration may vary from 80 to 480 km; other flocks are partially migratory, because they migrate with 24 to 32 km of their homesteads. There are hardly any flocks which can be considered stationary in the sense understood in those countries where flocks are confined to fenced pastures belonging to individual owners or companies. By and large the grazing hr of our sheep are also limited on account of the presence of wild animals, which put them at a disadvantage vis-a-vis the sheep in countries where they can graze throughout the 24 hrs of the day without any fear of wild animals. These conditions make our sheep long-legged, as they have to travel long distances to get their food. They can eat faster than exotic animals, which are accustomed to graze their pastures in peace. The differences in their grazing habits are conspicuous when they are grazed along with the local ones soon after their arrival in this country. Unless special efforts are made to help the exotic sheep to get adequate grazing, they are unable to compete with the local ones.

The availability of pastures which is seasonal depends upon the intensity of rainfall in different areas. Again, the rainfall is concentrated during a few months and is not evenly distributed over the year. Therefore the flocks of sheep migrate over extensive terrains, *e.g.* from the Rajasthan to the central forests of India. Sheep are constantly in search of fresh pastures, which may sometimes be along the edges of the forest lands and sometimes scrub along the main highways. The selection of the grazing areas by flock-owners is also related to the availability of irrigation facilities, etc. For instance, sheep from Rajasthan usually migrate to arid areas of Bundelkhand and western Uttar Pradesh after the *kharif* crops are over. There owing to the lack of irrigation, fields are left fallow and no cropping can be done during the *rabi* season. The Rajasthani sheep are found in large numbers folding over the fields.

9.2 Systems of Management

9.2.1 Traditional system

Sheep raising is mainly in the hands of the weaker sections of the community which either do not possess land or their landholdings are so small that crop cultivation does not provide remunerative employment all the year round. Further, in the major sheep rearing areas especially in north-western districts of Rajasthan grazing and stock watering resources are available only for a few months in a year, compelling shepherds to lead a nomadic life. The system of constant migration is one of the main reasons for the high percentage of illiteracy among these sections of the people. Even children of the family have no chances of education because they are also constantly on the move and are employed by their family for grazing the sheep flocks. Due to lack of education these sheep owners are not able to appreciate and adopt improved sheep husbandry practises brought to them by extension workers. Sheep management has thus remained in a neglected state. Migration and grazing practises have an impact on the present status of sheep husbandry in the country.

In the northern hill-region, migration of sheep usually starts in April-May and it takes about a month for the flocks to reach the Alpine pastures. The flocks graze on these pastures for about 5 to 6 months and then start returning to the foothills by September–October and reach the destination by November. The time taken to reach the alpine pastures largely depends upon weather conditions and the vegetation available on the route of migration. In the hill areas, one or two members of the family go with the flock during migration but the rest stay behind to look after agricultural operations. The economic condition of shepherds in this region is better than that of their counterparts in the plains. A fairly large percentage of sheep in Rajasthan is renominated on migratory basis. According to an estimate about 1.5 M sheep migrate from the State every year and out of this about 0.6 M sheep are on migration all the year round. Sheep migration is a regular practise with the sheep breeders of the western arid districts of Jaisalmer, Barrner, Jodhpur, Pali, Jalore, Nagaur and Bikaner. It is understood that prior to Partition, sheep from Jaisalmer and Barrner districts used to migrate to the canal areas of Sind during the months of March to June while the sheep from the eastern areas used to migrate towards further east or southwards to the bordering areas in the States of Uttar Pradesh, Madhya Pradesh and Gujarat. Sheep breeders from the districts of Jaisalmer, Barmer and Bikaner and parts of Jodhpur practise temporary migration extending over periods of 6 to 9 months. The flock owners usually start migration with their sheep after Diwali festival in November. Sheep flocks excepting those from Barmer and Jaisalmer districts usually migrate towards the south to Madhya Pradesh taking different routes and finally reach Nagda which is a central place for sale of wool. Sheep flocks also move towards east to Uttar Pradesh along the Chambal and Yamuna river belts. It has been estimated that about 0.2 to 0.3 M sheep from Jaisalmer and Barmer and parts of Jodhpur and Jalore migrate towards Gujarat and the breeders arrange for the sale of wool at Deesa in Gujarat. As the routes towards Madhya Pradesh are new and quite unfamiliar to the sheep breeders of Jaisalmer and Barmer they do not migrate long distances except under very acute conditions of drought and scarcity of fodder. This may perhaps be one of the reasons for heavy losses of sheep sustained by the sheep breeders of these districts during continuous drought periods. Sheep flocks from parts of Bikaner, Churu and Sikar districts usually migrate towards north to areas in Haryana, Punjab and Delhi. Sheep owners of Nagaur district practise migration all the year round because of heavy pressure on land and they depend on the forest areas in Rajasthan and Madhya Pradesh for the grazing of their sheep. Maintaining sheep on migration through most part of the year does not permit the sheep owners to obtain adequate income from their flocks. With such a system of constant movement of sheep over long distances with uncertain grazing and stock watering facilities, the sheep owners are unable to adopt scientific methods of breeding for genetic improvement of stock. Shearing of sheep in distant places and sale of wool at far away markets do not give them any chance for organized collection, processing and marketing of wool and the sheep breeders are unable to bargain for a remunerative price. These difficulties could be alleviated if the State Animal Husbandry/Sheep and Wool Departments set up service centres on the migration routes.

In the North temperate region sheep are grazed on Alpine pastures during most of the summer months; in autumn and winter seasons, they are grazed on harvested fields in the lower ranges of the hills and reserve forests. Usually 7 to 8 hr of grazing is practised in a day. However, grazing in winter does not meet the nutritional requirement especially for the animals in advanced pregnancy. This reduces lambing percentage and results in poor lamb survival and growth due to low mothering ability. Serious efforts are, therefore, called for in improving grazing/fodder resources during the winter. This is possible through raising a short term fodder crop after rice. It is estimated that even at present about 13% of the flocks are hand fed on some grains, conserved fodder or green leaves during winter. In the North western region, sheep are grazed on uncultivated lands during monsoon months. During post-monsoon period when *kharif* crops are harvested, they are grazed on stubbles. In November and December, non-migratory flocks are mostly grazed on uncultivated lands, whereas migratory flocks are grazed on harvested fields and in reserve forests on their migration routes. Top feeding by lopping tree branches and the feeding of pods are common. During extreme summer months, the flocks are grazed on harvested fields and in reserve forests on their migration routes. Top feeding by lopping tree branches and the feeding of ods are common. During extreme summer months, the flocks are grazed in the cool hr of the day and rested during the noon hr. In the peninsular region, there is hardly any migration of flocks. The flock size in this region is generally small and the sheep are mostly grazed on stubbles in the harvested fields or in forest or hilly areas. The sheep have thus to subsist at present mostly on stubble, tree loppings and grasses in denuded lands all over the country except in northern hill region where during summer they graze on relatively better pastures in the alpine region. Almost no attention has so far been paid to improve grasslands and to make provision for supply of drinking water to the flocks.

9.2.2 Extensive management

Under extensive system of management sheep are mainly raised under exclusive rangeland grazing with supplementation of top feed resources during critical lean season. Organized husbandry practises, nutritional inputs commensurating with various physiological stages and production traits and prophylactic and curative health coverage are almost negligible. The native sheep have poor production potential and efforts for their genetic up gradation by introduction of high producing breeds has not been largely successful under field condition, subsisting by grazing/browsing on variety of grasses, bushes and top feeds. The present grazing practises for sheep are indiscriminate and governed by compulsions of feed resource availability, taboos and tradition specific for the regions and socio-economic conditions of the farmers. Under the existing management practises of exclusive range land grazing the lambs hardly achieve 15 to 16 kg at 6 months of age whereas they marketed for slaughter, at a weight of 18–20 kg.

9.2.3 Semi-intensive management

The slaughter weight in farm is lower and the age at which target weight is achieved is far higher than expected. The system requires input of concentrate which although cost effective, yet has not been adopted due to socio- economic conditions of the farmer and the present marketing avenues. The work done in the country on grazing with supplementation have indicated that in addition to free grazing, supplementation of limited quantity of concentrate amounting to 1.5 to 2.0% of body weight will provide marketable finishing weight of 25 and 30 kg at 6 and 9 months of age respectively. The lambs under sole grazing on *Cenchrus* pasture with concentrate supplementation @ 1.5% of their body weight; achieved 26.7 kg of body weight at 6 months of age with about 100 g ADG.

9.2.4 Intensive management

The mutton lambs require intensive feeding to allow exploitation, of their growth potential. Maximum growth during preweaning period is very important for subsequent growth of the lamb. From birth to 3 weeks of age the lamb has non-ruminant stomach and feeding during this period should be mostly on milk through suckling or feeding of milk substitutes. From 3 weeks to 2 months of age, it is in transition phase, during which rumen development takes place. A creep feed should be fed to lambs as early as 2 weeks of age so that they would consume more energy and also the rumen development could be faster. In addition to *ad lib* feeding of cultivated fodder as green or hay, concentrate feed should also be given. The concentrate feed should contain higher proportion of grains like maize, barley and sorghum. It should also contain fish-meal or meat-meal at a level of about 5% along with mineral mixture and vitamins to ensure good growth. The digestible crude protein content should be about 14 to 16%. Composition of a typical creep ration is: maize or barley, 60.0; groundnut-cake, 15.0; fish and meat-meal, 5.0; wheat bran, 17.0; mineral mixture, 2.0; common salt, 1.0; and vitamin mixture, @ 25 g per quintal of feed.

The lambs on an average consume 200–250 g creep ration per head per day from 15th day of age to 90th day of age under *ad lib* feeding, and grow @ 125–150 g in native and 175–200 g per head per day in mutton cross-breds. After weaning there is a depression in growth for the first 2–3 weeks which should be avoided by ensuring good development of the rumen by appropriate feeding management during preweaning period. The lambs weaned completely at 90 days of age should be supplemented with 500 g concentrate mixture in addition to grazing for 8 hr on grass-legume pasture or fed *ad lib*, on cultivated leguminous fodder as green and or hay in the stalls up to the age of 6 months. Fed in this manner they can reach about 24–25 kg body weight in the indigenous breeds and 30 kg in cross-breds at 180 days of age. When a good grass-legume pasture is not available, weaned lambs should be kept under stall-feeding. Under the intensive feeding, the lambs are maintained totally under stall-feeding. They are offered *ad lib* complete feeds comprising concentrates and roughages. A complete feed with concentrate: roughage ratio of

50:50 is most economical for fat lamb production. Among roughages cowpea, berseem and lucerne hay-meals and *Zizyphus nummularia* (pala), *Prosopis cineraria* (khejri) leaf-meals are useful. Leaves of Ailanthus excelsa (ardu) can also be used. A feedlot gain of about 150 g in natives and 200 g per head per day in mutton cross-breds can be easily achieved by feeding a 50:50 concentrate roughage complete feed.

A package of practises for fat lamb production has been developed at the CSWRI, Avikanagar, in which the mutton ewes grazing on *Cenchrus ciliaris* pasture for 8 hr a day, were supplemented with 300 g of concentrate mixture per head per day during last 45 days of pregnancy, and with 400 g of concentrate mixture per head per day during the first 60 days of lactation. The lambs born weighed about 2.50–2.75 kg in natives and 3.00–3.50 kg in mutton cross-breds at birth. Lambs were allowed suckling and *ad lib* creep ration and a little green fodder in the form of lucerne, cowpea or pearlmillet from 10 days to 60 days of age. They were completely weaned thereafter. The lambs on an average, consumed 150–200 g of creep ration per head per day during this period and attained a weaning weight of about 12–13 kg in native and 14–15 kg in cross-breds. These lambs were then put on feedlot for 90 days and offered *ad lib* a complete feed comprising 50% concentrate and 50% roughage. They reached a live weight of 24–25 kg in native and 28–30 kg in cross-breds at the age of 150 days. The average dressing percentage of lambs slaughtered at this age was 50–52 and bone: meat ratio of 1:4.5 to 1:5.0. The composition of a typical feedlot ration is given below:

Ingredient	**Parts**	
	For lambs weaned at 90 days	**For early weaned lambs (60 days)**
Cowpea hay or meal or *Zizyphus nummularia* or *Prosopis cinereria* or *Ailanthus excelsa* leaf-meal	50	40
Maize or barley	30	40
Groundnut-cake	18	18
Mineral mixture	1	1
Common salt	1	
Vitablend AD3	25g/quintal of feed	20 g/quintal of feed

9.3 Handling and Care Animals

9.3.1 Lambing ewes

An ewe about to lamb prefers to leave the flock and will attempt to scrap a bed for herself. She will appear restless, the udder is often distended and external genitalia are in a flushed and placid condition. Normally a healthy ewe should not encounter any difficulty in parturition, still the following precautions may be taken during and after parturition.

(i) Expectant ewes should be separated from general flock at least 3 weeks prior to expected date of lambing and given extra feeding as per schedule.

(ii) The expectant ewe immediately prior to lambing should be removed to individual lambing pen 4" × 3" with small feeding and watering trough and is properly bedded and protected from direct wind.

(iii) The lambs and mother should stay for 24–28 hr in the individual pen.

(iv) The milk yield and mothering ability of ewe should be determined by examining udder and from the interest shown by ewe in her lamb and in case of poor milk yield/poor mothering ability, the lambs should be given supplementary milk feeding as per schedule.

(v) The lambs should be identified by ear tagging and tattooing and weighed within 12 hr of birth when dry. The weighment will be done close to 25 g. The naval cord should be severed, if not done automatically and painted with tincture iodine. The ewe should also be weighed at the time the lamb is weighed and care should be taken that placenta has been expelled before weight is recorded. In case of retention of placenta the medicines as recommended by Veterinarian should be administered.

(vi) Care may be given to ewes showing difficulty in parturition and the lamb, may be helped to stand for suckling in case necessary. Similarly help in cleaning the eyes and nostrils of lambs, in case mother is exhausted and is not taking care, may be rendered.

(vii) The ewes upto one month from lambing date would be kept separately in a group of 75–100 and grazed close to the corrals so that they can be brought during noon for suckling their lambs.

(viii) The lactating ewes should be given supplementary feeding as per schedule.

(ix) The young lambs upto one month should stay in the corrals and after 15 days of birth should be given some grain and green leafy feed (creep feeding). They should go out for grazing separately from their mothers after one month of age. The supplementary feeding of young lambs should be done as per schedule. The lambs should stay with their mothers only at night.

(x) Weaning should be done at 90 days of age and both mother and lambs should be weighed on the day of weaning.

(xi) It is usually seen that muconium does not pass in some lambs especially where mothers do not have good milk yield. In such cases about 5–10 cc of liquid paraffin may be administered per rectum or about ½ ounce of it may be given orally.

(xii) Keep a vigilant eye for dystokia or difficult birth due to malpresentation or other causes.

(xiii) Maiden ewes in poor condition or small framed ewes lambing to big rams will generally be finding difficulty on parturition.

(xiv) Seek prompt veterinary aid and advice from an experienced shepherd/ stockman.

(xv) Newborn lambs, after being licked by the mother, generally stand on their legs and start seeking for teats and suckle milk.

(xvi) If they are not able to do so after sufficient time, provide help to them in suckling first milk.

(xvii) Colostrum (1st milk) is very essential for health and survival of lambs.

(xviii) Save newborn lambs from cold, rain and winds.

(xix) Care for disowned or orphan lambs by resorting artificial milk feeding or arranging foster mothers.

(xx) Ligate, sever and antiseptically dress the naval chord of the lamb.

(xxi) Give a tea-spoon full of castor oil/liquid paraffin to the lamb to facilitate defecation and passing out of muconium easily.

(xxii) Don't handle lambs too frequently and let the ewes lick and recognize them properly.

(xxiii) Allow newborn lambs to remain with their mothers for a week or so.

(xxiv) Feed sufficient quantity of good quality hay and concentrates (if possible) to the lactating ewes for the increase in let down of milk.

(xxv) Provide plenty of clean fresh drinking water as the lactating ewes drink a surprising amount of water during lactation.

9.3.2 Lambs

Efforts should be made to have maximum care for optimum growth during early period of life of lamb.

Proper suckling by the young should be ensured. The growth of the lamb starts actually from the time of conception itself. This indicates that for effective lamb management and for reducing the lamb losses, care of ewe should start from the time of conception till the lambs are weaned. If reasonably well fed, ewe will form her lamb out of the feed she gets, but if she is half starved, she sacrifices her body weight to form the lamb. In younger ewes that are still growing, this sacrifice is not so natural nor so easily made and it is possible that either the lamb or the ewe may suffer. Udder of the ewe should be examined for blindness of teats and mastitis. The orphan and disowned lambs should be artificially fed on milk or foster mothers.

The most critical period in the life of lamb is during first 48 hrs of life. Immediately after birth the naval cord should be pinched 4" from the body. It should be painted with Tincture Iodine to avoid possibility of any infection. In the normal lambing, the mother usually cleans the lamb by licking. However, in the case of dystocia or poor mothering ability all mucous should be removed from around the mouth and then the lamb should be kept near the head of the ewe so that the ewe cleans up the lamb. The mucus from the lamb's body is applied on the nose of the ewe. This helps in owning the lamb. When the ewe owes the lamb, he should be left to her. The lambs and mothers should stay together for at least 72 hr in the individual pens. A healthy lamb would stand within 5 to 10 minutes of birth and start suckling between 10 to 20 minutes, at the most within one hr for a week the lamb needs assistance. Teat of the ewe should be put in the lambs mouth and milk should be stripped. This would enable to know if there are any plugged teats. Care should be taken to clean the udder to prevent infection. Colostrum is essential for cleaning the bowels and provides passive immunity. Through the colostrum the, lamb gets the antibodies which prevent the lambs from early infection as well as constipation. In the case of constipation, it would be better to give castor oil/paraffin. If pinning is there, it would be better to remove manually. If colostrum is not available, then the lamb is fed with warm milk containing castor oil/paraffin. The milk yield and mothering ability of the ewe should be determined by examining udder and from the interest shown by the ewe in her lamb and in case of poor milk yield/poor mothering ability, the lamb should be given supplementary milk feeding as per the following schedule.

Warm milk with cod-liveroil at the rate $^{1}/_{10}{}^{th}$ of the body weight.

First fortnight	–	6 times a day
Second fortnight	–	4 times a day
Upto 3 months	–	Twice a day

The following steps may be taken for increasing survivability of lambs and their better growth.

1) Ensure proper suckling of lambs. Examine udders for blindness of teats or mastitis.
2) Take care of indifferent mothers and arrange suckling of lambs by restraining such type of ewes.
3) Provide creep feed mixture to suckling lambs in addition to suckling of milk from 10th day to weaning stage.

4) Make available green leguminous grasses to lambs to nibble at during suckling period.
5) Perform lamb marking operation when lambs are between 24 weeks old.
6) Lamb marking embraces ear tagging/tattooing, tail docking and castration of male lambs if to be done at all.
7) Lamb marking site/place should be perfectly dry, clean and hygienic so as to minimize the risks of losses from tetanus, profused bleeding and blood poisoning.
8) Use sterilized clean knife and resort to proper ligation and antiseptic dressing to the needed lambs for tail docking operation.
9) Elastrator rubber band methods of tail docking and castration are very handy.
10) Tatooing forceps and ear-tagging forceps should be clean and sterilized at the time of use.

9.3.3 Pregnant, parturient and lactating ewes

Careful management of the pregnant, parturient and lactating ewes will have a marked influence on the percentage of lambs dropped and reared successfully. Following steps may be taken to afford proper attention to these animals.

1) Do not handle in-lamb ewes too frequently.
2) Separate them from the main flock and take effective care in relation to their feeding and watering.
3) Extra nutrition (flushing) during the later part of pregnancy will be beneficial for milk production of ewe, birth weight and growth of lambs.
4) Inadequate and poor nutrition may result in pregnancy toxaemia, abortions and premature birth of weak lambs.
5) Bring lambing ewes into lambing corals 4–6 weeks before parturition and provide maximum comfort.
6) Watch gestation length which ranges from 142 to over 150 days. Early maturing breeds have shorter gestation period.
7) Save parturient ewes from cold and chilly weather.

If the ewes are flushed and good grazing is available, no additional supplementation is necessary during early pregnancy upto about 14 weeks of gestation period. During advanced pregnancy (last 6 weeks) however, extra feeding is essential. During this stage, there is also depression in the intake capacity and feed digestibility. Hence highly digestible feed need to be fed during this stage.

9.3.4 Dry ewes

No extra care is required for these animals. Regular grazing for 8 to 10 hr on a good pasture is sufficient to maintain their weight/condition.

Flusing is conditioning of ewes for breeding. If the ewes are in low plane of nutrition prior to breeding, additional supplementation for about one month has beneficial effect in bringing the ewes into oestrus. Even without additional supplementation when there is flush green pasture, there is flushing effect. Supplementation of about 250 g of concentrate could bring about flushing of ewes quite well. If the plane of nutrition of the animal is good prior to breeding, flushing is not at all required.

9.3.5 Care of weaners and fattening stock

Complete separation of lambs from their mothers is called weaning. The practises and problems of weaning and care of weaners vary from place to place. The management of weaners play an important part in good sheep husbandry because these weaners will be the future breedable animals. The following steps will greatly help in proper care and management of weaners.

i) Weaning should preferably be done at 90 days.
ii) Avoid malnutrition, as it will result in stunted growth and susceptibility to worm infestation.
iii) Provide supplementary feeding and good clean pastures.
iv) Drench them regularly against various gastro-intestinal parasites as these are very prone to worm infestation.
v) Vaccinate them against enterotoxaemia, struck and black quarter diseases.
vi) Do not graze weaners in burry and seedy type of pastures which may cause skin irritation to lambs, damage to wool and cause ophthalmic diseases.
vii) Provide them shelter against vagaries of climate and predation.
viii) They should have easy access to fresh and clean water and nutritious green pastures.

This includes identification, tail docking and castration. Ear tagging for identification is done immediately after birth when except in their pens to encourage the lamb to start nibbling. The young lambs upto one month should stay in corrals from one month onwards, lambs should be given 50–100 concentrate. They should go out for grazing separately from their mothers after one month of age. They should stay with their mother at night. During the second month 100 to 150 g of concentrate should be given depending on their body size. It should be increased to 200 or 250 g in third month, after weaning 350 g of concentrate should be offered along with grazing for 6 to 8 hr in the lustrous pastures.

The mutton sheep require improved pastures, cultivated cereal fodders and legumes, grain and oilseed milling by-products as concentrates for intensive feeding. If sufficient land for growing cultivated fodders or pasture is not available, the sheep should be supplemented with concentrates to the extent of 300 g per head per day during last 45 days of pregnancy, with 400 g per head/day during first 60 days of lactation and with 250 g head/day for flushing. The breeding rams should be provided with 400 g per day concentrate mixture during the breeding season. Crop residues could also be improved in their nutritive value through treatment with certain fungi, ammoniation using urea, steam treatment with/without addition of urea and molasses, and pelleting. Roughage feeding should be done at night especially during summer to avoid heat increment due to rumen fermentation.

Mortality in young lambs and other stock

As a guideline, the following mortality figures should not be allowed to exceed under the ideal sheep management practises:

Birth to weaning	–	5 – 10%
Weaning to 6 months	–	5 – 8%
6 months to 1 year	–	3 – 5%
1 year and above	–	0 – 5%

9.3.6 Teaser rams

The number of vasectomized rams used is 1% of the original number of ewes *i.e.* for 5,000 ewes, 50 teasers and even when there are only 1,000 ewes left in the mob 50 teasers should still be used. All vasectomized rams are to be tested for brucellosis. Two teams of teasers are used and alternated every 3rd day. The time to change teasers is when a large percentage of the ewes that are presented for insemination have creamy vaginal mucus. This indicates that estrous has terminated and this means that teasers did not detect estrous at the onset, either because there were too many ewes or they had lost interest.

9.3.7 Breeding rams

Rams, should have access to pasture or irrigated plots and before they are let out on pasture, should be inspected daily for fly strike, screw worm and foot abscess. Rams at two tooth stage are separated from adults to prevent fighting and homosexuality.

When the actual insemination programme starts, the rams are still housed and fed as before and walked to the shearing shed and back daily, to ensure exercise. When in use at the shearing shed, they are penned separately awaiting their turn as semen donors to avoid fighting with other males and to save time in handling.

The training, actually does not begin till the programme starts and it seldom takes more than two days to get all rams to serve an ewe and subsequently into an A.V. The number of collections to be obtained in a day varies from ram to ram. From most of the rams semen can be collected three times a day producing average ejaculates of 1.0 ml, sperm concentration of 35×10^8 per ml and live counts 85–90% (Miller, 1975).

9.3.8 Management of ewes

Keast and Meeley (1949) and Dun *et al.* (1960) recommended overnight teasing and drafting raddled ewes off in the morning. This is quite manageable for small lots, but when large numbers have to be inseminated it takes lot of time as ewes cannot be presented for AI at proper time during oestrus. Under the circumstances the rams have to be worked too frequently to provide necessary semen. When large number of ewes are to be handled, harnessed vasectomized rams are released with the ewes at about 5 pm and the mob turned into a paddock. They are mustered into the yards by 8.00 am the following morning and marked ewes drafted off. With few modifications to most sheep yards, this can be done at the rate of 4,000 per hr.

Marked ewes are either inseminated to get the labour force acquainted with procedure and a few rams checked and trained or these are put aside and brought back into the mob at the end of the cycle. Generally the drafted ewes are inseminated on the acceptable type of oestrus mucus. Obviously those ewes which are drafted late in oestrus are rejected.

After the morning draft, the remaining flock is redrafted at 4 pm. These ewes which are called as "A ewes" have commenced oestrus between 9 am and 4 pm. The main flock is returned to the paddock with teasers and brought in for drafting again the following morning. The marked ewes which are called as "B ewes" have commenced estrous between 4 to 8 am.

The "A ewes" are inseminated between 7.00 to 8.00 am the following day. The "B ewes" are inseminated between 10 am and 3 pm the day they are drafted off.

All the "A ewes" are inseminated between 15 and 23 hr after the onset of oestrus.

All the "B ewes" are inseminated between 6 and 11 hr after the onset of oestrus.

When the rams are to be maintained they should be only for breeding purpose, otherwise they should be sold for slaughter. In some places in the field, farmers keep large number of rams in their flocks only for getting the wool clip. Some farmers have also religious sentiments not to slaughter their rams. This practise should be discouraged. Generally if good grazing is available, rams do well on grazing alone and no additional supplementation is required. During breeding season, concentrate supplement should be provided. In the absence of concentrate, supplement with good legume hay should be provided in the quantity twice the recommended concentrate allowance. During non-breeding season if the rams loose weight, fodder supplement should be provided. About 5 to 6% DCP in the fodder/ pasture is adequate to maintain the animal during non-breeding season.

9.4 Breeding Management

Sheep are bred either by natural mating or through artificial breeding. Rams of the indigenous breeds donate good quality semen under proper management conditions throughout the year. However the rams of temperate breeds if not protected from high temperature, high humidity and high solar radiation will not produce good quality semen during hot dry and hot humid seasons. Females of the tropics breed throughout the year. Temperate breeds which are affected by hr of day light and breed with declining day length. They come in heat in autumn from August to November, although some may breed upto February.

9.4.1 Natural breeding (individual, pen and flock mating)

The natural breeding is done either by flock mating, penmating and hand mating. In flock mating system, sheep can be bred naturally or through artificial insemination. In flock-mating system, breeding rams are usually turned out in the flock during the mating season at the rate of 2 to 3% of the ewes all through day and night. It is most widely practised in the flocks of all farmers. Semi flock breeding or pen breeding is done to conserve the energy of rams and give them rest. In this, the rams are turned out for service with the flock in the pen during night, and confined and stall-fed or grazed separately during the day time. Hand mating is practised when exotic purebred sires are used, or when it is considered desirable to extend the services of ram over much larger flocks. Sheep in heat do not manifest behavioural symptoms. Hence, the teaser rams are employed for detecting the ewes in heat. These ewes are then taken out of the flocks and bred to the designated sire of the flock. In pen and hand-mating systems and when teaser rams are used for heat detection, some dye mixed in grease or simple linseed oil is smeared on the brisket of the ram. This makes it possible to record the date when the ewes are bred and also to remove them from the breeding flock. The colour of the dye should be changed every 16 to 18 days so that the repeaters can be discovered, if the bred ewes have not been removed from the flock. This is termed as marking of ewes by breeding ram and marked ewes are considered as bred.

Artificial insemination has been already described earlier in Chapter 7.

9.4.2 Heat detection methods

Visual signs of approaching oestrus are a swelling and redness of the vulva and restlessness or nervousness indicating a desire for company, but the most obvious sign is riding and in turn being ridden. The breeding occurs only during estrous although the ram is capable of breeding at any time.

Overnight teasing and drafting ridden ewes off in the morning. This is quite manageable for small lots, but when large numbers have to be inseminated it takes lot of time as ewes cannot be presented for AI at proper time during oestrus. Under the circumstances the rams have to be worked too frequently to provide necessary semen. When large number of ewes are to be handled, harnessed vasectomized rams are released with the ewes at about 5 pm and the mob turned into a paddock. They are mustered into the yards by 8.00 am the following morning and marked ewes drafted off. With few modifications to most sheep yards, this can be done at the rate of 4,000 per hr.

Marked ewes are either inseminated to get the labour force acquainted with procedure and a few rams checked and trained or these are put aside and brought back into the mob at the end of the cycle. Generally the drafted ewes are inseminated on the acceptable type of oestrus mucus. Obviously those ewes which are drafted late in estrous are rejected.

9.4.3 Insemination procedures

Under the insemination technique there are a number of prerequisites such as equipment for insemination, restraint of ewes, number of ewes for insemination and adjustment of inseminations during better part of estrous period and whether heterospermic inseminations are desirable or not.

i) Equipments

A speculum, a head light torch and a syringe with insemination pipette is all the equipment required. There are three types of specula in use (1) the duck bill type, (2) metal barrel and (3) glass speculum. It is the usual experience that the duck bill type is easy for insertion and gives greater ease and freedom of movement. It can be easily dilated in the vaginal passage and removed in a closed condition after use. It is also easy for sterilization. A glass speculum is also used. A 2 ml glass syringe attached to inseminating pipette with a rubber connection is most convenient for insemination. A fresh insemination pipette should be used for each insemination.

ii) Restraint of ewes for insemination

With the help of two attendants, ewes can be inseminated at the rate of 100 per hr, if the operation is streamlined, using fresh neat semen. Various methods of holding ewes have been used and some methods used in Australia between fifties and sixties are:

1. Ewes held upside down in a cradle.
2. Ewes placed on a battery of bails on a raised platform.
3. Ewes held on a rail, as for lamb marking.
4. The operator working in a pit and a ewe manoeuvred upto a hock bar at the edge of it.

Other methods used in Asia and in South America, include the use of various cradle and jacking devices to lift the hind quarters. These are not easy to operate. The method used in Australia, at present, is for the attendant to saddle the ewe facing the tail and with a hand under each flank to throw her hindquarters over a rail 24" above floor level. Attendant keeps the ewe's hind legs extended by pressing into her stifles. If the ewe's front feet are on the floor, she will not struggle. After insemination the ewe is released through a gate in the pen.

A false floor about 2 feet high is put into the catching pen to enable the inseminator to have the ewes presented at eye level.

When the ewe is placed over the rail, an assistant inserts glass speculum whilst the inseminator loads his syringe. The inseminator then holds the speculum, locates and positions the cervical opening and inseminates the ewe. The semen is deposited in the cervix.

A simple wooden crate with sloping platform is considered for restraint. The rear portion is raised to a height of 3 feet so that the inseminator can conveniently inseminate, in standing position.

9.5 Handling and Management of Sheep Products

9.5.1 Wool

Skirting is done after shearing. It is removing the objectionable parts such as tags, leg pieces, neck pieces, bellies, locks and stained portions from the body of the fleece.

After skirting the actual grading of the wool is done by a trained classes. The method of classing in the principle wool-producing countries differs considerably. The grading or classing of fleece is done on visual appraisal of length, fineness (handle or feel). Grading of Indian wools was initiated under FAO/UNDP project based on wool quality, length, colour and vegetable fault (content of vegetable matters, *e.g.* burr, seed). About 90 types were prescribed besides lower lines. The Indian Standard Institution (IS 2900:1979) evolved a grading system based on micron value. It prescribes almost 120 grades. Thereafter style grading was introduced in Rajasthan which cuts the grades to 32 only. However, none of the above grading systems is practised in the country to any significant extent. The results led to the following recommendations: (a) wool from different agro-climatic regions and breeds may be marketed as such; (b) at the time of shearing fleece should be properly skirted; (c) fleece after skirting may be classed into at the most 2 to 3 grades, like fine, medium and coarse; and (d) bales from individual lots should be sampled, and these samples be subjected to objective quality assessment through laboratory testing for yield, length, average fibre diameter and vegetable matter content.

9.5.2 Meat

In our country the word mutton is used in a very broad sense to include goat meat also. However, mutton by definition is the flesh together with the associated tissues such as blood vessels, nerves, glands, fat and bones from carcasses of sheep older than 12 months of age. Sometimes, mutton from sheep carcasses of 12–24 months of age is called as yearling mutton whereas that from carcasses older than 24 months of age is called mature mutton. Meat from sheep carcasses younger than 12 months of age is called lamb. Flesh from goats should be called chevon (pronounced as 'shevon' which means a kid).

Mutton and chevon are substituted for each other once in a while by meat traders in places where there are local preferences and price differential. Although it is very difficult for a layman to differentiate sheep and goat flesh, the following points might be useful: Lamb is pale pinkish with evenly distributed firm white fat. Mutton is slightly darker than lamb and many a times there is a well developed and thick fat cover. Mutton may also have marbling in it. Both lamb and mutton may have a few wool fibres sticking here and there. Chevon is dark red with coarse texture and a sticky sub-cutaneous connective tissues layer which may have adherent sheep hairs.

For the production of wholesome meat with good keeping qualities, sound husbandry methods and cleanliness of the animal during final stages before marketing are necessary. Feed additives and antibiotics should be withdrawn at appropriate period before slaughter. Sheep may often become dirty due to diarrhoea or during wet weather. Judicious drying off of wet sheep on a clean straw for 12 hr before slaughter is of great value. Detection and segregation of sick sheep before transporting them to a slaughter point will avoid transportation and subsequent losses. During transportation from a farm to abattoir, every possible care should be taken to avoid injury, lameness, suffocation or transit fever. Sheep should never be lifted by wool during loading and unloading as this will cause bruises leading to carcass trimming losses. Water must be provided to sheep and goats during a journey of 36 hr or over. A minimum floor space of 213 cm^2 per sheep must be ensured during transportation. Sheep normally loose up to 3–6 kg of their body weight during transportation depending on the weather and duration of the journey, due to sweating, exhaustion, and excretion of urine and faeces. Stress during transportation and before slaughter affects carcass yields, microbial contents of the tissues, onset of rigor mortis and the keeping quality of flesh and preserved meat. Therefore, sheep should be rested at least for 16 hr prior to slaughter with continuous access to water.

9.5.3 Milk

Since sheep are small, many sheepmen prefer to have the ewes stand on a milking platform. This also gets the milch animal up off the soiled floor. On the other hand, in India the people are used to squatting down when they milk ewes/does so there is probably not enough to be gained to warrant the cost of constructing a milking stand. Large commercial dairies in European countries construct a masonry platform 38 to 45 cm above the floor level. The ewes/does place their heads in a stanchion and while 10 to 20 ewes/does are being milked either by hand or a milking machine, the ewes/does are eating their concentrate happily. By the time the 10 to 20 ewes/does are milked, they too have finished their grain and are ready to be released for the next group of ewes/does. In India the habit is to milk the ewes/does by hand from the left side. This is preferable to milking from the rear.

The secretion of milk takes place day and night but it is slowed up by the pressure of the accumulated milk in the udder. So milking the animals at regular intervals gives the best results. Feeding should also be done at the same time each day. When possible it is preferable that the same person manages the ewes/goats.

Milk is a food and should be treated with the utmost cleanliness and sanitation. Keep the wool clipped from around the flank and udder to prevent dirt from collecting, which might fall into the milk. Brush the ewe daily or at more frequent intervals. Wash the udder or wipe it with a damp cloth before milking. The person doing the milking must have clean hands and clothes. Milking utensils must be clean. Milk in a clean environment away from the ram.

9.5.4 Edible and non-edible offal

For effecting an improvement in quality and to increase the quantity of casings collected for processing, the first essential step is improvement in the conditions prevailing in the slaughter houses. Modernization of slaughter houses and provision of a by-product wing in them are absolutely essential. However, pending modernization an interim measure to improve the quality of the casings immediately would be to make provision for adequate water supply in the existing slaughter houses so that the guts are cleaned properly within the precincts, and with the least possible delay.

In many slaughter houses, guts are not removed soon after the slaughter of animals. This results in deterioration of the quality of guts. As such, it should be made obligatory on butchers to remove guts within prescribed hr. The by-products wing of each slaughter house should have a processing unit for guts under hygienic conditions. Guts may then be sorted out according to calibre, grade etc. Thus making it easy for the exporter as well as the processor to do only the final grading. This would fetch a better price for the graded product.

9.5.5 Skins and hides

Hides of all animals which die cannot be fully claimed as arrangements for timely flaying of dead animals are not available in many parts of the country. Loss of hides of animals lost in floods, famines and those dying in remote areas has been estimated to range between 4 and 10% in different parts of the country. In the case of skins, the wastage due to non-collection is negligible, the major proportion of skins being recovered from slaughtered stock. The estimated annual loss of skins due to non-collection from dead stock has been reckoned at 1 to 2%.

Though India stands second in the world in the production of hides and fifth in the matter of production of skins, it cannot forge ahead in capitalising on its large production unless stress is laid on quality right from the initial stages of production. As such, improved methods of flaying should be introduced in the slaughter houses and better flaying should be encouraged by payment of premia to good flayers. The all India Khadi and Village Industries Commission has set up a number of flaying centres in the rural areas through the State Khadi and Village Industries Boards. There is however an urgent need for establishing more village flaying centres for skilled flaying of fallen animals. Demonstration-cum-training centres should be established in important places for imparting training in curing, tanning and rational utilization of hides. Facilities need to be suitably strengthened to carry out grading of hides and skins according to Agmark standards. Cold storage facilities for preserving raw hides and skins also need to be provided, wherever possible. Before export, a system of compulsory pre-shipment inspection should be introduced. This measure is of vital importance in securing quality.

CHAPTER 10

PRODUCTION PERFORMANCE OF SHEEP

India possesses 40 breeds of sheep of which 5 can be classified into medium to fine wool breeds, 24 into coarse carpet quality wool breeds and rest into hairy meat type breeds. The average wool and meat yield per sheep in India is extremely low as compared to the same from sheep in some agriculturally advanced countries. The problem of increasing productivity of sheep in India centres around a) improving the yield of carpet quality wool and making it more uniform, b) improving the mutton type yield of our native mutton breeds especially in the South and c) developing new fine wool and mutton breeds through crossing our native breeds with exotic fine wool and mutton breeds to suit the different agro-climatic regions of the country.

Knowledge about the consequences of cross breeding especially the heterosis exhibited by crosses, level of exotic inheritance which can be sustained and the native and exotic breeds producing crosses coming closest to the desired goal, is necessary. The information available on exotic and indigenous sheep indicate that except for fertility, all characters concerned with wool production and wool quality as well as growth and carcass yield are highly heritable and most of the genetic variability is due to the additive effects of genes.

To improve the genetic potential of the indigenous sheep, attempts through selection and through crossbreeding have been made in the country. The Central Sheep and Wool Research Institute (CSWRI), Avikanagar, Rajasthan has been making efforts to improve the productivity of Indian sheep through crossbreeding of indigenous breeds with exotic fine wool and mutton breeds at the main and regional stations and at the units under the All India Co-ordinated Research Project, (AICRP) on sheep breeding located in different regions of the country.

10.1 Studies on Purebred Performance

Purebred performance of important native breeds of sheep in the country has been studied and it is observed that the performance of our native sheep for body weight gains, reproduction, survival and greasy wool production is relatively poor as compared to the sheep in the developed countries. But keeping in view the conditions of feeding and management in which they are maintained they are not inferior as it is apparent from the face value of the records. Among the breeds of North-western region, Magra can be considered to be the best whereas in the Southern peninsular region, Madras Red has been found to be superior to all other breeds in the region. Based on the estimates of genetic and phenotypic parameters studied for different economic characters, it is observed that 6 month body weight is highly heritable and has positive genetic correlation with market weight and adult body size and selection based on this weight should improve market weight and the ewe productivity. Similarly, selection based on an index combining greasy fleece weight and medullation percentage, latter being given a negative weightage will improve the greasy wool production and quality in carpet wool breeds.

10.2 Studies on Crossbreeding for Improving Wool Production and Quality

Crossbreeding of indigenous breeds of various regions in the country, namely Gaddi in temperate region, Chokla, Nali, Malpura, Jaisalmeri and Patanwadi in Northwesterly region, and Nilgiri in the Southern hills with exotic fine wool breeds, namely Rambouillet and Russian Merino has been tried. The results indicate major improvement in greasy fleece production at half-bred level although there is improvement in fleece quality with increase in Rambouillet/Merino inheritance beyond 50%. It has thus been recommended that the coarse wool breeds may be crossed with Rambouillet Russian Merino and the level of exotic inheritance be limited to 50% in North-western region. In the Northern-temperate region and Southern hills, however, the inheritance level can be increased to 75% because the animals with this inheritance do not create any management problem and the improvement in wool to the finer quality which is evident at this level can be exploited to fill the gap of fine wool in the country. Using this technology a new strain of sheep, *viz,* 'Avivastra' have been evolved at CSWRI. This strain has been tested under conditions of grazing on reseeded *cenchrus* pasture without any feed supplementation and has shown excellent performance with respect to reproduction, lamb and adult survival, greasy wool production and quality. Avivastra produce 2.5 kg of annual greasy fleece weight of 21 μ fibre diameter and 21% medullation and gives about 21% higher income from the sale of wool alone compared to Chokla.

Under the AICRP on sheep breeding two new synthetics/strains of fine wool-one with Chokla and the other with Nali have been generated at CSWRI.They produce more than 2.5 kg greasy fleece annually. The Chokla synthetic produce 7.89% more wool showing 16.39 and 52.94% improvement in fibre diameter and medullation percentage respectively over the Chokla. The Nali synthetic produce marginally more wool and showing 19.93% improvement in fibre diameter and 62% in medullation percentage over the Nali. Reproduction and survivability in both the synthetics and natives are not different. The synthetics produced the apparel type wool of 20–2 μ fibre diameter, with 0 to 5% medullation and 5 cm staple length. Another fine wool strain *viz.* Nilgiri synthetic has been evolved in Southern hills at Sandynallah with annual production capacity of 3 kg having 18 μ average fibre diameter and 0% medullation.

10.2.1 Wool production

a) Apparel wool production

For improving apparel wool production, research has been in progress under a number of ICAR sponsored sheep improvement programmes at CSWRI and under the AICRP on sheep breeding which have involved evaluation of different exotic fine wool breeds and native breeds in crosses, level of exotic inheritance and effect of interbreeding crossbreed. Of the two exotic fine wool breeds, Rambouillet and Russian Merino in crosses, little differences were seen in more extensive breeding experiments. However, in case of indigenous breeds the superior carpet wool breeds showed improvement both in production and quality parameters but brought the lesser wool reasonably closer to apparel wool whereas the crosses with coarse and hairy breeds showed a larger improvement in greasy wool production and quality and brought the wool towards superior carpet quality at a different level. There was little increase in wool production with the increase in level of exotic inheritance and the half-bred had the maximum improvement. However, there was improvement in the quality as reflected by a decline in average fibre diameter and medullation percentage with increase in exotic inheritance. It was concluded that the half-breds involving exotic fine wool breeds and indigenous superior carpet wool breeds both in produc tion and adaptation as reflected by their grazing behaviour, capacity to travel long distances and survival, are most suitable. The CSWRI has evolved two new high producing strains-Avikalin, a superior carpet wool strain and Avivastra, a superior apparel wool strain from half-breds of Rambouillet with Malpura and Chokla respectively. Similarly, under the AICRP on sheep breeding, fine wool synthetics involving Chokla and Nali, Pattanwadi and Nilgiri as indigenous breeds and Rambouillet and Merino as exotic fine wool breeds have been developed.

b) Carpet wool production

For improving carpet wool production, results available so far indicate that in case of coarse and hairy breeds, where the wool is not useful for quality carpets, the crossbreeding with superior fine wool breeds should be undertaken and the exotic inheritance be stabilized at 50%. It should be followed by selection for increasing carpet wool production and maintaining quality towards superior carpet wool (average fibre diameter 30 μ and 30% partially medullated (heterotypic) fibres. The CSWRI has evolved a new superior carpet wool producing strain Avikalin through crossbreeding Malpura with Rambouillet stabilizing the exotic inheritance level at 50%. The superior carpet wool breeds such as Chokla, Nali, Magra, Patanwadi should be improved through selection is based on 6 monthly greasy fleece weight and against medullation percentage. In other coarse and hairy breeds, where crossbreeding with exotic fine wool is not feasible, grading with better carpet wool breeds from Rajasthan namely, Nali and Magra be undertaken. The results of such grading have shown improvement in wool production and quality as reflected from an increase in staple length and decline in percentage of medullated fibres.

10.3 Sheep for Mutton Production

The researches on crossbreeding conducted under the AICRP on sheep breeding for mutton units located in different regions of the country namely CSWRI, Avikanagar, Rajasthan, Central Goat Research Institute, Makhdoom, Uttar Pradesh, Sheep Breeding Research Station, Palmnar, Andhra Pradesh and Mahatma Phule Krishi Vidyalaya, Rahuri, Maharashtra involving different sheep breeds of the region with exotic mutton breeds, namely Suffolk and Dorset indicate that half-breds are superior to other grades with $^{3}/_{4}$ and $^{5}/_{8}$ inheritance in respect of performance as well as survivability. Pure indigenous breeds, though inferior in performance than crosses, have better survivability and lambing percentage. Suffolk crosses have higher mortality and low fertility and are superior to Dorset crosses in body weight gains. The half-breds can attain 30 kg live weight at six months of age when fed under individual feed lot trials.

Based on the performance of crossbred at 50 and 75% levels of exotic inheritance and combining ability with two exotic namely Suffolk and Dorset utilized in the crossbreeding programme, three mutton synthetics (strains), one at CSWRI and two at Palmnar unit under the AICRP of Sheep Breeding have been evolved. The synthetics/strains show superiority in feed lot grains, efficiency of feed conversion and to a small extent in dressing percentage on live weight basis over the indigenous breeds evolved. They attained live weight of 30 kg at six months of age under intensive feeding conditions, and are 60% more economical in terms of meat production under the conditions of intensive feeding. Keeping in view the research results so far available, it has been recommended that to augment body weight and the overall meat productivity, the native breeds with no wool or with very coarse wool may be crossed with Dorset. The exotic level of inheritance may be limited to 50% only.

10.4 Sheep for Pelt Production

Breeding Karakul sheep for pelts has also given encouraging results. Performance of Karakul sheep imported from USSR. has been tested at the Division of Carpet Wool and Karakul Pelt Production at Bikaner of CSWRI which is located in the hot arid region and also at Sheep Breeding Farm, Kumbhathang, Leh (Ledakh), Jammu and Kashmir which is located in the cold arid region as purebreds and has been found to be satisfactory as reflected from the body weight, overall survival, conception rate, greasy fleece weight and quality of pelts. The results of crossbreeding Karakul with some indigenous coarse carpet wool breeds have also been encouraging. The pelts produced from half-bred involving Malpura and Sonadi indigenous breeds of Rajasthan have shown marked improvement and are comparable to Karakul pelt types. This technology can profitably be utilized by farmers as a new dimension for increasing their income.

10.5 Studies on Canary Colouration

Biological causes of canary colouration (a major problem in the North-western region) have been investigated. It is observed that it is a sequel to adaptive mechanism in hot and humid climate conditions and thermoregulatory mechanism requiring dissipation of body heat through cutaneous evaporative cooling. Alkaline suint under the conditions of high temperature and humidity reacts with wool fibre and causes the yellow colouration. For the elimination of this defect it has been recommended to practice a shearing regimen with an interval of four and eight months. The sheep may be shorn in June and again in September, so that, the June to September clip which is canary coloured can be curtailed to a minimum quantity.

CHAPTER 11

SHEEP PRODUCTION AND LAND DEGRADATION

India has approximately 13.8% of the world's cattle, with a collective herd of 185 million head. India also has 120 M goats, 98 M water buffalo, 63 M sheep (FAOSTAT – website year, 2006). Cultivation of marginal areas and overgrazing of pastures in arid regions of India have resulted in degradation of land. A quarter of the whole area needs urgent attention for soil conservation. Animal wealth in India has increased manifold and the animal husbandry practises have changed to a great extent following the introduction of newer technologies particularly for crossbreeding and up-gradation of indigenous breeds. More recently, with the liberalization of trade after the advent of WTO's SPS agreement, the chances of ingress of exotic diseases in to the country have increased. For ensuring the maintenance of disease free status and to be compatible with the standards laid by the Office International des Epizooties (OIE) – World Animal Health Organization, major health schemes have been initiated to support the animal health programmes in the states.

Land degradation refers to a decline in the overall quality of soil, water or vegetation condition commonly caused by human activities. The Vegetation Management Act 1999 states that the phrase includes soil erosion, rising water tables, the expression of salinity, mass movement by gravity of soil or rock, stream bank instability and a process that results in declining water quality.

Degradation is also considered to include a change in the ground cover to less palatable species, or a change from predominantly perennial grasses to predominantly annual grasses. Environmental dilapidation is brought about by pollution especially in urban areas, which not only experience a rapid growth of population due to high fertility rates, low mortality and increasing rural-urban migration, but also due to the rampant industrialization.

Major ecological and socio-economic crisis are perpetrated by land/soil degradation. Direct impact of agricultural development on the environment arise from farming activities, which contribute to soil erosion, salinity/brackishness of land and loss of nutrients. The Green Revolution has been accompanied by over exploitation of land and water resources and use of fertilizers and pesticides have increased manifold. In the race to urbanize virgin territory, there has been rampant violation of the land laws.

The livestock sector is the biggest single anthropogenic user of land using 26% of the earth's surface not covered by ice (Steinfeld *et al.*, 2006). It is estimated that about one third of arable crops grown are fed to livestock, primarily monogastrics. The change in diet of the people in the rapidly growing economies of China and India is driving significant expansion of intensive pig and poultry units demanding more arable-cop based feeds. This is a significant driver in the increase in food grain production through productivity increases and expansion of cropping areas into grazing and forest lands. Delegards *et al.* (2007) had predicted surplus crop feeds at stable prices which would drive livestock production. This has not happened, as the energy cost of crops are not sustainable.

About 60% of the world's pasture land (about 2.2 Mkm^2), just less than half the world's usable surface is covered by grazing systems. Distributed between arid, semi-arid and sub-humid, humid, temperate and tropical highland zones, this supports about 360 M cattle (half of which are in the humid savannas), and over 600 M sheep and goats, mostly in the arid rangelands (Steinfeld *et al.*, 2006).

Grazing systems supply about 9% of the world's production of beef and about 30% of the world's production of sheep and goat meat. For an estimated 100 M people in arid areas, and probably a similar number in other zones, grazing livestock is the only possible source of livelihood.

Livestock can improve soil and vegetative cover and plant and animal biodiversity, for *e.g* by removing biomass, which otherwise might provide the fuel for bush fires, by controlling shrub growth and by dispersing seeds through their hoofs and manure, which can improve plant species composition. In addition, trampling can stimulate grass tillering, improve seed germination and break-up hard soil crusts (Steinfeld *et al.*, 2006).

However, many people associate grazing animals with overgrazing, soil degradation and deforestation. To them livestock keeping in arid regions of the tropics provokes images of clouds of dust and an advancing desert. The two most quoted sources are the Global Assessment of Soil Degradation (Oldeman *et al.*, 1990), which estimates that 680 Mha of rangeland have become degraded since 1945, and Dregne *et al.* (1991) who argue that 73% of the world's 4.5 billion hectares of rangeland is moderately or severely degraded. In humid areas, livestock are associated with ranch encroachment and deforestation of tropical rainforests and competition with wildlife.

However the concept of degradation implies a point of view and a value judgment. For example prolonged heavy grazing by cattle and sheep undoubtedly contributes to the disappearance of palatable species to them and the subsequent dominance by other, less palatable, herbaceous plants or bushes, which in turn are more palatable to goats and other browsing species. Excessive livestock grazing by heavy cattle also causes soil compaction and erosion, decreased soil fertility and water infiltration and a loss in organic matter content and water storage capacity. On the other hand, total absence of grazing also reduces biodiversity because a thick canopy of shrubs and trees develops which intercepts light and moisture and results in over-protected plant communities which are susceptible to natural disasters.

The environmental challenge is thus to identify the policies, institutions and technologies which will enhance the positive and mitigate the negative effects of grazing in order to achieve the product "economic, social and environmental well being" required by man.

Livestock are often blamed for degradation of the environment that is caused by other users. For *e.g.* a general pattern in the exploitation of new land is that "waste land" or forests are first burned and grazed. Successive grazing and burning will change the tree-bush vegetation into a tree-grass vegetation. This opens ways for the harvesting of (fire) wood (charcoal burning) followed by cleaning the land for crop cultivation. After a number of years of cultivation, land is left fallow or abandoned because soil fertility has dropped and land has become vulnerable to erosion and degradation. When livestock are grazing this fallow land, it is blamed for all the degradation while in fact intensive wood harvesting, the cleaning of land and repeated crop cultivation in earlier phases have contributed more to the degradation.

Grazing of fallow vegetation can stimulate the rehabilitation of the vegetative cover and composition (cycling of nutrients and spreading of seeds) from animal dropping. In general the density of palatable species on fallow land is low. Therefore they are often grazed by sheep and browsed by goats. Their selective grazing behaviour can result in a delay or even inhibit further regeneration of the vegetation. This can be a desired result when the objective of farmers is to cultivate the fields again after a couple of years of fallow. If the objective is to rehabilitate degraded land through natural regeneration or reforestation, the grazing by cattle and sheep and the browsing by goats of these areas become issues for local level management and control (Steinfeld *et al.*, 2006).

Estimates of Steinfeld *et al.* (2006) indicate that the developed world uses half of all the water used by livestock in the world, while, for example, Africa only uses 14% of the world's water. As water gets scarcer it will have to be used more efficiently. However at present using it to support livestock in extensive systems is entirely rational.

There are many livestock-environment hot spots, which need careful attention for sustainable management, otherwise both environment/ecology and the livestock production systems would collapse. The important livestock-environment hot spots are (i) land degradation particularly of semi-arid region (ii) extensive grazing and large-scale forest degradation, and loss in biodiversity; (iii) animal waste production exceeds the absorption capacity of land and water. Livestock waste emits greenhouse gases such as methane and nitrous oxide, contributing to global warming; (iv) groundwater contamination, and pollution; (v) involution of mixed farming system; (vi) slaughter houses.

A Task Force to evaluate the impact of sheep and goat rearing in ecologically fragile zone was appointed by the Government of India in 1987 under the Chairmanship of Prof. C.H. Hanumantha Rao, Institute of Economic Growth, Delhi University. The Task Force concluded as follows:

1. Sheep and goats did not pose a threat to the ecology as was generally believed. The negative effects, if any, have been vastly exaggerated.
2. Small ruminants made a very important contribution to the survival of the economically weaker section of the society and met the essential requirements of industry, associated with wool, leather and meat, besides earning valuable foreign exchange through exports.
3. The unproductive cattle, intensive agriculture on submarginal land and excessive stocking rate of livestock etc. should be critically examined to improve the situation.
4. Forest departments and panchayats should play a role for protecting and improving the grazing land resources.

5. In the programme on social forestry and expansion of forest, enough attention has not been given to fodder trees. Fodder trees provide a better autotroph and heterotroph relation under the ecosystem over only timber trees.
6. Small ruminants needed to be given due priority under Central as well as State plans so as to raise per animal production.
7. The existing marketing system for the sheep and goat products entailed considerable exploitation of the rural poor. As a result, economically weaker sections could do very little to improve production or generate further employment in rural areas.

CHAPTER 12

WOOL PRODUCTION AND QUALITY

12.1 Greasy Fleece Production

The greasy (unwashed) fleece weight is usually considered as a measure of the wool yield of a sheep. Though the clean fleece yield is decisive form in an economic point of view, but phenotypic correlation between the greasy fleece kept under same environmental conditions is too high (around 0.9) thus the former can be considered satisfactory. A genetic correlation of 0.7 between the greasy and scoured is another measure. Fineness is measured as average fibre diameter and recorded usually on a lanameter or a projection microscope. Such measurements are made at constant temperature and humidity since wool is highly hygroscopic and the diameter of the same fibre will change with the change in temperature and humidity. The fineness and the length are the major factors contributing to the manufacturing quality. There is a high correlation between the spinning count (number of spins of 560 yards in length which can be spun from 1 lb of scoured wool) and fibre fineness. In Indian wools, measure of medullated fibres (as a percent of total fibres by weight of the total fibres) is also an important character. Uniformity in fineness is usually considered interms of coefficient of variation of average fibre diameter.

12.2 Factors Affecting Wool Production

Level of wool production and fleece structure may be affected by various climatic factors and seasonal changes which can influence the shedding and regeneration of follicles, fibre diameter and fibre length growth rate (Bhasin and Desai, 1965). In cold countries sheep is shorn annually. In India, it is normally shorn twice a year but in hot states they are shorn thrice a year. Shearing even in warm climate results in change in the micro-climate of sheep. Shearing in a warm environment might stimulate wool growth or in cold conditions might bring about a stress response from adrenal cortex. Injections of adreno-corticotrophic hormone (ACTH), cortisone and the application of cortisone ointments depress wool growth.

12.2.1 Non-genetic

a) Humidity and atmospheric temperature

It has been found that humidity differences of 25–30% and temperature differences of upto 30 °C had no effect. On the other hand there is a marked correlation between the rate of wool growth and mean atmospheric temperature amongst maiden Corriedale ewes fed with a uniform maintenance diet throughout the year.

b) Seasonal rhythm

The seasonal rhythm persists amongst Corriedale ewes fed a ration of constant composition at rates which maintained the effective body weights of the ewes constant. Marston (1955) found five fold differences in the rate of wool production of fine wool sheep between lush pastures of spring and the poor pasture of winter. The fundamental causes of the seasonal rhythm of wool growth remains obscure.

c) Photoperiodicity

The ewes produced lesser wool in winter than in summer. Nutrition, pregnancy and lactation can override or magnify the seasonal effect, but it can be isolated and measured, provided these factors, are adequately controlled. The photoperiodicity rhythm may be one factor and it is probable that the seasonal rhythm in temperature may be another.

d) Hormones

Speculation of the mechanisms involved here is perhaps unwarranted at this stage though the probable connection between hormone output and temperature and photoperiod rhythms may be mentioned. Atmospheric temperature is known to affect the output of the thyroid. It has been shown that thyroxine administration and deprivation influences wool growth.

The thyroid gland has a permissive effect on wool follicle development and wool growth. Thyroidectomy reduces wool growth to about 40–50% of normal. administration of thyroxine to normal sheep stimulates wool growth.

Dry ewes, given a single 60–100 mg implant responded by an increase in wool production by 15%.

Oral administration of L-thyroxine to Suffolk sheep brought about an increase in wool production per unit area of skin due to an increase in fibre length, growth rate and also caused an increase in milk secretion during the second to eighth week of lactation in Black Leicester × Scotch Black face ewes.

Cortisol

The main glucocorticoid secreted by the adrenal cortex of the sheep is cortisol and the plasma level of cortisol in the sheep is exceptionally low compared to other species. Administration of cortisol or ACTH to normal sheep suppresses wool growth and complete inhibition can be produced even though the plasma cortisol level remains below than that of other species for which the plasma levels are known. Raised levels of plasma cortisol cause "breaks" and "tenderness" of fleece. Stress conditions such as injury, disease, starvation and extremes of environmental temperature stimulates adrenal cortical secretion and a break in the fleece following such conditions is probably mediated via the adrenal cortex.

Growth hormone

Growth hormone having direct action, does not appear to have a permissive effect on wool growth in that it does not permit growth in hypophysectomized sheep, more over it does supplement the response to thyroid hormone in the sheep. Other pituitary hormones such as prolactin, gonadotrophic hormones or oxytocin do not appear to have any significant permissive or regulatory effect on wool.

e) Physiological state

It has significant effect on wool growth. In general, entire males produce more wool than castrated males which produce more than females. There is no real evidence, however, to suggest that these differences are determined by anything other than differences in size and nutritional state. The reproductive cycle in females can reduce wool growth significantly. During late pregnancy, depression of wool growth rate in the range of 20–40% has been found and even in breeds which exhibit a marked seasonal depression, pregnancy can depress the rate still further. In lactating ewes, wool growth is also reduced by upto 30% or more, despite the accepted increase in voluntary intake associated with lactation.

The average reduction in clean fleece weight between handicapped animals (twins and progeny of young ewes) and unanticipated (singles and the progeny of adults) animals born in the same year was 150 g. The period most sensitive to nutrition deficit is from 30 days before to 35 days after birth (Fraser, 1954).

f) Age

Rate of wool production in most breeds increases to a maximum between four and five years and then declines, often rapidly.

g) Nutrition

The most important factor affecting adult wool growth however, is the current level of nutrition (Fraser, 1954; Kapoor *et al.,* 1972). Increase in energy intake, except at very low levels of protein content in the diet, can have an immediate effect on wool growth detectable within a few days. It is generally accepted that wool growth rate is dependent on energy rather than protein intake. Wool protein contains a high proportion of the high sulphur amino acid, cysteine and it has been shown that variation in the availability of the sulphur amino acids to the follicle can affect both fibre growth rate and fibre composition.

It is not possible to estimate the actual amount of energy which produces wool. The amount of metabolizable energy which is drawn upon for wool production is likely to exceed materially the combustible energy for wool fibre and of the secretions which accompany it, which is 3% of the metabolizable energy derived from rations (Marston, 1955). Deficiency of minerals such as calcium and vitamin D are associated with depressed wool growth. These act indirectly possibly by affecting health and appetite.

Carbohydrate supplementation to sheep increases body weight and fleece weight (Fraser, 1954). The addition of carbohydrate to diet converts a negative nitrogen balance into a positive one (Krishnan, 1939). Carbohydrate is necessary for mitosis to take place and there is importance of glycogen in the follicle for wool growth.

12.2.2 Genetic

There is variation in wool production in various breeds. In five wool breeds, winter growth rate has been found to be around 85% of that in summer, whereas in British hill breeds the winter rate may be as low as 30% of that in summer. The wool quality of Deccani and their crosses with Dorset and Merinos, Patanwadi and

Deccani crosses has been studied by Daflapurkar *et al.* (1979) and Deshpande and Sakhre (1987).

The importance of differences between strains and probably breeds of sheep is indicated by recent work with Merinos, indicating small but significant differences in efficiency of conversion of feed to wool.

The behavioural differences between individual sheep subjected to competition for limited feed, are another source of variation. Feeding groups of sheep once per week gave higher mean body weights and fleece weights than feeding them equal amounts on daily basis. Difference of social order within a flock or group of sheep also determines wool production.

12.3 Wool Quality

The Sheep farmers/producer of wool should have a working knowledge of the wool quality. They should be able to determine the value of the wool raised so as to look out for their own interests. Otherwise they will find themselves at the mercy of the wool buyers at market time.

The various factors that are taken into account for judging the quality of wool are:

12.3.1 Fibre fineness

The diameter or thickness of wool varies. The thinner and finer the fibres, the more valuable is the wool. The unit for measurement of thickness is a μ or (1×10^{-4}) of a centimeter. The wool fibre has a diameter varying from 12 to 80 μ. Fine wools range from 15 to 25 μ in diameter. The fleeces obtained from the Merino breed are of this quality.

12.3.2 Staple length

The length of the fibre is important in determining the use that may be made of it in manufacture. Long wool is more valuable than short wool.

12.3.3 Medullation

A true wool fibre has no stiff central core in it. As a rule, the finer the wool and more elastic and uniform the fibres are, the better it is considered for spinning purposes. Apart from its being of a fine type, it is of utmost importance that throughout the fleece, there should be as much uniformity as possible in the fineness of the fibres. There should also be uniformity in the diameter of the individual fibre so that each one is even throughout its entire length. Wools from improved breeds do not show variations in excess of 50% from the finest to the coarsest, while in the case of unimproved sheep, the coarsest fibre may be 5 or 6 times as thick as the fine ones.

12.3.4 Crimp or curliness of fibres

Wool, as it grows on sheep, develops a wavy structure which is called the crimpiness. In fine wools, these curls are larger in number and more regular in a unit length of fibre than in coarse fleeces.

Crimp is often emphasized by some as being a very important feature of wool. The number of crimp per inch is usually associated with fineness of fibre, but this is not necessarily always the case. Wool fineness cannot be judged on the basis of crimp alone. Some Australian wool has a "bold" crimp but is finer than other wool that has a closer crimp. Crimp is related to elasticity and some other features, but its chief value is probably in the fact that an even, distinct crimp indicates a well-grown, sound fibre of uniform diameter and length. The other features of good wool are enhanced and made more evident by such crimpiness.

12.3.5 Strength of fibres

The stronger the fibres and the less they break, the higher is the quality of the wool. Strength is another matter that influences the value of wool. Strong durable fabrics can be made only from wool that is well grown and sound. The strength or soundness of the wool fibres is largely dependent upon the health and nutrition of animals. Certain other environmental factors may affect the strength of the fibres. Sheep that remain in good health and that have regular nourishment, both with respect to quality and amount, are able to maintain a uniform growth of fibre and hence have sound fleeces. If the nourishment is suddenly reduced or radically changed so as to affect the health of the sheep, the fibres are weakened and hence reduced in value. If the condition of health is seriously affected, the fibres may be so severely weakened that the sheep will lose the entire fleece. This loss occurs a few weeks after the onset of the sickness. Most feeds fed in reasonable amounts do not adversely affect growth or soundness of the fibres. Any feeds fed so as to disturb the health will result in weakened fibres. Ewes nursing lambs, especially if on dry feed, do not grow as strong wool as at other times.

12.3.6 Lustre, colour and elasticity

The more lustrous and shining the wool, the whiter its colour; and the more elastic the wool, the more it is prized, because it produces smooth and shining yarn and cloth, and can be dyed better.

The analysis of fabrics is beyond the abilities and knowledge of most consumers. In order to protect the textile quality of the fabric, the law now requires that fabrics be labeled with the kind of fibres which they contain. Wool which has not previously been used is designated as new or virgin wool; that which has been previously used may be designated as reused or reworked wool. If other fibres are used with the wool, the fabric labeling act requires that they be designated also. While there has been some opposition to this law, it has been beneficial to both growers and consumers. The act does not prohibit the use of any fibres which the manufacturers may want to use, but it does require that the fibre content of the material be given. Styles and the need for economy have great effect upon the types of fabrics made as a result of the demands of consumers. The very highest grades of reused wool may be as good or even better than very poorly grown low grade new wool. For the most part, virgin wool is superior, and fabrics made from the best grades of such wool are the most expensive and serviceable.

Foreign wool and foreign-made goods are sometimes spoken of as vastly superior to those of domestic origin. Environment and breeding have effects upon wool, but it has never been proved that all foreign wool is better than much of the domestic wool. The best foreign wool may be better than the best domestic wool, but at least a part of the expressed superiority is fancied. There are differences in the way the wool is prepared for market and in the shrinkages, but the differences in the actual fibres are not as evident as is often implied. Some of the wool, especially that from Australia and New Zealand, when of a certain fineness, may be superior in length, softness, and uniformity of fineness, but little if any is superior to domestic wool in colour and strength. It is unlikely that wools from other countries are better than similar wool grown by good producers in this country. Workmanship in manufacturing is a matter apart from the wool fibre's inherent characteristics.

CHAPTER 13

MEAT PRODUCTION

13.1 Meat Production, Consumption and Marketing Trends

With a formidable 28% of large ruminant population and around 25% of small ruminant population of the world, India should have dominated the meat trade. However, ironically this country produces less than 1% of the total of 225.9 Mt meat production in the world. Of the annual meat production of around 1.2 Mt a little over 50% is from small ruminants, a quarter from large ruminants and an equal amount from poultry and pigs. Inspite of the vast reservoir of ruminant population, the meat exports from this country are only Rs. 100 crores annually.

Mutton Production from sheep from 1999–2003 is given in Table 13.1.

Table 13.1. Meat Production from Sheep

Year	No. of animals slaughtered (in 000 Nos.)	Average yield (in kg)	Meat production (in 000 tonnes)
1998–1999	18,118	10.10	183.05
1999–2000	16,814	11.06	185.99
2000–2001	16,768	11.45	191.94
2001–2002	19,456	11.53	224.26
2002–2003	21,811	11.63	253.67
2003–2004	17,625	12.22	216.00
2004–2005	18,928	11.89	225.00
2005–2006	19,110	12.59	241.00

Source: State Animal Husbandry Department

Although the percentage increase in meat production has been 23.85, the contribution to total rneat production has decreased by 2.86 during the last 20 years. The total meat production from all meat animals has increased from 1.10 Mt in 1985 to 6.03 Mt in 2004. During the same period the contribution of sheep and goat has relatively decreased from 52.61 to 43.44%; the average carcass weight remained at 9 kg. These changes indicate that the meat production from sheep and goat has not been coping with the increased demand and the relative role of sheep and goat enterprises towards meat production is getting decreased. This kind of situation might be due to the changes taking place in Indian agriculture (Table 13.2). A decrease in permanent pasture land certainly indicates the decreased role of sheep, if the present system of extensive rearing has to be continued. However, improved meat production from sheep and goat is possible by taking advantage of the increased crop production through adoption of semi-intensive rearing. In the case of sheep, percentage of total food to dam is 72 against 28% to progeny. This indicates a differential system of rearing which may be more economical. The dam may be maintained mostly on grazing while the post-weaned lambs may be reared under semi-intensive or even under intensive system if the economics of such a system permits.

Large areas of arid zones are being utilised for sheep and goat farming and mostly marginal and landless labourers live on these enterprises. Besides the domestic need, there exists a vast potential of export of live animals as well as meat from sheep and goat to Gulf countries, having proximity to these countries, and established preference for Indian carcasses by virtue of their leanness. Hence economic efficieney, adaptability and social acceptability of different livestock enterprises have to be evaluated for assessing the future production potential of different livestock enterprises and targeting of meat production.

Table 13.2. Some Changes in Indian Agriculture (1975–85)

Itern of change	% change
Irrigation	+ I8.19
Amble land	+ 00.68
Pernlanent pasture	– 06.67
Cereals-area	+ 04.04
Yield (kg/ha)	+ 23.71
Production	+ 28.77

13.2 Components of Meat Production

13.2.1 Prolificacy

Sheep in India and other parts of tropics in the world, breed throughout the year, but are seasonal breeders in most areas of the United states, and sub-tropical and temperate areas the breeding season is in the fall. Some breeds, such as the Merino and Dorset Horn, may be bred under some conditions to produce two lamb crops per year. Most breeds produce just one crop per year, although rams produce sperm throughout the year. Some rams are suceptible to high summer temperatures, and may be infertile or of low fertility in late summer during the first part of the breeding season.

The number of lambs raised per ewe is one of the most important factors deter nining the efficiency of production. Lamb production varies a great deal under different conditions and with different breeds. Ewes under farm conditions are usually more prolific than those produced on the range. This is probably due to a higher level of nutrition generally found in farm flocks. Twinning in sheep often is desirable, because a ewe that weans twins produces from 10 to 15 kg more lamb than the ewe that weans a single lamb.

Fertility in sheep is lowly heritable, with an average heritability and repeatability estimate of 7 to 13%. These estimates are in agreement with those for other classes of livestock. This indicates that fertility in sheep is not greatly affected by additive gene action and could be improved very little by selection. Most of the phenotypic variation, therefore, is due to environmental factors and attention to these should improve the lamb crop.

13.2.2 Growth rate and feed conversion efficency

The age at which lambs are weaned varies under different conditions, but the age of 180 days is often used for selection purposes.

Lambs can be weighed as they reach 180 days, or can be corrected to this age. the correction is done by multiplying the average daily gain from birth by 180 and adding this product to the birth weight.

The age of the ewe may have considerable influence on the weaning weight of her lambs. Two-year-old ewes wean lambs that are from 2 to 4 kg lighter than those from mature ewes. Production of ewes usually increases to four or five years of age. Probably the most important adjustment is that for weaning weights of lambs from ewes that are two years of age. This adjustment can be made by comparing the production of the two-year-old ewes with that of mature ewes in the same herd and then adding the difference to the weaning weights of lambs from the younger ewes. If this is not possible, an adjustment may be made by adding 3.1 kg to the weight of the lambs.

13.2.3 Type and conformation

Desirable type and conformation have also recieved attention in sheep as in other classes of farm animals. With this species however, attention must also be paid to selection for wool production in addition to mutton quality and rate and efficiency of gains.

Animals possessing vesry obvious defects, such as over-shot jaws, undershot jaws, black wool, wool-blindness, skin folds, shallow bodies, and poor mutton qualities, should be culled from the flock. If animals with these defects are eliminated from the breeding flock and selections are made on the basis of body weight and quantity and quality of wool, especially in the selection of rams, perhaps this will be sufficient attention to desirable type.

13.3 Factors Affecting Mutton Production

13.3.1 Body weight

Body growth affects the composition of the lamb. It has been observed by many research workers (Reid *et al.*, 1968) that the 63, 95 and 88% of variation in ash, protein and fat, respectively are due to changes in body weight at slaughter.

13.3.2 Heredity

The rate of growth varies from breed to breed, even if given same type of fodder and concentrate to all the breeds. The growth rate is affected by the adult body weight (Dackson, 1974).

13.3.3 Sex

The sex hormones of males and females affect their growth rate. The ewe differ from the ram in that it has slower growth rate, a more early maturing carcass and reaches a lower mature size due to the effect of oestrogen in restricting the growth of long bones of the body.

13.3.4 Castration

The castration of male decreases their growth rate and they mature at an early age. The appreciation of the central role of the sex hormones in growth and development as manifested in the consequences of castration have led to extensive study of the effects of the administration of natural or synthetic hormones to meat animals in order to influence performance.

13.4 Specification for Mutton (sheep) and Chevon (goat-meat) (AGMARK Standard)

1. *General*: Mutton and chevon shall be obtained from healthy animals and slaughter in slaughtered houses. The animals shall be subjected to ante-mortem and post-mortem inspections by Veterinarians designated and/or of recognized local agency in respective states by the Agricultural Marketing Adviser. It shall not be treated with colours, dyes, additives, preservatives and chemicals.

2. *Special Types*: Mutton shall be of the following types:

   I. Type A : Fresh, Chilled Carcasses
   II. Type B : Fresh, Frozen Carcasses
   III. Type C : Fresh, Chilled (Bone-in)
   IV. Type D : Fresh, Frozen (Bone-in)
   V. Type E : Fresh, Chilled-Deboned (Boneless)
   VI. Type F : Fresh, Frozen-Deboned (Boneless)

The AGMARK grades and standards formutton is given in Table 13.3.

Table 13.3. AGMARK Grade and Standards of Mutton

Commercial name	Grade designation	Special characteristics	Bacteriological standards
1	2	3	4
Mutton	Standard Grade	The dressed carcasses or meat shall have/be	The sample per lot drawn and tested shall fulfil the following requirements:
		1. Boneless cuts to be entirely free from bone pieces, wood dust, metal pieces or other undesirable matter. Not exceeding 100 per gm	1. Total plate count not exceeding 10 microganisms per gram. 2. *Escherichia coli*
Chevon	Standard Grade	2. Free from slime, discoloration, malodour and structural alterations	3. *Salmonella*-Absent
		3. Lean	
		4. Firmness in consistency, *i.e.* will not pit on pressure	

CHAPTER 14

MILK PRODUCTION

14.1 Production Level

Sheep have been used for milk production since ancient times. World production of sheep milk in 1979 (FAO) was 7.328 Mt as compared to 6.958 Mt of goat milk. However, in India, though production of goat milk is 0.65 Mt, production of sheep milk is negligible inspite of there being over 40 M sheep as compared to 70 M goats. This is especially striking when we see country like Turkey which has 41.5 M sheep producing 1.08 Mt of sheep milk per annum. Thus sheep production for milk has been a greatly neglected sector of animal husbandry in India. The matter of milk yield in sheep has received scant attention in India, but some studies that have been made indicate that the quality of milk with respect to its fat content varies widely among individuals within breeds and this seems to be more pronounced than variations between breeds. There is also a great variation in the milk of an individual at different periods during lactation and vary probably during different lactation periods.

France is the leading country in the world as far as milk production from sheep is concerned. In Southern France in 1951, 33 M litres of sheep milk were produced from 600,000 ewes, while in 1978, 59 M litres of milk were produced from the same number of ewes. This was because the annual milk yield per ewe increased from 55 litres to 100 litres (Flament and Morand-Fehr, 1982).

14.2 Milk Composition

Data indicate ewe's milk to have a considerably higher average fat content than Goats' milk (Table 14.1). Ritzman showed that 158 samples from a great variety of ewes ranged from 2.4 to 12.1% in fat content and averaged 6.0%. But observations of the growth of lambs support the statement that the amount of milk yielded by the ewes is of more consequence than its quality if this is not too far from average.

Table 14.1. Average Composition in g/lt

	Dry matter	Fat	Lactose	Proteins		Ash
				Casein	Albumin	
Goat	115 – 135	25 – 45	40 – 45	24 – 30	4 – 6	7 – 9
Sheep	170 – 185	55 – 70	43 – 50	45 – 50	8 – 10	9 – 10

14.3 Milking Frequency

Milk yield in Deccani flock ranges from 0 to 1000 ml per day (milked twice), the average appearing to be around 300 ml. Jaisalmeri ewes yield about half litre of milk daily which is used for ghee making (Mittal and Shivprasad, 1987).

Gatenby (1986) has reported milk yields of sheep in the tropics and sub-tropics, which is reproduced in Tables 14.2 and 14.3.

Table 14.2. Milk Yields of Selected Breeds in India

Breed	Yield kg/day	Period measured (days)
Avikalin	0.45 – 0.47	92
Chokla	0.46 – 0.51	90
Jaisalmeri	0.40	90
Malpura	0.50 – 0.60	90
Merino	0.70	112
Nilgiri	0.65 – 0.67	112

Table 14.3. Milk Yield of Selected Breeds of Sub-tropics

Breed	Country	Yield kg/day	Period measured (days)
Assaf	Israel	1.34	156
Assaf	Israel	2.23	100
Awassi	Iran	1.16	207
Corriedale	Kenya	1.36	70
Merino	Egypt	0.85	126
Sardinian	Italy	0.86	44

14.4 Utilization of Milk

Sheep are kept by the underprivileged section of our society so that any improvement in milk yield of sheep will benefit a large deserving population. An increase in milk production will also enhance lamb survival and growth and provide additional milk for home consumption and perhaps for sale.

The increase in weights of lambs is closely proportional to yield and consumption of milk, and these increases may vary as much as 75% between lambs receiving copious amounts and those receiving very small quantities. The feeding and management of the ewes will have a great bearing on the quantity of milk they yield, but good feeding and management will not result in a high yield from an ewe having no inherent capacity for high production.

CHAPTER 15

UTILIZATION OF BY-PRODUCTS

15.1 Hides and Pelts

15.1.1 Hides

Though India stands second in the world in the production of hides and fifth in the matter of production of skins, it cannot forge ahead in capitalising on its large production unless stress is laid on quality right from the initial stages of production. As such, improved methods of flaying should be introduced in the slaughter houses and better flaying should be encouraged by payment of premia to good flayers.

As against the declining trend in the exports of hides and skins, leather and leather manufactures have been emerging in a significant manner in Indian exports. The average annual exports of this item improved from Rs. 63 crores during the triennium ending 1968–69 to Rs. 146 crores during the triennium ending 1973–74 with the peak having been reached in 1973 when exports amounted to Rs. 175 crores. AGMARK Grade Designations and Definition of Quality of Skins and Hides is given in Tables 15.1 and 15.2.

15.1.2 Pelts

The pelts are produced either through the slaughter of lambs with in 24–48 hr of birth or through killing of unborn lambs removed through slaughter or through abortion of the mother around 130–140 days of pregnancy. The latter pelts, known as "broad (fat) tailed Persian" are twice more valuable than the best types produced

Table 15.1. Grade Designations and Definition of Quality of Hides (Kids, Buffaloes and Calves) Produced in India

Grade designation	Constitutional	Branding iron marks	Hair side pox, scab, sores and skin disease marks	Marks due to injuries	Scratches
1	2	3	4	5	6
A	Hides from young or middle aged animal with no rib marks or signs of emaciation.	Not allowed	No pox, score or marks of skin diseases allowed but one small scab and a small sore may be allowed if not on the butt.	Not allowed	One small superficial scratch scratch not affecting the grain grain allowed if within 4" of the outer edge of the neck or belly portion only.
B	Slight rib marks allowed	Not more than one brand mark allowed which may only be on the belly neck or cheek	No pox or skin disease marks allowed. Two small scabs or sores allowed if not on the butt In the case of kids and buffalo hides these may be on the butt provided they are within 2" of the line of the spine.	Permitted on belly $	Superficial scratches not affecting the grain allowed.
C**	Rib marks freely allowed	Atleast, one half of the hide shall be free from brand marks	Atleast half the hide shall be unaffected	Allowed if on neck belly or below hip $ joint.	Reasonably free from deep scratches.

Table 15.1. (*Contd.*)

General characteristics					
Flesh side		State or condition	Trimmed weight prior to curing		
Flayer's marks and cuts	Warbles and warble holes*		Weight groups	Limits of weight buffalo hides	Kids
7	8	9	10	11	12
No deep cuts allowed. Not more than 4 shallow cutts other than on the butt permitted $	One open and not more than 3 blind warbles allowed	They hides shall be reasonably clean and free from flesh, fat and extraneous matter. Horns ears feet (from above the fetlocus) and tail (beyond 4 inched) shall be removed			
Shallow cuts allowed and one deep cut allowed if not on the butt	Not more than 5 open and 8 blind warbles allowed	-do-	Heavy (H)	60 lb and over	28 lb and over
Shallow cutts allowed a reasonable number of deep cuts or holes permitted if not on the batt	Not more than 10 open and a reasonable number or blind warbles allowed	-do-	Medium (M) Light (L) Calf (C)	40 to 60 lb 25 to 40 lb under 25 lb	18 to 28 lb 8 to 18 lb Under 8 lb

* Warble holes are holes or spots which can be opened with a blunt skewer

** But hides weighing over 12 lb. shall not be graded higher than C. Young bull hides upto 12 lb. may be placed in B grades but not in A

$ In case of heavy and medium weight buffaloes only "One sore or cut mark on the hair of flesh side within two inches of the spine" may be permitted

Table 15.2. Grade Designations and Definition of Quality of Skins (Goat and Sheep) Produced in India

Definition of quality

Special characteristics respecting State of condition

Grade designation	Constitutional	Hair side		
		Mange, scab and skin disease marks	Marks due external injuries	Pox, scab and other diseases marks
Primes	Plump and thick skins without signs of emaciation	Not allowed	Not allowed	Not allowed
Seconds	Without signs of emaciation less plump and thinner than primes but not so thin as to become papery on drying	No more than four allowed	No more than four allowed	No more than foure allowed

General characteristics

Flesh side Flayer's cuts and marks	State or condition	Measurements groups	Trimmed measurements Limits of measurement! Length along back!!	
			Goat (cm)	Sheep (cm)
6 Not allowed except for 1 or 2 cuts on the belly portion*	The skin shall bear no branding from marks and shall be reasonably free from bloods dirt and other extraneous matter Head and feet shall be removed	Extra heavy	≥ 102	≥ 91
		Heavy	91 – 102	
A reasonable number allowed on the belly portions* but not more than 4 cuts or holes (with a combined total length of 51 mm may be on the remainder of the skin		Medium	84 – 91	81 – 89
		Light	71 – 84	69 – 79
		Kids lambs	56 – 71	56 – 69

* Provided that the total number of marks and cuts, apart from those on the belly portion*, shall not exceed four or cover a total area of more than 6.5 sq. cm.

** The skin having been evenly folded lengthwise and a measurement made at right angles from the fold to a point midway between the force and hind legs any part of the skin lying more than seven-eights of this distance from the central fold shall constitute the belly portion.

* ! The measurements given are for dry salted skins. A shrinkage allowance of 2–1/2% shall be added in the case of wet salted and 5% in the case of raw skin.

* !! The skin having been evenly folded lengthwise and laid flat the length along back shall be measured from the root of the tail to the point where the skin of the front part of throat is the widest when measurement at right angles to the fold.

from slaughter after birth as they have better ornament, are lighter and more lustrous. The lamb pelts sell from 15–20 dollar per pelt in international market depending upon their size and quality. The quality of pelt is generally determined by the ornament (type of curls, their size and tightness), lustre and its weight. The present trend is to get light skins with short hair. The pelts are classified into various types and into quality grades within each type; the types are jacket, ribbed, flat and caucasian.

In order to obtain the best quality pelts so as to fetch the maximum price, one to two day old lambs should be slaughtered and the pelt removed. Beyond 2 days of age, the hair grow longer decreasing the quality of the pelt. This pelt of lamb serves as the raw material and it has to pass through many processing techniques before being actually converted into the final product. As the quality of the end product depends to a large extent on the quality of the raw material, It is very essential to protect the raw pelts of the freshly slaughtered lambs from putrification so that the best quality pelts could be produced. For this purpose the raw pelts are subjected to primary processing known as "curing". Though there are many methods-of "curing" or preserving the raw skins, the "dry salting" method as practiced in the USSR is the simplest, practically feasible and most economical in the conditions prevailing in India. As the primary processing is to be done immediately after slaughtering, the raw Karakul pelts are dry salted at the sheep farm itself and hence the complete knowledge of this method is necessary for every sheep farmer rearing Karakul sheep.

15.2 Dead Carcass Utilization

Failure to make use of the carcasses of fallen animals is responsible for an enormous wastage of otherwise useful material.

Rendering plants fully equipped to process animal carcasses should be established on a subdistrict, district or regional basis and that incentive payments need to be made to those who bring fallen animals to these plants. Each state should establish a suitable number of carcass utilization and hide flaying centres with a view to prevent the colossal waste that is occurring at present.

The huge financial loss sustained due to faulty methods of flaying, curing of hides and non-utilization of the by-products of fallen animals has been observed. This needs to be checked by proper utilization by establishing modern utilization centres.

15.3 Slaughter By-Products

a) Bones

A major portion of the bones collected in India is utilized for production of crushed bones and bone grists and a small quantity is used for the manufacture of bonemeal. There were 100 bone crushing mills and about 360 bone digesters in the country at the end of the Fourth Plan. Most of the mills crush bones primarily with the object of exporting crushed bones and bone grists. The bone digesters are working on a cottage industry basis set up in various states with the aid of the Khadi and Village Industries Commission or the State Governments for converting locally available raw bones into bonemeal for utilization as fertilizer. New bone digesters will have optimum use if set up only in areas capable of utilizing the bonemeal, preferably in remote places unconnected by rail or road. Cooperatives of bone collectors should be provided with bone digesters on rent. Encouragement should be given by the State Governments for setting up factories for the manufacture of gelatin, glue and Neat's foot oil.

b) Animal fats

Full and rational utilization of animal fats which are available in large quantities from fallen and slaughtered animals is highly important. This would not only benefit the livestock producers but would also help in saving foreign exchange worth crores of rupees, which are being spent at present on the import of animal fats. As large a quantity as 61 Mkg of animal fats valued at Rs. 8.9 crores were being imported in a year (1972–73).

Slaughter houses which are being modernized should have a by-product plant within their precincts or in close proximity so that all available fats from slaughtered stock could be processed. Efforts should also be made for efficient and quick recovery of fats from the dead animals. Since the proportion of fallen stock is very much higher in the case of bovines, a chain of carcass idolization centres should be established in areas of concentrated bovine population.

c) Blood

At present blood is being collected only in a few slaughter houses of the country. In developed countries, blood finds several important uses. It is consumed by human beings in the shape of black puddings and blood sausages, and blood albumen is employed as a substitute for eggs in the ice cream manufacture and in bakeries. It is used extensively for industrial purposes, as a fertilizers and is incorporated as an important constituent for the manufacture of stock feeds. In India, only blood collected from sheep, goats and pigs is utilized for human consumption. It is consumed by the weaker sections of the society after frying it with spies or boiling it with rice. Occasionally it is mixed with wheat flour in the preparation of chapaties.

Collection of blood from all the slaughtered animals is highly important. When incorporated in the livestock feeds, it would provide a valuable source of animal proteins and as a fertilizer it would enrich the soil. Uncollected blood in a slaughter house becomes a serious sanitary problem. It quickly clots, choking drains, septic tanks etc. and rapidly decomposes serving as an ideal medium for bacterial growth. Blood collection on efficient lines will be possible only in modern slaughter houses as collection has to be done speedily and without dilution with water. Otherwise, processing would be prolonged making moisture removal highly expensive.

d) Horns and hoofs

Horns and hoofs constitute a very small portion of animal by-product but because these are rich in keratin and have considerable value as fertilizer after conversion into meal, these have much economic value. The horn core is particularly rich in ossein which is used in developed countries for manufacture of gelatine. The export trade in buffalo horns and antlers fetched Rs. 6.2 lakhs in 1972–73. It is desirable that wastage in the collection of horns and hoofs is reduced as much as possible and export trade in these items is increased. Further, horns and hoofs left in the country should be processed for the manufacture of gelatine and the unnoticed portion converted into meal for use as fertilizers.

e) Meat unfit for human consumption

The annual production of useless meat, *i.e.*, meat condemned for human consumption and meat which remains adhered to the bones and other tissues is estimated to be of the order of 23,000 tonnes. Since useless meat is an excellent source of nitrogen in poultry feed and fertilizer for tea and coffee plantations, such meat should be converted into meat meal and should not be wasted as is being done at present.

15.4 Sheep Manure and Composting

Sheep dropping improve the fertility of soil considerably and penning of sheep in harvested fields (sheep folding) brings in additional income to the flock owners.

CHAPTER 16

SHEEP RECORDS

It is essential to maintain the necessary records at an organized sheep farm to know about inputs and outputs. This will help in working out economy of sheep production per adult unit and develop a profit and loss statement.

The recording system is required to be simple, accurate capable of collecting the required information and finally kept upto date. The data collected should not be kept but should be used for proper analysis. In sheep growth rate, feed conversion, wool quality, dressing percentage are highly heritable characters. Prolificacy, milking capacity and fleece weight are characters of low heritability. More profits can be obtained if more attention is paid to characters of high heritability. Having decided his breeding programme, a farmer must then choose his parent stock.

Good record should identify the offspring with the parents. They should also give date of birth, sex and final disposal of individual. This information together with desirable confirmation is the basis upon which selection for flock improvement should be made.

There should be a system for ear marking, tattooing and or tagging. In case of short ear breeds, tattooing on inside of thighs or tails be practised. A combination of tattooing and tagging is desirable.

Following data/observations are of great help at an organized farm:

i) Livestock inventory

This register gives information about the individual identification, parentage and date and reason for sale or culling.

ii) Wool weight and quality

Wool evaluation should be limited to greasy fleece weight of each clip, staple length at a particular site, average fibre diameter, medullation percentage, visual assessment of fineness and making careful examination of undesirable characters such as black fibres or coarse fibres.

iii) Growth records

Body weights at birth 3, 6, 9 and 12 months of age are important. Birth weight is related with vigour, survivability and growth rate. Weaning weight is used to measure total lamb production of an ewe. Six month weight is correlated with market and yearling weight and is highly related to adult weight and ewe productivity. Body weights of each lamb are required to be correlated to age, sex, type of birth, season and age of dam.

iv) Prolificacy

Twin producing ewes are more profitable than single producing ewes excepting in areas where grazing conditions are poor.

v) Health

Information on causes of death and incidence of diseases helps in planning prophylactic health programme.

In addition to above data, the following records should be kept properly.

1) Livestock account record
2) Breeding record
3) Lambing record
4) Shearing/wool yield record
5) Mortality record
6) Sale of animals and wool record
7) Purchase of livestock medicines and equipments record.

16.1 Guidelines for Recording of Data

All recordings should be done by the farm manager or technical officers available to him.

16.1.1 Body weights

The body weight should be recorded at birth and after every 7 days upto 28 days and thereafter every 15 days upto 268 days and thereafter every 30 days upto 358 days. The earlier weights (10 kg) should be recorded closer to 25 g and weights upto weaning (20 kg) closer to 50 g and later weights closer to 100 g.

16.1.2 Body measurements

The body measurements, *viz.* heart girth, length and height at withers closer to 1 cm should be recorded at birth, weaning, 6, 9 months and 1 year with the help of a metallic tape marked in cm and after placing the lamb on a flat even surface with up right standing position.

16.1.3 Fleece yield and quality

i) Immediately after birth fleece colour and the extent of body covered with the colour should be recorded.

ii) The record of 6 monthly greasy fleece weight of all animals should be made closer to 25 g. The exact age at shearing should also be recorded.

iii) Samples of fleece from approximate 1" × 1" area (in case density is not to be determined otherwise exact area should be marked and complete sample taken) from mid-side should be clipped with a sharp scissor as closer to the skin as possible and used for recording clean yield percent, average staple length, average fibre diameter, medullation percentage and density.

iv) The intensity and extent of yellow (canary) coloration may also be recorded at the time of shearing.

16.2 Culling and Disposal

i) All lambs which are stray (of unknown parentage) highly deviating from the characters of a genetic group, in colour and quality of wool and the lambs which are extremely poor in growth (less than 2 standard deviation below the weaning weight) would be culled at weaning.

ii) All male lambs should be ranked on the basis of 6 months body weight and those below the mean should be culled.

iii) Out of the rest 50%, the requisite number in each genetic group should be retained on the basis of greasy fleece weight and fleece quality (average fibre diameter). While retaining efforts should be made to have largest numbers of sires represented.

iv) All ewes not breeding/lambing in two consecutive seasons should be culled (this may not be strictly followed in exotics if otherwise the condition of the animal is good).

v) All ewes beyond the age of 7 years or those that become gummers earlier should be culled.

vi) All rams that have been used for 2 consecutive breeding seasons and have been mated to at least 10 ewes of each genetic group should be removed.

vii) All ewes showing extremely poor mothering ability should be culled.

viii) The culled/extra animals should be disposed off to the State Governments for breeding purpose if suitable or through the conductor on basis of their live weight. The information about the extra males suitable for breeding should be circulated among all the development agencies.

ix) The disposal of culled animals, through the contractor should be the responsibility of farm manager. He will obtain the necessary sanction for disposal and writing off of these animals, giving age, live weight and reason for culling.

CHAPTER 17

DISEASES AND THEIR CONTROL

Morbidity and mortality are the two important factors resulting in heavy losses in sheep production and improvement programmes. Prevention is always better than cure as it is a lot cheaper. This has a special significance with sheep as they seem to respond less to treatment when sick than other livestock species. Disease in sheep can be broadly categorized as non-infectious and infectious.

17.1 Non-Infectious Diseases

80.2% deaths in lambs have been estimated due to non-infectious causes. Starvation, primarily from mis-mothering and behavior, nutritional and environmental stress, reproductive problems and predation are the major causes reported.

17.1.1 Drenching pneumonia

It is one of the most common and important pathological conditions in sheep. It is characterized clinically by increased respiration, coughing and abdominal breathing. A toll of 20–40% of the mortality at Central Sheep and Wool Research Institute, Avikanagar have been due to this conditions, although mainly due to bacterial or viral causes but occasionally one can come across with "aspiration" or "drenching" pneumonia during mass drenching operation. The best remedy is to avoid force drenching and allow the animal to deglutition the drench itself. If some fluid has entered in respiratory tract, lower the animal's head immediately and slap few times on its head.

17.1.2 Ruminal tympany

"Bloat" or over-distension of left flank either due to gas alone or with froth. This is also generally encountered in "greedy feeders" either when lush pasture is available or in farms where concentrate feed is offered. Tying a "stick" in the mouth as a bit is most practical and can be done immediately. Oral administration of sweet oil with turpentine oil or at times with formalin is advised.

17.1.3 Acute impaction of the rumen

The ingestion of large amounts of highly fermentable carbohydrate feeds causes an acute illness due to the excess production of lactic acid in the rumen. Clinically the disease is manifested by dehydration, blindness, recumbency, complete ruminal stasis and a high mortality rate.

Antihistamine such as avil, normal saline solution 500 ml I/V, sodabicarbonate 4 gm orally is advised.

17.1.4 Intussusception

It is a common occurrence in sheep due to several reasons such as nodular worms, change in feed and local intestinal problems.

The animal looks dull, off feed, kicking at the belly with no rise of temperature and frequent straining with no defecation, an excitement with symptoms of colic at later stages and the animals become recumbent and dies. No rational treatment except surgery is possible. Mild purgative with astringents may be of little avail.

17.1.5 Deficiency diseases

Young sheep grazing on drought stricken pasture can suffer serious depletion of reserves of the minerals and vitamins. In adult sheep on a deficient diet for few months before hepatic stores are depleted and disease becomes evident. Copper and cobalt deficiency in sheep are characterized by anorexia and wasting. Growth and wool production are severely retarded. Wool may be tender or broken. Fine wool becomes limp and glossy and loses crimp developing straight steely appearance. Anaemia, diarrhoea and unthriftiness occurs in conditions of extreme cases. Required dosage of copper or cobalt sulphate treatment causes rapid disappearance of the symptoms.

i) Diseases caused by the deficiency of Phosphorus, Calcium and vitamin D

The daily requirements of calcium, phosphorus and vitamin D for an adult sheep is about 2.5 g, 1.5 g and 300 to 500 units respectively. The lamb may develop rickets and in adult sheep it may result in osteomalacia. Supplementing the diet with mineral mixture is essential.

ii) Vitamin A deficiency

Vitamin A deficiency occurs in sheep on dry range country side during the periods of drought. Clinical vitamin A deficiency does not occur commonly because hepatic storage is usually good if period of deprivation is not sufficiently long for these stores to reach a critically low level. Young sheep grazing on drought stricken pasture can suffer deficient diet for 18 months before hepatic stores are depleted and the disease becomes evident. The syndrome includes night blindness, corneal keratinization, pityriasis, defects in hooves, loss of weight and infertility. Congenital defects are common in the offspring of deficient dams. For treatment, place the animals on green pasture and supply vitamin A in feed.

17.1.6 Pregnancy toxaemia of sheep

Amongst domestic farm animals the metabolic disease achieve their greatest importance in milch and pregnant ewes. Pregnancy toxaemia of sheep is called as Ketosis of pregnant ewes and is highly fatal. Hypoglycaemia and hyperketonemia are the primary metabolic disturbances in ovine ketosis. The most important etiological factor in pregnancy toxaemia is a decline in the plane of nutrition and short period of starvation (40 hr) during the last two months of pregnancy particularly large lambs may be affected. It is primarily a disease of intensive farming system and is relatively rare in extensive grazing units.

The earlier sign is separation from the flocks and apparent blindness, constipation, grinding of teeth later marked drowsiness, tremors of the head, twitching of lips, champing of jaws and salivations, circulating, sheep goes position of limb and elevation of chin (star grazing posture), in co-ordination and falling when attempting to walk. A smell of ketones may be detectable in the breath. In three to four days, profound depression, comma and death. For treatment 100 ml of 50% glucose I/V may be given. Supply of molasses in ration is also recommended. During the last two months of pregnancy a provision of concentrate at the rate of 400 g per day; increasing to 800 g per day in last two weeks be made.

17.1.7 Poisoning caused by

a) The chlorinated hydrocarbon insecticides

This group includes DDT (Dichloro Diphenyl Trichloroethane), benezene hexachloride (BHC), its pure gamma isomer lindane, aldrin, dieldrin, chlordane, toxaphane, methocychlor, DDD, insecticides to control external parasites on sheep or to control insect infestation of pasture, range forage plants and crop plants. Inhalation, ingestion, aspiration and percutaneous absorption are possible portals of entry into the body. Most of the insecticides accumulate in the fat depots, may be excreted in the milk in dangerous amounts. Fat depots are a potential source of danger in that sudden mobilization of the blood stream and the appearance of signs of toxicity. The toxic effects produced includes increased excitability and irritability followed by muscle tremor, weakness and paralysis and teeth grinding is noticed. Complete anorexia occurs constantly.

Short acting barbiturate (pentobarbitone), later followed by a dose of chloral hydros by mouth is given as a treatment regimen. When the insecticide has been applied on the skin washing with detergent is advisable. It is usually recommended that calcium borogluconate should be given I/V possibly together with glucose saline.

b) Organo-phosphorus insecticides

The group includes some of the most dangerous chemicals which are highly toxic but they are used and recommended for control of insects attacking plants, available for control of internal parasites of sheep. Some of the compounds are chloration, carbophenothion, demton, dasnon, dimethylaparation, trichlorphon, dioxalthion, malathion, darathion, scharadan etc. For blow flies, keds, lice and ticks diazon, cobal, malathion are used. The recommended concentrations of dips sprays are 0.04% diazon, 0.25% for coral, 0.5% for malathion. As anthelmintics rueleam (Asuntol), coral (Bayer 21/199) and neguvon (Bayer L13/59), have been used. The clinical signs of poisoning are profuse salivation, muscle stiffness dyspnoea, with open mouth breathing diarrhea and muscular tremors. Such cases can be treated by giving. atropin sulphate 2.0 mg/kg body weight : $^1/_3$ to be giving I/V in 10 ml distilled water and rest S/C.

c) Snake bite

Venomnous snakes fall in to 2 classes: the elapine snakes which include cobra and the viperine snakes which comprise Russels viper, Elapine snakes have short fangs and tend to chew their victims. Their venom is neurotoxic and kills the victim by paralysing the respiratory centres. Viperine snakes attack once and then withdraw Their venoms mainly haemolytic causing pronounced local damage.

Experimental work suggest an order of decreasing sensitivity is horse, sheep, ox, goat, dog, pig and cat. Sheep is usually bitten on udder or scrotum. The presence of hair may abscure the typical fangs marks though a close examination should reveal the point of entry. Prolonged pain, muscular weakness, impaired vision, nausea, paralysis are generally exhibited. Shock in all its classical elements accompanies all severe snake bites.

If antivenin is not available and the bite is located in an area where tourniquatus cannot be applied excision of an area of skin, 7 cm in diameter, including the associated sub-cut tissue may be life saving. Normally no treatment is possible due to sudden death in grazing areas.

17.1.8 Electricution

One of the natural calamity where death ensues in few seconds. Common causes are short circuit in electric line, leakage of current in wire fencing.

The animals are generally found dead with eyes wide open and pupil dilated. The anus may also left dilated with some faecal material voided out. On autospsy haemorrhages of varying size are observed under the skin and viscra. Heart may show streaks of haemorrhages on myocardium with pin point haemorrhanges. The blood may appear black and liquefied. This needs to be differentiated from lightening stroke when there is no evidence of lightening fall and no signs of searching and sinking of skin and hairs on the body of the animal.

17.1.9 Wounds

During monsoon season, a large number of animals suffer from wounds at various sites. Around the ear, sternum, under surface of fore and hind legs are usual places. The main reason may be the awns of *Aristida* and *Hetropogan* grasses, which initially break the continuity of skin which is attacked by flies making the wound infected and maggoty. Antiseptic dressing with fly repellents, *i.e.* turpentine with chloroform is applied over the wound and adjacent area. As the weather changes the incidence is also reduced.

Grass seed infestation *viz. Hetropogan controtus* and *Aristida* sp. play a vital role in certain areas in the production of stress in young lambs with consequent ill effects. Abcess formation specially on various places at head and legs are not uncommon. It causes conjunctivitis, opacity of cornea and blindness.

17.1.10 Dystokia (caesarean operation)

Common causes are insufficient opening of cervical canal, heavy lambs specially crossbreds, position of embryo and uterine torsion. The pregnant ewe should not have fever and the operation should be conducted at a clean place on clean table.

Skin of the near about area is to be cleared and shaved with soap water and on cleaned area tricture iodine or similar antiseptic preparation should be applied.

2% novocain is the best anesthesia which can be used. Left or right flanks can be used. Right side is better because, rumen is not an obstacle and uterus will be found immediately after opening 20–25 cm long incision leaving 3 fingers from tubercoxae directing towards knee joint. The impregnated horn should be lifted towards the opening and uterus should be opened taking care that no embryonal water should go in the peritioneal cavity. Alive lamb should be cared by an helper, who should clean it. The umbilicus should be cut nearnest to endometrium and uterus should be closed keeping sufficient antibiotic, specially terramycin and closed through catgut. Skin should be sutured separately with muscles. Muscles can be sutured with intermittent sutures putting sufficient antibiotic to avoid any sepsis. For 3–4 days, antibiotic should be given parenterally also.

17.2 Infectious Diseases

In our country, much efforts of the workers on the animal health are devoted to the control of serious epizootic diseases like sheep pox, anthrax, black quarter, foot and mouth disease, hemorrhagic septicaemia. These diseases, apart from causing economic loss to the breeders, have been largely responsible for poor health of livestock, which in turn, effects the economy and cause serious human food shortage. It is therefore, imperative that utmost efforts be made for their prompt control and eradication from the country. It is important for any profitable enterprise that chances of loss are eliminated or reduced to a minimum. In livestock farming the major factors of loss are the diseases; infections, infestations and clinical ailments. Whereas clinical ailments can be done to prevent them, the infections and infestation affect a large number of stock, often fatally and it would amount to reducing the chances of loss or completely eliminating them if preventive measures are taken. These measures are collectively known as prophylactic practices. As prophylaxis means preventing, the measures should be taken prior to the probable time of occurrences and therefore different measures have to be taken at different periods of time in the space of a year.

The term vaccination pertains to the introduction into the hosts body of an antigenic substance from the parasite (the whole of a virulent parasite) which will enable the host to resist subsequent infection by a virulent parasite, by means of humeral (antibodies) and cellular agents of immunity.

The substance that is introduced into the host's body is called the vaccine. It may contain a part or whole of the organism in either dead or alive state. Most of the viral vaccines contain living organisms whose infective potential has been reduced and are called attenuated organisms.

17.2.1 Blackleg

Blackleg is an acute, infectious disease caused by *clostridium chauvoei* a gram positive spore farming rod shaped bacterium and characterised by inflammation of muscles, severe toxaemia and high mortality (approaches 100%). The spores are highly resistant to the environment. All age groups are susceptible in sheep. Increased protein feeding of sheep increases their susceptibility to blackleg. It is a soil borne infection. The portal of entry is through the alimentary mucosa after ingestion of contaminated feed. In sheep, the disease is always a wound infection. Infection of skin wounds at shearing and docking and of the naval at birth may cause the development of local lesions, When the blackleg lesions occur in the limb musculature in sheep, there is stiff gait and the sheep is disinclined to move due to severe lameness. Discolouration of the skin may be evident. There is a high fever, anorexia, depression and death occurs very quickly. Subcutaneous oedema is marked around the head. Gas is present in the affected region. In all the cases of suspected blackleg, smears of affected tissue should be made and material collected for bacteriological examination. Treatment with penicillin is logical. Blackleg antiserum is unlikely to be of much value in treatment unless very large doses are given. Annual vaccination with black quarter vaccine (polyvalent) should be done. The vaccine is to be inoculated subcutaneously 2 ml and same dose may be carried out after 10 days.

17.2.2 Enterotoxaemia (pulpy kidney)

This is an acute disease of sheep of all ages but primarly of lambs. It affects animals in high state of nutrition on a lush feed, grass or grain. It causes heavy losses particularly in flocks managed for fat lamb mutton production. Morbidity rates vary a great deal but seldom exceed 10%. The mortality rate approximates 100%. It is caused by *Clostridium perfringens* type D. It normally inhabits the alimentary tract of sheep. Under certain conditions the organism proliferate rapidly in the intestines and produce lethal quantity of toxin. In lambs, the course of illness is very short often less than 2 hr and never more than 12 hr and many are found dead without manifesting early signs. Acute cases may show little more than severe cloric convulsions with frothing at the mouth and sudden death. Cases which survive for a few hours show a green, pasty diarrohoea, for staggering, recumbency, opisthotonus and severe cloric convulsions. Adult sheep show staggering and knuckling, chewing of jaws, salivation and rapid shallow, irregular respiration, muscle tremors, grinding of teeth and salivation. A history of sudden death of several big single lambs justifies a tentative diagnosis of enterotoxaemia. The postmortem findings of a distended pericardium and congested kidney may be confirmed by identifying type DC1 *perfringens* in the intestinal contents by neutralization tests with mice. Animals that die suddenly may show no lesions whatsoever. Small haemorrhages may be present under the epicardium or endocardium. There is commonly present a haemorrhagic inflammation in the mucous membrane of both the abomasum and duodenum. Softening of kidney as a postmortem change that is not apparent until 3 to 4 hr after death. Putrefaction of the carcass takes place early. Hyperimmune serum is an efficient short term prophylaxis. Sulphadimidine is reported to be effective.

There are two major control measures available, reduction of the food intake and vaccination. A method of reducing the grain intake of lambs self-fed in feed lots is to mix clemental sulphur. (7.5 g per day per lamb) in feed. Entertoxaemia vaccine is a culture of a highly toxigenic strain of clostridium type D grown in anaerobic medium rendered sterile and atoxic by the addition of solution of formaldehyde in such a manner that retains its immunizing properties. In sheep 2.5 ml are injected sub-cutaeneously and is repeated after 14 days with same quantity. Lambs and sheep vaccinated during the previous year require only one injection in the subsequent year. Immunity conferred by this vaccine is for 1 year.

Entertoxaemia caused by *Clostridium perfringens* Type B and C

Infection with *Cl. perfringens* Types B and C results in severe enteritis with diarrhoea and dysentery in lambs. These include lamb dysentery (*Cl. perfringens* Type B), struck (*Cl. perfringens* Type C), and haemorrhagic enterotoxaemia (*Cl. perfringens* Type C). In affected groups of lambs due to Type B the morbidity rate may reach as high as 20 to 30%. The mortality rate approaches 100%. Struck caused by *Cl. perfringens* type C affect adult sheep, particularly when feed is abundant. The characteristic effect of the beta-toxin, the important toxin produced by *Cl. perfringens* Types B and C is the production of haemorrhagic enteritis and ulceration of the intestinal mucosa. The lesions described for " Struck" in adult sheep and the haemorrhagic enterotoxemia in young lambs are similar. The syndrome is of 2 weeks old. In the more common acute form, there is severe abdominal pain; recumbency, failure to such and passing of brown fluid faeces some times containing blood. Death usually occurs within 24 hr. The diagnosis is made by testing the intestinal contents for toxins and the isolation of the organisms. Multicomponent clostridial vaccine is inoculated 5 ml sub-cutaneously and repeated after 14 days.

17.2.3 Tetanus

It is an acute infectious disease manifested by toxic convulsions of the voluntary muscles. Among the domestic animals the sheep occupies the position next to horse in susceptibility. In sheep it more commonly follows the routine operations such as shearing, docking, castration and even vaccination. *Clostridum tetani* forms spores which are capable of persisting in soil for number of years. The spores are resistant to many standard disinsfection procedures including steam heat at 212 °F (100 °C) for 30 to 60 minutes but can be destroyed by heating to 239 °F for 20 mintues. *Cl. tetani* organisms are commonly present in the faeces of animals, especially horses. The portal of entry is usually through deep puncture wounds A high incidence to tetanus may occur in lambs following castration, shearing and docking. Docking by the use of elastic band ligatures appears to be especially hazardous. In sheep the first symptom is usually stiffness of the limbs, sway back attitude, lock jaw, opistholonous is common. The tail stands out in a rigid position. Death occurs by asphyxiation due to fixation of the muscles of respiration. The main principles in the treatment of tetanus are to eliminate the causitivie bacteria, neutralize residual toxin, relax the muscles to avoid asphyxia. Tetanus antitoxin is usually administered but is of little value when signs have appeared. To control it lambs are usually given 100 to 150 units of antitoxin subcutaneously after docking.

17.2.4 Pasteurellosis

Pasteurellosis is usually pneumonic in form although a septicaemic form is not uncommon in lambs. Morbidity and mortality rate reaches up to 40%. Death losses in feeder lambs are usually of the order of 5% but may be as high as 10% to 20%. In sheep *Pasteurella haemolytica* is often the primary cause. The transmission occurs by the inhaltaion or ingestion of the infected material. In outbreaks, sheep show high rise of temperature, mucopurulent discharge from eyes and nose, coughing depression, froth in the mouth, anorexia, increased pulse and respiration. The diagnosis is made in the isolation of the organisms from the heart blood. Little information is available on the treatment of Pasteurellosis in sheep. Administration of sodium sulphadimiadine gives excellent results with a 1 g per 7 kg body weight. Hyperimmune serum is similarly effective. Vaccination of sheep with Haemorrhagic septicaemia adjuvant vaccine is recommended. The dose of the vaccine is 3 ml intramuscularly. The vaccine is expected to give protection against disease for a period of one year. A slight swelling may occur at the site of inoculation and a slight rise in temperature which may persist for 2–3 days. The animals should not be sent out for grazing for 2 to 3 days following vaccination.

17.2.5 Paratuberculosis (Johne's Disease)

Paratuberculosis is a chronic wasting disease, characterised by progressive emaciation and a thickening and corrugation of the wall of the intestine. The morbidity rate is difficult to estimate because of uncertainty of diagnosis but the animal mortality rate in infected flocks may be as high as 10%. The disease may cause severe economic embarrassment in infected flocks. The reclamation value of clinically affected animals is usually negligible because of severe emaciation. The causative organism is *Mycobacterium paratuberculosis*, an acid fast organism. The progress of the disease is always slow. For this reason, it is most often observed in older animals. The incubation period may be as much as a year. Progressive loss of weight results in extreme emaciation. Lesions include thickening of intestinal mucosa, enlargement of lymph nodes. Diagnosis is made on the isolation of organism. Infected sheep react to intradermal injection of Johnin which is useful as an aid in diagnosis of clinical shedders. A complement fixation test has also been used as a diagnostic tool. No treatment is successful. Vaccine from Iceland may be of great value.

17.2.6 Sheep pox

Sheep pox is highly contagious. Although in most cases spread appears to occur by contact with infected animals and contaminated articles, spread by inhalation may also occur. It is characterised by development of vesicles and pustules on the skin and internal lesions. It is the most serious of all the pox diseases in animals, often causing death in 50% of affected animals. Major losses may occur in each new crop of lambs.

The etiological agent is a filterable virus. The virus is quite resistant, having been known to remain viable for 2 years in sealed tubes. The incubation period is of 2 to 14 days. In lambs the malignant form is quite common. There is high rise of fever, eyelids are swollen, discharge from the eyes and nose anorexia, by the second day eruption appears. There is a formation of papules with a depressed centre, later vesicles develop and finally pustules. These may develop on the wooly and on non-wooled areas. The pustules become dry with the course of time and fall off, excessive salivation occurs due the lesions in the mouth. There is a fluctuation in the temperature due to invasion of the pustules by secondary organisms. The course of the disease is 3 to 4 weeks, during which time the sheep become emaciated and may shed their wool. In fatal cases, death occurs in about 1 week after the 1st sysmtoms appear. In the malignant form pox lesions extend into mouth, pharynx, larynx and vagina. Lesions may also appear in the trachea with an accompanying catarrh and pneumonia. Lesions occasionally reach the abomasum and are accompanied by a haemorrhagic enteritis. The disease can be diagnosed on the clinical picture and with immunodiffusion technique. Vaccination at the proper time is the best control.

17.2.7 Foot and mouth disease

It is an extremely contagious acute disease by a virus. It is characterised by the development of vesicles in the oral cavity and in the interdigital space. The mortality is usually low (3%). The economic loss is chiefly due to loss of condition of the affected animal. When the disease breaks out in susceptible animal, it spread very rapidly. There are seven immunologically distinct types of foot and mouth disease virus. Type O, A and C widely distributed and found in Europe and South America. Type S, A, T, 1, 2 and 3 are found only in Africa and Type Asia I is widely distributed in Asia. The disease is transmitted by contact with diseased animal. The incubation period is short and less than 24 hr. There is high rise of temperature with acute painful stomatitis. Affected sheep show only small lesions in the mouth but all four feet may be badly affected with severe lameness. The field diagnosis of disease in sheep is based on the presence of vesicular lesions on the mucosa of the mouth and on the skin of the interdigital area of the feet. The lesions are found in the mouth, pharynx, oesophagus and forestomach. Antibiotics are recommended to check the secondary infection. As a prophylactic, foot and mouth vaccine is recommended. It is concentrated quadrivalent (Q, A, C, Asia I strain). It is inoculated subcutaneously 5 ml in sheep.

17.2.8 Contagious ecthyma

It is a viral disease and characterised by the formation of papules and pustules and the piling up of thick crusts. It is caused by filterable virus. The virus gains entrance through unobserved wounds on lips. The incubation period is dependant on the amount of virus introduced. The lesions are mostly found on the commissure of the lips. The lesions are covered by the scabs. The course of the disease is one to four weeks. The scabs become dry and drop off and the tissues return to normal except in the most severe cases, where scars remain. The diagnosis is made on the basis of type of lesions where scars remain. Antibiotics are used to check the secondary invadors. Good results have been obtained by applying vaseline containing 3% phenol.

17.2.9 Blue tongue

It is an infectious but non-contagious exotic disease of sheep. It is characterised by hyperaemia and ulcerative inflammation of the mucous membrane of the mouth, nose and gastro-intestinal tract. Blue tongue has been considered as an important "emerging disease". The causal agent of blue tongue is a filtrable virus. It is present in the blood and visceral organs of affected animals. Natural transmission takes place through insect vectors, *viz* culicoides, aedes and sheepked, *Melophagus ovinus*. The incubation period is less than a week. Rise in body temperature upto about 106 °F is the common initial symptom. The disease has its three clinical forms, *viz.* abortive, acute and sub-acute. The abortive form is mostly passed unnoticed. In the acute form there is fever lasting for 5 to 6 days with nasal discharge frothing and marked salivation. The nasal and oral mucosa are highly congested and may appear cynotic. Lips, gums, dental pad and tongue are often swollen and odematous. There may be excoriation of the epithelium of inner parts of lips, gums, cheeks and tip of tonuge. Examination of mouth cavity may reveal necrotic ulcers, elongated and lenticular in shape. There is purplish discolouration of the skin of interdigital space and occasionally pasterns and coronets show swelling. Sheep is unable to stand or move. The subacute form has the similar sympotms but less severe. The morbidity rate may be 50% or more. The mortality rate varies widely. It is diagnosed on the basis of clinical symptoms and lesions. Antibiotics can be used to check the secondary organisms. It can be prevented by enforcing certain prophylactic measures such as use of insecticidal large scale immunization and quarantine.

17.2.10 Rinderpest

Rinderpest is an acute, highly contagious disease of sheep and characterised by high fever and corrosive lesions confined largely to the mucosa of the alimentary tract. The disease occurs in plague form and is highly fatal. In India it has been endemic for centuries and still causes heavy losses. A number of report on rinderpest in sheep have been made from India. Many outbreaks of rinderpest in sheep in India between 1936 and 1950 with heavy mortality amounting to as much as 90% in some instances. There were 18,273 sheep affected with 14,084 deaths, a mortality of 77%. It is caused by a filtrable virus. Chief spread is through the contact with diseased animal. The entrance is believed to take place through the digestive tract. There is high fever, depression, anorexia, with accelerated respiration, nasal discharge, coughing and congestion of conjectiva. Profuse diarrhoea develops, emaciation also develops rapidly. In fatal cases death occurs in from 3 to 9 days after the 1st appearance of symptoms.

Lesions are present in the mucosa of the abomasum which is characteristically red and swollen and show multiple small submucosal haemorrhages. In addition to the erosions in the mouth, the pharyx and oesophagus are congested. Diagnosis is made on the basis of clinical history and postmortem findings. When outbreak occurs, all affected and in contact animals are vaccinated with goat adapted modified virus vaccine. There is no medicinal treatment. Serum 40–80 ml S/C be given in the febrile stage.

CHAPTER 18

ECONOMICS OF SHEEP FARMING

Sheep with its multi-facet utility for wool, meat, milk, skins and manure form an important component of rural economy particularly in the arid, semi-arid and mountainous areas of the country. It provides a dependable source of income to the shepherds through sale of wool and animals. The advantages of sheep farming are:

i) Sheep do not need expensive buildings to house them and on the other hand require less labour than other kinds of livestock.
ii) The foundation stock is relatively cheap and the flock can be multiplied rapidly.
iii) Sheep are economical converter of grass into meat and wool.
iv) Sheep will eat varied kinds of plants compared to other kind of livestock. This makes them excellent weed destroyer.
v) Unlike goats, sheep hardly damage any tree.
vi) The production of wool, meat and manure provides three different sources of income to the shepherd.
vii) The structure of their lips helps them to clean grains fallen or lost at harvest time and thus convert waste feed into profitable products.
viii) Mutton is one kind of meat towards which there is no prejudice by any community in India and further development of superior breeds for mutton production will have a great scope in the developing economy of India.

18.1 Scope for Sheep Farming and its National Importance

The country has 61.0 M sheep as per 2003 livestock census and ranks sixth in the world. The statewise sheep population are given in Table 18.1. During 2001–02

Table 18.1. Statewise Sheep Population in India–2003

Sl No.	States	(in thousands) Total
1	Andhra Pradesh	21376
2	Arunachal Pradesh	19
3	Assam	170
4	Bihar	1062
5	Chattisgarh	121
6	Goa	0
7	Gujarat	2062
8	Haryana	633
9	Himachal Pradesh	926
10	Jammu and Kashmir	3411
11	Karnataka	7256
12	Kerala	4
13	Madhya Pradesh	546
14	Maharashtra	3094
15	Manipur	6
16	Meghalaya	18
17	Mizoram	1
18	Nagaland	4
19	Orissa	1620
20	Punjab	220
21	Rajasthan	10054
22	Sikkim	6
23	Tamil Nadu	5593
24	Tripura	3
25	Uttar Pradesh	1437
26	Uttaranchal	296
27	West Bengal	1525
Union Territories		
28	Andaman and Nicobar Islands	0
29	Chandigarh	0
30	Dadra and N Haveli	0
31	Daman and Diu	0
32	Delhi	3
33.	Lakshadweep	0
34.	Pondicherry	2
	ALL INDIA	61468

Source: 17th livestock census, 2003

wool production stands at the modest level of 50.70 M, skin with wool sheep 524 Mt, sheep and goat meat 700500 Mt in 2001–02, sheep fresh 52,380 Mt. The export earnings from different woollen products during 2003 was Rs. 25,773 million.

The contribution of sheep to total meat production in the country is around 14%. The contribution of sheep through export of meat is 8% of the total export value of agricultural and processed food products. Live sheep are also exported for meat purpose. Sheep skin in the form of leather and leather products is also exported.

Sheep make valuable contribution to the livelihood of the economically weaker sections of the society. Amongst the livestock owners the shepherds are the poorest of the poor.

Realizing the importance of sheep in agrarian economy, the Central Government had established the Central Sheep and Wool Research Institute (CSWRI) at Avikanagar in Rajasthan. Number of sheep breeding farms were established during various plan periods throughout the country for evolving (i) new fine wool breeds for different agro-climatic regions capable of producing 2.5 kg of greasy wool per annum. (ii) new mutton breeds capable of attaining 30 kg live weight at 6 months of age under intensive feeding conditions. Sheep development activities were initiated under DPAP, MFAL and SFDA programmes. Intensive Sheep Development Projects (ISDP's) were introduced in many of the sheep rearing districts. Setting up of Wool Boards in important wool producing states was envisaged. States of Jammu and Kashmir and Karnataka have already set up these Boards. Some of the States have set up Wool Development Corporations/Federations. The states having such organizations are given in Table 18.2.

Table 18.2. State Sheep and Wool Boards/Federations/Corporations

Sr.No.	Name and address
1.	APCO Wool-3-5-770, Weavers Bhawan, Narayan Guda, Hyderabad500 029 (AP)
2.	Gujarat Sheep and Wool Development Corporation Limited "Shreekunj", Opp. Navarangpura Telephone Exchange, Ellisbridge, Ahmedabad 380 006 (Gujarat)
3.	Himachal Pradesh State Co-operative Wool Procurement and Marketing Federation Limited, Pashudhan Bhawan, Boileauganj, Shimla 171 005 (HP)
4.	Jammu and Kashmir State Sheep and Sheep Products Development Board, Kartholi, Bari Brahamana, Jammu 181 133 (J & K)
5.	Karnataka Sheep and Sheep Products Development Board, No.58, IInd Main Road, Vyalikaval, Bangalore 560 003 (Karnataka)
6.	Maharashtra Mendhi Va Sheli Vikas Mahamandal Limited, Mendhi Farm, Gokhale Nagar, Pune 411 016 (Maharashtra)
7.	Rajasthan State Cooperative Sheep and Wool Marketing Federation Limited, Gandhi Nagar, Tonk Road, Jaipur (Rajasthan)
8	Uttar Pradesh Poultry and Livestock Specialties Limited, Directorate of Animal Husbandry, Badshah Baug, Gorakh Nath Road, Lucknow (UP)

18.2 Financial Assistance Available from Banks/NABARD for Sheep Farming

NABARD is an apex institution for all matters relating to policy, planning and operation in the field of agricultural credit. It serves as refinancing agency for the institutions providing investment and production credit for agriculture and rural development. It promotes development through a well organized Technical Services Department at the Head Office and technical cells at each of the Regional Offices.

Loan from banks with refinance facility from NABARD is available for starting sheep farming. For obtaining bank loan, the farmers should apply to the nearest branch of a Commercial, Co-operative or Regional Rural Bank in their area in the prescribed application form which is available in the branches of financing bank. The Technical Officer attached to or the Manager of the bank can help/ give guidance to the farmers in preparing the project report to obtain bank loan.

For sheep development schemes with very large outlays, detailed reports will have to be prepared. The beneficiaries may utilize the services of NABARD Consultancy Services (nabcons) having good experience in developing livestock projects for preparation of the project report to avail the bank loan for the items such as purchase of breeding animals, construction of sheds, purchase of equipments etc. The cost of land is not considered for loan.

18.3 Scheme Formulation

A scheme can be prepared by a beneficiary after consulting local technical persons of State Animal Husbandry Department, DRDA, Sheep development Corporation, Sheep Co-operative society/Union/Federation and Commercial Farmers. If possible, the beneficiaries should also visit progressive sheep farmers and Government/ Agricultural University sheep farms in the vicinity and discuss the profitability of sheep farming. A good practical training and experience in sheep farming will be highly desirable. The sheep Co-operative societies established in the villages as a result of efforts by the Sheep Development Department of State Government/ Sheep Development Board would provide all supporting facilities, particularly marketing of live animals and wool. Nearness of the sheep farm to such a society, veterinary aid and breeding centre should be ensured.

The scheme should include information on land, livestock markets, availability of water, feeds, fodder, veterinary aid, breeding facilities, marketing aspects, training facilities, experience of the farmer and the type of assistance available from State Government, Sheep Society/Union/Federation.

The scheme should also include information on the number of and types of animals to be purchased, their breeds, production performance, cost and other relevant input and output costs with their description. Based on this, the total cost of the project, margin money to be provided by the beneficiary, requirement of bank loan, estimated annual expenditure, income, profit and loss statement, repayment period, etc. can be worked out and included in the scheme.

The scheme so formulated should be submitted to the nearest branch of bank. The bank's officers can assist in preparation of the scheme or filling in the prescribed application form. The bank will then examine the scheme for its technical feasibility and economic viability.

A) Technical feasibility: this would briefly include

(a) Nearness of the selected area to veterinary aid, breeding and wool collection centre and the financing bank's branch.

(b) Availability of good quality animals in nearby livestock markets. The distribution of sheep breeds in India are given in Table 18.3, while the wool production and quality in Table 18.4.

(c) Source of training facilities. The major institutions providing training in Sheep farming are given in Table 18.5.

(d) Availability of good grazing ground/lands.

(e) Availability of green/dry fodder, concentrate feed, medicines etc.

B) Economic viability this would briefly include

(a) Unit Cost-the average cost of Sheep units for some of the States is given in Table 18.6.

(b) Input cost for feeds and fodders, veterinary aid, insurance, shearing etc.

(c) Output costs *i.e.* sale price of animals, wool, penning etc.

(d) Income-expenditure statement and annual gross surplus.

(e) Cash flow analysis.

(f) Repayment schedule *i.e.* repayment of principal loan amount and interest.

Other documents such as loan application forms, security aspects, margin money requirements etc. are also examined. A field visit to the scheme area is undertaken for conducting a techno-economic feasibility study for appraisal of the scheme. The economics of sheep farming is given in Table 18.7.

Table 18.3. Region-wise Distribution of Sheep Breeds in India

North western arid arid and semi-arid region	Southern peninsular region	Eastern region	Northern temperate region
Chokla	Deccani	Chottanagpuri	Gaddi
Nali	Bellary	Shahabadi	Rampur
Marwari	Nellore	Balangir	Bushair
Magra	Mandya	Ganjam	Bhakarwal
Jaisalmeri	Hassan	Tibetan	Poonchi
Pungal	Macheri	Bonpala	Gurez
Malpura	Kilakarsal		Kashmir Merino
Sonadi	Vembur		Changathangi
Patanwadi	Coimbatore		
Muzzafarnagri	Nilgiri		
Jalauni	Ramnad white		
Hissardale	Madras red		
	Tiruchi black		
	Kenguri		

Table 18.4. Wool Production and Quality in Different Regions of India

S.No.	Particulars	North	North western	Southern temperature	Eastern penninsular
1.	Sheep population (M)	20.36	3.45	19.80	4.6
2	Percent contribution	42.23	7.15	41.07	9.54
3	Wool production (Mkg)	25.11	4.03	7.68	1.57
4	Percent contribution to total	65.40	10.50	20.00	4.10
5	Per capita production (kg)	1.23	1.16	0.38	0.34
6	Fineness (μ)	30.45	22.3	40.60	50 – 60
7	Medullation (%)	30.80	5.15	60.80	80 – 90
8	Burr content (%)	2 – 5	2 – 8	Below 5	1 – 3
9	Yield (washed) (%)	80 – 90	50 – 60	80 – 90 (except Nilgiri)	85 – 90

Source: CSWRI, Avikanagar

Table 18.5. Training Institutions in Sheep Farming

S.No.	Institutions
i)	Central Sheep and Wool Research Institute, Avikanagar (Rajasthan)
ii)	Central Sheep and Wool Research Institute (sub-station) Bikaner, (Rajasthan)
iii)	Central Sheep and Wool Research Institute (sub-station) Garsa (HP)
iv)	Sheep and Wool Training Centres established by Department of Animal Husbandry in respective states

Table 18.6. Economics of Sheep Farming (20 + 1 Units)

Techno-economic parameters followed

1. The present unit cost of one year Ewes and ram is taken at Rs. 1200 and Rs. 1600 respectively.
2. Lambing interval is taken as 12 months with lambing percentage of 75 and sex ratio of 50:50.
3. Mortality is considered as 10% and 5% among lambs and adults.
4. All female lambs are retained in the flock and males are sold at 8 to 9 months of age.
5. The culling rate among ewes is 20% and above from third year onwards
6. Grazing charges @ Rs. 4/- per adult per year.
7. Cost of concentrate feed per pregnant Ewe for 30 days @ 250 gms/animal costing Rs 5/kg.
8. Insurance is 4% per year and cost of veterinary aid is Rs. 10 and 5 per adult and young animal per year.
9. Shearing will be done twice a year. The adult sheep will yield 1.2 kg wool per year and the lamb will yield 600 gm per year. The shearing charges are Rs. 2.00/kg and the price of wool is Rs. 40 per kg.
10. Sale price is Rs. 800 per ram lamb, Rs. 1000 per adult ewe and Rs. 1200 per adult ram. The sale value of closing stock is considered for working out cash flow analysis.
11. Penning charges are Rs. 8 per adult animal per month for 6 months in a year.
12. The economics have been worked out basing on the opening stock.

A. Capital Investment	I Year	II Year	III Year	IV Year	V Year	VI Year
Cost of 20 ewes	24000	–	–	–	–	–
Cost of 1 ram	1600	–	–	1600	–	1600
Cost of housing and other equipmentsb	4000	–	–	–	–	–
Total capital investment	29600	–	–	1600	–	1600
B. Working cost (variable)						
Feeding cost						
Grazing	84	84	84	84	84	84
Concentrate	768	768	768	768	768	768
Veterinary aid	210	210	210	210	210	210
Shearing charges	42	346	342	346	342	346
Total variable cost	1104	1408	1404	1408	1404	1408
C. Working cost (fixed)						
Insurance cost	1024	1024	1024	1024	2592	2592
Depriciation @10% on total capital investment	–	2960	2644	2400	2160	1943
Total fixed cost	1024	3984	3668	3424	4752	4535
D. Income						
Income from sale of male lambs	–	6400	5600	6400	5600	6400
Income from sale of ram	–	–	1200	1200	1200	1200
Income from sale of culled ewes	–	–	4000	4000	4000	4000
Income from sale of lamb ewes	–	–	2400	2400	2400	2400
Income from sale of wool	1176	1392	1344	1392	1344	1392

Table 18.6. (*Contd.*)

D. Income	I Year	II Year	III Year	IV Year	V Year	VI Year
Income from penning charge	1008	1008	1008	1008	1008	5040
Total income	2184	8800	15552	16400	15552	20432
E. Total expenditure	31728	5392	5072	6432	6156	7543
Total income	2184	8800	15552	16400	15552	20432
Profit	-29544	3408	10480	9968	9396	12889
F. Bank loan required						
Farmer's equity @ 25% of capital investment		7400				
Working capital for 6 months			1064			
Amount required as loan			23264			
Interest rate per annum			12%			
G. Birth and death register						
Item description	**I Year**	**II Year**	**III Year**	**IV Year**	**V Year**	**VI Year**
Ewe	20	–	20	20	20	20
Ram	1	–	1	1	1	1
Birth						
Male	–	8	7	8	7	8
Female	–	8	7	8	7	8
Culling of ewes	–	–	4	4	4	4
Culling of rams	–	–	1	1	1	1
H. Income						
Income from sale of male of lambs	–	6400	5600	6400	5600	6400
Income from sale of ram	–	–	1200	1200	1200	1200
Income from sale of culled ewes	–	–	4000	4000	4000	4000
Income from sale of lamb ewes	–	–	2400	2400	2400	2400
Income from sale of wool	1176	1392	1344	1392	1344	1392
Income from penning charge	1008	1008	1008	1008	1008	1008
Total income	**2184**	**8800**	**15552**	**16400**	**15552**	**16400**
Gross Profit [income-(fix cost + var. cost)]	-14256	-3159	10480	11568	9396	10457
Repayment of loan ($^1/_5$th annually)	4653	4653	4653	4653	4652	–
Time to pay back loan	5 years					
I. Cash flow statement						
	I Year	**II Year**	**III Year**	**IV Year**	**V Year**	**VI Year**
Gross income	2184	8800	15552	16400	15552	16400
Loan outstanding	23264	18611	13958	9305	4652	–
Interest	2233	2233	1675	1117	558	–
J. Profitability						
Loan instalment	4653	4653	4653	4653	4652	–
Net surplus	-4702	1914	9224	10630	10342	16400
Net income/month	-392	159	769	886	862	1367

Table 18.7. Economics of Sheep Farming (100 + 5 Units)

Techno-economic parameters followed

1. The present unit cost of one year Ewes and ram is taken at Rs. 1200 and Rs. 1600 respectively.
2. Lambing interval is taken as 12 months with lambing percentage of 75 and sex ratio of 50:50.
3. Mortality is considered as 10% and 5% among lambs and adults.
4. All female lambs are retained in the flock and males are sold at 8 to 9 months of age.
5. The culling rate among ewes is 20% and above from third year onwards
6. Grazing charges @ Rs. 4/- per adult per year.
7. Cost of concentrate feed per pregnant Ewe for 30 days @ 250 gms/animal costing Rs. 5.00/kg.
8. Insurance is 4% per year and cost of veterinary aid is Rs. 10 and 5 per adult and young animal per year.
9. Shearing will be done twice a year. The adult sheep will yield 1.2 kg wool per year and the lamb will yield 600 gm per year. The shearing charges are Rs. 2.00/kg and the price of wool is Rs. 40 per kg.
10. Sale price is Rs. 800 per ram lamb, Rs. 1000 per adult ewe and Rs. 1200 per adult ram. The sale value of closing stock is considered for working out cash flow analysis.
11. Penning charges are Rs. 8 per adult animal per month for 6 months in a year.
12. The economics have been worked out basing on the opening stock.

A. Capital investment	**I Year**	**II Year**	**III Year**	**IV Year**	**V Year**	**VI Year**
Cost of 100 ewes	120000	–	–	–	–	–
Cost of 5 ram	8000	–	1600	1600	1600	1600
Cost of housing and other equipments	20000	–	–	–	–	–
Total capital investment	148000	–	1600	1600	1600	1600
B. Working cost (variable)						
Feeding cost						
Grazing	420	420	420	420	420	420
Concentrate	3600	3600	3600	3600	3600	3600
Veterinary aid	1030	1360	1360	1360	1360	1360
Shearing charges	206	346	342	346	342	346
Total variable cost	5256	5726	5722	5726	5722	5726
C. Working cost (fixed)						
Insurance cost	5120	5120	5120	5120	5120	5120
Depriciation @10% on total capital investment	–	14800	13320	11988	14759	9313
Total fixed cost	5120	19920	18440	17108	19879	14433
D. Income						
Income from sale of male lamb	–	26400	27200	26400	27200	26400
Income from sale of ram	–	–	1200	1200	1200	1200
Income from sale of culled ewes	–	–	20000	20000	20000	20000
Income from sale of lamb ewes			10400	11200	10400	11200
Income from sale of wool	5880	6624	6672	6624	6672	6624
Income from penning charge	5040	5040	5040	5040	5040	5040
Total income	10920	38064	70512	70464	70512	70464

Table 18.7. (*Contd.*)

	I Year	II Year	III Year	IV Year	V Year	VI Year
E. Total expenditure	158376	25646	25762	24434	27201	21759
Total income	10920	38064	70512	70464	70512	70464
Profit	-147456	12418	44750	46030	43311	48705
F. Bank loan required						
Farmer's equity @ 25% of capital investment			37000			
Working capital for 6 months			5188			
Amount required as loan			116188			
Interest rate per annum			12%			
G. Birth and death register						
Item description	**I Year**	**II Year**	**III Year**	**IV Year**	**V Year**	**VI Year**
Ewe	100	–	20	20	20	20
Ram	5	–	1	1	1	1
Birth						
Male	–	33	34	33	34	33
Female	–	33	34	33	34	33
Culling of ewes	–	–	20	20	20	20
Culling of rams	–	–	1	1	1	1
H. Income						
Income from sale of male lambs	–	26400	27200	26400	27200	26400
Income from sale of ram	–	–	1200	1200	1200	1200
Income from sale of culled ewes	–	–	20000	20000	20000	20000
Income from sale of lamb ewes	–	–	10400	11200	10400	11200
Income from sale of wool	5880	6624	6672	6624	6672	6624
Income from penning charge	5040	5040	5040	5040	5040	5040
Total income	10920	38064	70512	70464	70512	70464
I. Profitability						
Gross profit [income-(fix cost + var. cost)]	-14256	-14492	46350	47630	44911	50305
Repayment of loan ($^1/_5$th annually)	23238	23238	23238	23238	23238	–
Time to pay back loan	5 years					
J. Cash flow statement						
Gross income	10920	38064	70512	70464	70512	70464
Loan outstanding	116188	92950	69712	46474	23236	–
Interest 13943	11154	8365	5577	2788	–	–
Loan instalment	23238	23238	23238	23238	23238	–
Net surplus	-26260	3672	38909	41649	44486	70464
Net income/month	-2188	306	3242	3471	3707	5872

18.4 Sanction of Bank Loan and its Disbursement

After ensuring technical feasibility and economic viability, the scheme is sanctioned by the bank. The loan is disbursed in stages against creation of specific assets such as construction of sheds, purchase of equipments and animals. The end use of the loan is verified and constant follow-up is done by the bank.

18.5 Lending Terms – General

18.5.1 Unit cost

Each regional office (R.O) of NABARD has constituted a State Level Unit Cost Committee under the chairmanship of RO in-charge and with the members from developmental agencies, commercial banks and cooperative banks to review the unit cost of various investments once in six months. The same is circulated among the banks for their guidance.

18.5.2 Interest rate for ultimate borrowers

Banks are free to decide the rate of interest within the overall RBI guidelines. However for working out financial viability and bankability of the model project we have assumed the rate of interest as 12% p.a.

18.5.3 Margin money

NABARD has defined the farmers into three different categories and where subsidy is not available. The minimum down payment to be contributed by the beneficiaries are given in the following table.

Sr.No.	Category of farmer	Beneficiary's contribution (%)
A	Small farmers	5
B	Medium farmers	10
C	Large farmers	15

18.5.4 Security

Security will be as per NABARD/RBI guidelines issued from time to time.

18.5.5 Repayment period of loan

Repayment period depends upon the gross surplus generated. The loans will be repaid in suitable half yearly/annual installments usually within a period of about 5–6 years with a grace period of one year.

Table 18.8. Economics of Sheep Farming (200+10 Units)

Techno-economic parameters followed

1. The present unit cost of 1 year ewes and ram is taken at Rs.1200 and Rs.1600 respectively.
2. Lambing interval is taken as 12 months with lambing precentge of 75 and sex ratio of 50:50.
3. Mortality is considered as 10% and d5% among lambs and adults.
4. All female lambs are retained in the flock and males are sold at 8 to 9 months of age.
5. The culling rate among ewes is 20% and above from third year onwards
6. Grazing charges @ Rs. 4/- per adult per year.
7. Cost of concentrate feed per pregnant Ewe for 30 days @ 250 gms/animal costing Rs. 5.00/kg.
8. Insurance is 4% per year and cost of veterinary aid is Rs. 10 and 5 per adult and young animal per year.
9. Shearing will be done twice a year. The adult sheep will yield 1.2 kg wool per year and the lamb will yield 600 gm per year. The shearing charges are Rs. 2/kg and the price of wool is Rs. 40 per kg.
10. Sale price is Rs. 800 per ram lamb, Rs. 1000 per adult ewe and Rs. 1200 per adult ram. The sale value of closing stock is considered for working out cash flow analysis.
11. Penning charges are Rs. 8 per adult animal per month for 6 months in a year.
12. The economics have been worked out basing on the opening stock.

A. Capital investment	**I Year**	**II Year**	**III Year**	**IV Year**	**V Year**	**VI Year**
Cost of 200 ewes	240000	–	–	–	–	–
Cost of 10 ram	16000	–	3200	3200	3200	3200
Cost of housing and other equipments	40000	–	–	–	–	–
Total capital investment	296000	–	3200	3200	3200	3200
B. Working cost (variable)						
Feeding cost						
Grazing	840	840	840	840	840	840
Concentrate	7200	7200	7200	7200	7200	7200
Veterinary aid	2100	2760	2760	2760	2760	2760
Shearing charges	420	684	692	684	692	684
Total variable cost	10560	11484	11492	11484	11492	11484
C. Working cost (fixed)						
Insurance cost	10240	10240	10240	10240	10240	10240
Depriciation @10% on total capital investment	–	29600	23976	21578	19421	17478
Total fixed cost	10240	39840	34216	31818	29661	27718

Table 18.8. (*Contd.*)

D. Income	I Year	II Year	III Year	IV Year	V Year	VI Year
Income from sale of male lambs	–	52800	54400	52800	54400	52800
Income from sale of ram	–	–	2400	2400	2400	2400
Income from sale of culled ewes	–	–	20000	20000	20000	20000
Income from sale of lamb ewes	–	–	10400	11200	10400	11200
Income from sale of wool	5880	6624	6672	6624	6672	6624
Income from penning charge	5040	5040	5040	5040	5040	5040
Total income	10920	64464	98912	98064	98912	98064
E. Total expenditure	316800	51324	48908	46502	44353	42402
Total income	10920	64464	98912	98064	98912	98064
Profit	-305880	13140	50004	51562	54559	55662
F. Bank loan required						
Farmer's equity @ 25% of capital investment		74000				
Working capital for 6 months			10400			
Amount required as loan			232400			
Interest rate per annum			12%			
G. Birth and death register						
Item description	**I Year**	**II Year**	**III Year**	**IV Year**	**V Year**	**VI Year**
Ewe	200	–	40	40	40	40
Ram	10	–	2	2	2	2
Birth						
Male	–	66	68	66	68	66
Female	–	66	68	66	68	66
Culling of ewes	–	–	40	40	40	40
Culling of rams	–	–	2	2	2	2
H. Income						
Income from sale of male lambs	–	52800	54400	52800	54400	52800
Income from sale of ram	–	–	2400	2400	2400	2400
Income from sale of culled ewes	–	–	40000	40000	40000	40000
Income from sale of lamb ewes	–	–	20800	22400	20800	22400
Income from sale of wool	11760	13248	13344	13248	13344	13248
Income from penning charge	10080	10080	10080	10080	10080	10080
Total income	21840	76128	141024	140928	141024	140928
I. Profitability						
Gross Profit [(income-(fix cost+var. cost)]	-14256	-29014	95316	97626	99871	101726
Repayment of loan ($^{1}/_{5}{}^{th}$ annually)	24718	24718	24718	24718	24718	24718
Time to pay back loan	5 years					

Table 18.8. (*Contd.*)

J. Cash flow statement	I Year	II Year	III Year	IV Year	V Year	VI Year
Gross income	21840	76128	141024	140928	141024	140928
Loan outstanding	232400	185920	139240	92560	45880	–
Interest	27888	22310	16709	11107	5506	–
Loan instalment	46480	46680	46680	46680	46480	–
Net surplus	-52528	7138	77635	83141	89038	140928
Net income/month	-4377	595	6470	6928	7420	11744

Table 18.9. Economics of Sheep Farming (400+20 Units)

Techno-economic parameters followed

1. The present unit cost of 1 year ewes and ram is taken at Rs. 1200 and Rs. 1600 respectively.
2. Lambing interval is taken as 12 months with lambing percentage of 75 and sex ratio of 50:50.
3. Mortality is considered as 10% and 5% among lambs and adults
4. All female lambs are retained in the flock and males are sold at 8 to 9 months of age.
5. The culling rate among ewes is 20% and above from 3rd year onwards
6. Grazing charges @ Rs. 4/- per adult per year.
7. Cost of concentrate feed per pregnant ewe for 30 days @ 250 gms/animal costing Rs. 5.00/kg
8. Insurance is 4% per year and cost of veterinary aid is Rs.10 and 5 per adult and young animal per year.
9. Shearing will be done twice a year. The adult sheep will yield 1.2 kg wool per year and the lamb will yield 600 gm per year. The shearing charges are Rs. 2.00/kg and the price of wool is Rs. 40 per kg.
10. Sale price is Rs. 800 per ram lamb, Rs. 1000 per adult ewe and Rs. 1200 per adult ram. The sale value of closing stock is considered for working out cash flow analysis.
11. Penning charges are Rs. 8 per adult animal per month for 6 months in a year.
12. The economics have been worked out basing on the opening stock.

A. Capital investment	I Year	II Year	III Year	IV Year	V Year	VI Year
Cost of 400 ewes	480000	–	–	–	–	–
Cost of 20 ram	32000	–	6400	6400	6400	6400
Cost of housing and other equipments	80000	–	–	–	–	–
Total capital investment	592000	–	6400	6400	6400	6400
B. Working cost (variable)						
Feeding cost						
Grazing	1680	840	840	840	840	840
Concentrate	7200	7200	7200	7200	7200	7200
Veterinary aid	4200	5520	5520	5520	5520	5520
Shearing charges	840	1368	1384	1368	1384	1368
Total variable cost	13920	14928	14944	14928	14944	14928
C. Working cost (fixed)						
Insurance cost	20480	20480	20480	20480	20480	20480
Depriciation @10% on total capital investment	–	59200	51840	46656	41990	37791
Total fixed cost	20480	79680	72320	67136	62470	58271

Table 18.9. (*Contd.*)

D. Income	**I Year**	**II Year**	**III Year**	**IV Year**	**V Year**	**VI Year**
Income from sale of male lambs	–	105600	108800	105600	108800	105600
Income from sale of ram	–	–	4800	4800	4800	4800
Income from sale of culled ewes	–	–	20000	20000	20000	20000
Income from sale of lamb ewes	–	–	10400	11200	10400	11200
Income from sale of wool	5880	6624	6672	6624	6672	6624
Income from penning charge	5040	5040	5040	5040	5040	5040
Total income	10920	117264	155712	153264	155712	153264
E. Total expenditure	609760	92368	93024	87824	83174	78959
Total income	10920	117264	155712	153264	155712	153264
Profit	-598840	24896	62688	65440	72538	74305
F. Bank loan required						
Farmer's equity @ 25% of capital investment		144000				
Working capital for 6 months			16880			
Amount required as loan			448880			
Interest rate per annum			12%			
G. Birth and death register						
Item description	**I Year**	**II Year**	**III Year**	**IV Year**	**V Year**	**VI Year**
Ewe	400	–	80	80	80	80
Ram	20	–	4	4	4	4
Birth						
Male	–	132	136	132	136	132
Female	–	132	136	132	136	132
Culling of ewes	–	–	80	80	80	80
Culling of rams	–	–	4	4	4	4
H. Income						
Income from sale of male lambs	–	105600	108800	105600	108800	105600
Income from sale of ram	–	–	4800	4800	4800	4800
Income from sale of culled ewes		–	80000	80000	80000	80000
Income from sale of lamb ewes	–	–	44800	41600	44800	41600
Income from sale of wool	23520	26496	26688	26496	26688	26496
Income from penning charge	20160	20160	20160	20160	20160	20160
Total income	43680	152256	285248	278656	285248	278656
I. Profitability						
Gross profit [income-(fix cost+var. cost)]	-14256	-49276	198624	197232	208474	206097
Repayment of loan ($^1/_5$th annually)	24718	24718	24718	24718	24718	24718
Time to pay back loan	5 years					
J. Cash flow statement	**I Year**	**II Year**	**III Year**	**IV Year**	**V Year**	**VI Year**
Gross income	43680	152256	285258	278656	285248	278656
Loan outstanding	448880	359104	269328	179552	89776	–
Interest	53866	43092	32319	21546	10773	–
Loan instalment	89776	89776	89776	89776	89776	–
Net surplus	-99962	19388	163153	167334	184699	278656
Net income/ month	-8330	1616	13596	13944	15392	23221

18.6 Package of Common Management Practises Recommended for Sheep Farmers

Modern and well established scientific principles, practises and skills should be used to obtain maximum economic benefits from sheep farming. Some of the major norms and recommended practises are given here under.

18.6.1 Selection and purchase of animal

i) It is necessary to select suitable improved breed of sheep available in particular area.

ii) Crossbred sheep are available for purchase from State Government/ Government of India sheep breeding farms.

iii) Ewes can be purchased in regular sheep markets or from breeders in villages, while male sheep (rams) of exotic/crossbred from Government farms.

vi) A certificate regarding age and health of sheep is obtained from the veterinary assistant surgeon.

vii) The animals purchased have to be identified by fixing ear tags.

viii) Sheep should be vaccinated for important diseases like sheep-pox and entero-toxaemia.

ix) The price of sheep depends on breed, age and health status.

x) An entrepreneur should have a unit of 20–30 ewes and one ram.

18.6.2 Pregnant, parturient and lactating ewes

Careful management of the pregnant, parturient and lactating ewes will have a marked influence on the percentage of lambs dropped and reared successfully. So, the following steps may be taken to afford proper attention to these animals.

i) Do not handle the pregnant ewes too frequently.

ii) Separate the advanced pregnant ewes from the main flock and take effective care in their feeding and management.

iii) Extra feed during the later part of pregnancy (3–4 weeks before parturition) will be beneficial for the condition of the pre-parturient ewes which will help in improving milk production of ewes, birth weight and growth of lambs.

iv) Inadequate and poor nutrition may result in pregnancy, toxaemia, abortions and premature births of weak lambs.

v) Bring lambing ewes into lambing corals 4–6 days before parturition and provide maximum comfort. If possible, provide soft, clean bedding and individual lambing pens.

vi) Watch gestation length which ranges from 142 to over 150 days. Early maturing breeds have slightly shorter gestation period.

vii) Save parturient ewes from cold and chilly weather.

18.6.3 Care at lambing/parturition

An ewe about to lamb prefers to leave the flock. She is restless, the udder is often distended and external genital are in a flushed and flaccid condition. Generally in a healthy ewe, parturition is normal. Still the following precautions may be taken during and after parturition.

i) Keep a vigilant eye for dystokia or difficult birth.

ii) Maiden ewes in poor condition or small framed ewes mated to big rams will generally have difficulty in parturition and will have to be assisted.

iii) Seek prompt veterinary aid and advice from an experienced shepherd or stockman in case of dystokia.

iv) Newborn lambs, after being licked by the mother generally stand on their legs and start seeking for teats and suckle milk. If they are not able to do so after sufficient time, provide help to them in suckling colostrum (the 1st milk) which is very essential for health and survivability of lambs.

v) Save newborn lambs from cold, rain and winds.

vi) Resort to artificial milk feeding or arrange foster mothers to disowned or orphan lambs. Goats can serve as excellent foster mother but ewes which have lost their lamb early after birth may also be utilized.

vii) Ligate, sever and antiseptically dress the naval cord of the lamb.

viii) Give a teaspoonful of castor-oil or liquid paraffin to the lamb to facilitate defecation and passing out of muconium easily.

ix) Do not handle lambs too frequently immediately after birth and let the dams lick and recognize them properly.

x) Allow newborn lambs to be with their mothers all the 24 hours for 1st week or so.

xi) Feed sufficient quantity of good quality hay and concentrates (if possible) to the lactating ewes for meeting nutritional requirements for early lactation.

xii) Provide plenty of clean fresh drinking water as the lactating ewes drink surprisingly higher amount of water during lactation.

18.6.4 Care of lambs

The lamb should be taken care of the maximum extent during early period of life. This will also ensure better survival. The following steps may be taken for ensuring better growth and survival.

i) Ensure proper suckling of lambs. Examine udders for blindness of teats or mastitis.

ii) Take care of indifferent mothers and arrange suckling of lambs by restraining such type of ewes.

iii) Provide creep feed (good quality hay with or without concentrate mixture) to suckling lambs in addition to suckling of milk from tenth day to weaning age.

iv) If possible, make available green leguminous fodder or fresh tree leaves to lambs to nibble during suckling period.

v) Lambs may be ear tagged or tattooed on the ear for identification (tattooing forceps and ear tagging forceps should also be cleaned and sterilized at the time of use). Tail docking and castration may also be done in 1st week or so by placing elastrator (strong rubber band) at the intervertebral space and not on the vertebra.

vi) Alternatively use sterilized and clean knife for castration and docking and resort to proper ligation and antiseptic dressing at the roof of scrotum with testicles before it.

vii) During castration keep the lambs on perfectly dry, clean and hygienic site so as to minimize the risks of losses from tetanus.

18.6.5 Weaning and care of weaners

The management of weaners plays an important part in good sheep husbandry. The following steps are important in proper care and management of weaners.

i) Weaning should preferably be done at 90 days, although in breeds with low milk production or where re-breeding is desired it can be done around 60 days.

ii) Supplementary feeding and good clean pastures for growing weaners should be provided. The nutrient requirements for growing lambs is given in Tables 8.5 and 8.6 and requirements of different nutrients for maintenance of adult sheep is given in Table 8.4.

iii) Weaned lambs should be drenched against gastro-intestinal parasites by first month; and vaccinated against enterotoxemia and sheep-pox.

iv) Weaners should not be grazed on poor burry and thorny types of pasture since it could cause skin irritation, injury to the eyes and damage to wool.

v) They should be protected against vagaries of climates and predation.

18.6.6 Housing and shelter management

i) Normally sheep do not require elaborate housing facilities but minimum provisions will definitely increase productivity, especially protection against inclement weather conditions (sun, rain and winds) and predation. Shed could be provided with gunny bags or temporary or removable protections made of thatching material and bamboos. The roof of the shed should be made of the asbestos sheet supported by tubular or angular steel, but wooden rafters and thatching material could also be used.

ii) Exotics should be provided 0.9–1.1 m^2 and native and crossbred sheep, 0.8–0.9 m^2 space per head. Sheds measuring 18 m × 6 m can accommodate about 120 sheep.

iii) A chain link fencing or thorny bush enclosure of 12 m × 6 m can be provided for night padlocking of sheep on each side of the shed.

18.6.7 Culling

Culling of sheep is very important for the development of a good flock. It helps to remove undesirable animals and breeding from those which are most approximating the ideal sheep. About 10–20% culling should be practised annually to develop a good flock. The flock size should be maintained by replacing culled ewes by ewe lambs born in the flock.

18.6.8 Maintenance of records

It is very essential to maintain the necessary records at an organized (experimental or commercial) sheep farm to know about the inputs and outputs. This helps in working out the economy of sheep production per unit of area and per animal. The following records should be maintained: livestock strength, breeding, lambing, shearing, wool production, mortality, purchase of animal feeds, medicines, sale of animals and wool.

18.6.9 Identification

The shepherds commonly practise notching or punching holes in the ears for identification of lambs. Tattooing is also satisfactory but is more expensive. Metal or plastic ear tags with stamped letters and numbers are most suitable although they are relatively expensive and heavy for the ears of the smaller native sheep. These ear tags are applied with the help of a clincher.

18.6.10 Dipping

To control the ectoparasites, the sheep should be dipped a few weeks after shearing when they have grown sufficient new wool to hold the chemical substance. There are standard designs for sheep dips and there are many products effective against ectoparasites. A footpath may also be provided at the entrance of the farm to prevent the spread of contagious diseases like foot-and-mouth and foot rot.

18.6.11 Health management

The health management calendar for vaccination against major diseases may be followed for better health of stock.

18.6.12 Shearing

i) Shearing is done mechanically either with clippers, a pair of scissors or by power operated machines depending upon the size of operations.

ii) Most flocks are usually shorn twice a year, *i.e.* March–April after the winter and September–October after the rains. In some states like Jammu and Kashmir and Rajasthan sheep are shorn thrice a year.

18.7 Marketing of Wool and Animals

i) The wool and meat sector in the country is small and widely scattered. This is the sector, where organized sector and decentralized sector run complimentary to each other towards meeting the requirements of all sections of the domestic market as well as export. The sector is very loosely knit and heterogenous. Taking the due cognizance of these complexities in the wool sector, the Government of India has constituted the Central Wool Development Board in the Ministry of Textiles for the overall growth and development of this sector. The State Governments have also set up separate corporation/federation to encourage the sector.

ii) Shepherds generally market wool and animals, through rural agents on the basis of rough estimates of weights. The animals are also sold in village weekly markets, where shepherds are exploited to a great extent.

iii) State Governments through sheep corporation, board and federation have also set up sheep co-operative societies which also purchase wool and animals on weight basis and shepherds are paid better prices.

REFERENCES

Abboud, S.Y. (1989). Genetic and phenotypic parameters associated with growth of purebred and crossbred lambs. Dissertation Abstr. International, B. Sci. and Engg. 50: 38, c.f. *Anim. Breed.* Abstr. (1991) **39**: 1029.

Aboul-Naga, A.M., Aboul-Ela, M.B., Mansour, H. and Gabr, M. (1989). Reproductive performance of Finnsheep and crosses with sub-tropical breeds under accelerated lambing. *Small Rum. Res.* **2**: 143–150.

Abulkhaliq, A.M., Harvey, W.R. and Parker, C.F. (1989). Genetic parameters for ewe productivity traits in the Columbia, Suffolk and Targhee breeds. *J. Anim. Sci.* **67**: 3250–3257.

Acharya, R.M. (1978). Breeding strategy for sheep in India. A discussion paper. Mimeographed. CSWRI, Avikanagar (Rajasthan).

Acharya, R.M. (1981). Status if sheep production in India. Indian Society of Sheep and Goat production and utilization, Jaipur.

Acharya, R.M. (1982). Sheep and goat breeds of India. FAO Animal Production and Health Paper 30. FAO, Rome.

Acharya, R.M. (1990). Sheep Production. *In*: "Handbook of Animal Husbandry". Publications and Information Division, ICAR, New Delhi. pp. 43–118.

Acharya, R.M. and Arora, C.L. (1971). Genetic basis for improving Indian sheep for wool production. Wool and Woollens of India, pp. 38–40.

Acharya, R.M. and Arora, C.L. (1972). Evolving fine wool breeds in India. Wool and Woollens of India, pp. 15–21.

Acharya, R.M. and Arora, C.L. (1976). Improving sheep for fine wool. *Indian Farmers' Digest* **X(3)**: 19–22.

Acharya, R.M., Malhi, R.S. and Arora, C.L. (1972). Studies on sampling procedure for assessing wool quality in Indian sheep: Determination of body region for sampling – A preliminary note. *Indian J. Anim. Sci.* **42**(**12**): 1043–1044.

Acharya, R.M., Arora, C.L., Malik, R.C. and Sahni, M.S. (1972). Genetic aspects of performance of Indian breeds of sheep and their crosses with exotic fine wool breeds. Review of research in sheep and wool production, 1–48; CSWRI, Avikanagar (Rajasthan).

Acharya, R.M., Arora, C.L., Singh, L.B. and Amin, M. (1980). Note on preliminary observations on the performance of pure-bred Karakul sheep and its crosses with carpet wool breeds in the hot, and cold arid and hot semi-arid climates. *Indian J. Anim. Sci.* **50**(**1**): 79–82.

Agar, N.S. (1968). The adaptive significance of blood potassium and haemoglobin types in sheep. *Experientia* **24**: 1274.

Agar, N.S. and Evans, J.V. (1969a). Immunological studies of sheep and goat haemoglobins. *Aust. J. Exp. Biol. Med. Sci.* **47**: 401–403.

Agar, N.S. and Evans, J.V. (1969b). Haemoglobin variants in Indian sheep. *Anim. Prod.* **11**: 247–250.

Agar, N.S., Rawat, J.S. and Roy, A. (1969). Blood potassium and haemoglobin polymorphism in Indian sheep. *J. Agric. Sci.*(Camb) **73**: 197–202.

Agar, N.S. and Seth O.N. (1971). Haemoglobin polymorphism in some sheep breeds of Himalayan region. *Am. J. Vet. Res.* **32**: 360–362.

Ahmed, S.I. (1955). Effect of glycine on storage of ram semen. *J. Agric. Sci.* **46**: 164–67.

Amble, V.N. and Malhotra, J.C. (1966). Review of crossbreeding of sheep for wool production in India. Seminar on Animal Breeding, Haringhata/Calcutta. ICAR, New Delhi. pp. 219–239.

Amble, V.N. and Malhotra, J.C. (1968). Statistical studies on Rambouillet × Rampur Bushair cros of sheep at Pipalkoti, U.P. India. *J. Vet. Sci.* **38**: 101–111.

Andreevskii, V.Ja. (1940). Pricizy proconosti spermy u baronov (Causes of semen abnormality in rams). *Trud. Lab. iskusst. Osemen.* Zivoth. (Mosk.), **1** : 36–45 (*Anim. Breeding Abstr.* **13**: 36).

Armstrong, D.G. and Gilbert, H.J. (1985). Biotechnology and the rumen: A mini review. *J. Sci. Food Agric.* **36**: 1039–1046.

Arora, C.L. (1970). Genetics of haemoglobin, transferrin and potassium type in sheep and their relationship with production and reproduction traits. Ph.D. Diss., Haryana Agril. Univ, Hisar.

Arora, C.L. (1982). Selecting on biochemically polymorphic traits for improving economic characters in Indian sheep I. Potassium types. *Livestock Adviser* **VII**: 15–20.

Arora, C.L. (1983). Research highlights: 1971–83. CSWRI, Avikanagar (Rajasthan).

Arora, C.L. (1984). Selecting on biochemically polymorphic traits for improving economic characters in Indian sheep II-Haemoglobin types. *Livestock Adviser* **IX**: 9–18.

Arora, C.L. (1988). Sheep breeding in India. *In*: "Animal Productivity", pp. 471–480. Oxford and IBH Publishing Co. Pvt. Ltd., New Delhi.

Arora, C.L. and Acharya, R.M. (1971a). Transferrin polymorphism in Indian sheep. *Indian J. Exp. Biol.* **9**: 209–211.

Arora, C.L. and Acharya, R.M. (1971b). Genetic analysis of variation in whole blood potassium in Indian sheep. *Indian J. Exp. Biol.* **9**: 212–213.

Arora, C.L. and Acharya, R.M. (1972a). A note on the association between transferrin types and production traits in Indian sheep. *Anim. Prod.* **15**: 93–94.

Arora, C.L. and Acharya, R.M. (1972b). A note on the haemoglobin and potassium types in Nali breed of Indian sheep and their relationship with body weight and wool yield. *Anim. Prod.* **15**: 95–97.

Arora, C.L. and Acharya, R.M. (1972c). Factors affecting growth, carcass yield and wool yield in Indian breeds of sheep and their crosses. *Indian Vet. J.* **49(6)**: 578–584.

Arora, C.L. and Acharya, R.M. (1976a). Regionalisation of sheep in India. *Indian Farmers' Digest* **X(3)**: 26–28.

Arora, C.L. and Acharya, R.M.(1976b). Improvement of Mandya sheep in Karnataka : A review. *Indian J. Anim. Sci.* **46**(**7**): 340–345.

Arora, C.L. and Acharya, R.M. (1980). Genetic improvement of sheep in the arid and semi-arid regions of India. *Annals of Arid Zone* **19**(**4**): 487–492.

Arora, C.L., Acharya, R.M., Badashiya, B.S. and Dass, N.C. (1975a). Characterization of Chokla breed in Rajasthan and future prospects for its improvement. *Indian J. Anim. Sci.* **45(6)**: 345–350.

Arora, C.L., Acharya, R.M., Badashiya, B.S. and Dass, N.C. (1975b). Characterization of Malpura breed in Rajasthan and future prospects for its improvement. *Indian J. Anim. Sci.* **45(11)**: 843–848.

Arora, C.L., Acharya, R.M. and Kakar, S.N. (1970). Distribution of blood potassium and haemoglobin types in Indian sheep. *Indian J. Exp. Biol.* **8(4)**: 335–336.

Arora, C.L., Acharya, R.M. and Kakar, S.N. (1971a). A note on the association of haemoglobin types with ewe and ram fertility and lamb mortality in Indian sheep. *Anim. Prod.* **13**: 371.

Arora, C.L., Acharya, R.M. and Kakar, S.N. (1971b). New transferrin type(s) in sheep. *Curr. Sci.* **40**: 661–662.

Arora, C.L., Acharya, R.M., Singh, Ramadhar and Buch, S.D. (1975). Characterization of Nali breed in Rajasthan and future prospects for its improvement. *Indian J. Anim. Sci.* **45(11)**: 849–855.

Arora, C.L., Acharya, R.M., Sangolli, B.V. and Nagaraja, M.S. (1977). Characterization of Sonadi breed in Rajasthan and future prospects for its improvement. *Indian J. Anim. Sci.* **47(3)**: 120–125.

Arora, C.L. and Arora, R.L. (1978). Biochemically polymorphic characteristics in sheep and their relationship with production and reproduction. National Seminar on progress of research in animal genetics and breeding during a decade. NDRI, Karnal, June 12-15, 1978.

Arora, C.L. and Batta, J.N. (1983a). Production and quality of wool in Gaddi and its crosses with Rambouillet and Soviet Merino. *Indian J. Anim. Sci.* **53(9)**: 1028–1030.

Arora, C.L., and Batta, J.N.(1983b). Growth performance in Gaddi and its half-breds with Rambouillet and Soviet Merino. *Indian J. Anim. Sci.* **53(10)**: 1147–1150.

Arora, C.L., Balgopal, P.N., Shukla, S.K. and Bapna, D.L. (1981). Note on the combining ability of Rambouillet and Russian Merino rams with Chokla and Nali ewes with respect to wool production and quality. *Indian J. Anim. Sci.* **51(4)**: 498–499.

Arora, C.L., Kakar, S.N. and Acharya, R.M. (1970). Haemoglobin polymorphism in some breeds of Indian sheep and their crosses. *Indian J. Anim. Prod.* **1**: 91–93.

Arora, C.L., Kakar, S.N. and Acharya, R.M. (1970). A new haemoglobin type in Indian sheep. *Indian J. Biochem.* **7(1)**: 77–78.

Arora, C.L. and Krishnamurthy, U.S. (1984). Growth and wool production and quality upto six months of age in Nilgiri and its halfbreds with Rambouillet and Soviet Merino. *Indian J. Anim. Sci.* **54(8)**: 824–828.

Arora, C.L., Mishra, R.K. and Gupta, D.C. (1983). Variation in the growth traits of Nali, Chokla and their half-breds with Rambouillet and Merino sheep. *Indian Vet. J.* **60**: 111–117.

Arora, C.L., Reddy, K.K. and Gupta, G.C. (1986). Evaluation of breeding data in crossbreeding. *Indian J. Anim. Sci.* **56(3)**: 367–69.

Arora, C.L. and Swain, N. (1989). Relative performance of sheep and goats under rangeland management system. *In*: "Rangelands Resource and Management", pp 436–442. Range Management Society of India, IGFRI, Jhansi.

Arora, R.L. (1979). Studies on inheritance of isozymes haemoglobin and erythrocite potassium in sheep and their relationship with production and reproduction traits. Ph.D. Thesis, Agra. Univ. Agra.

Arora, R.L. and Arora, C.L. (1980). Note on reproduction and performance of Dorset and Suffolk breeds of sheep. *Indian J. Anim. Sci.* **50(3)**: 287–290.

Aswad, A., Abdou, M.S.S., Al-Bayaty, F. and El-Sawaf, S.A. (1976). The Validity of the 'Ultra-sonic Method' for pregnancy diagnosis in ewes and goats. *Zentralblatt fur Veterinarmedizin A* **23**: 467–474.

Aswad, A., Rebesko, B., Fathalla R. and Fahri, R. (1974). Sinhronizacija porodaja (Parturition synchronisation). *Veterinarski Glasnik* **28**: 609–614.

Ayalon, N. and Shemesh, M. (1979). Use of milk progesterone measurement for determining pregnancy and ovarian activity in cattle and sheep. *Rafuah Veterinarith* **36**: 79.

Balaine, D.S., Singh, Jarnail, Acharya, R.M. and Kanaujia, A.S. (1971). Post-weaning feed lot performance of Nali and Lohi sheep, and their crosses with Nellore and Mandya. *Indian Vet. J.* **48**: 50–58.

Balaine, D.S., Wadyal, S.S., Acharya, R.M. and Biswas, D.K. (1970). Studies on wool traits of Nali and Lohi sheep. *Indian J. Anim. Prod.* **1**: 37.

Banerji, R. and Arora, C.L. (1982). Note on certain inherited blood characteristics and their combinations in indigenous, exotic and cross-bred sheep. *Indian J. Anim. Sci.* **52(11)**: 1114-1116.

Basuthakur, A.K. and Acharya, R.M. (1981). *Eds.* 1st Proceeding of the Society and National Seminar on Sheep and Goat Production and Utilization. Indian Society for Sheep and Goat Production and Utilization, Jaipur.

Bell, T.D. (1945). Breeding studies in Rambouillet ewes and rams. Pr. Bull. N. Mex. Agric. Exp. Sta. No. 1003: 2, (*Anim, Breeding Abstr.* **14**: 154).

Berger, Y. and Ginisty, L. (1980). Bilan de 4 annees d'etude de la race ovine djallonke Cote d'lvoire (Results of a 4-year study of the Djallonke breed of sheep in the Ivory Coast), *Revue d'Elevage et de Medicine Veterinaire des Pays Tropicaux* **33**: 71–78.

Bhadula, S.K. and Bhat, P.N. (1981). Genetic and non-genetic factors affecting pre- and post-weaning gains and feed efficiency traits in Muzaffarnagri and its halfbreds with Corriedale. *Anim. Breed. Abstr.* **49**: 830.

Bhaik, N.L. and Kohli, I.S. (1980). Reproductive performance of Chokla and Magra ewes in Mbtropical area of North Western arid zone of Rajasthan. 2. Parturition and involution of uterus. *Indian Vet. J.* **57**: 327–33.

Bhaskar, B., Krishnamurthy, U.S. and Rahtnasabapathy, V. (1976). Genetics of blood potassium and haemoglobin in Mandya sheep. *Cherion* **5(1)**: 8–13.

Bhasin, N.R. and Desai, R.N. (1965). Studies on factors affecting the characters concerning quality of wool fibre in Chokla flock of sheep. *Indian Vet. J.* **42**: 782.

Bhat, P.N. (1978). A note on the inheritance of haemoglobin types in Muzaffarnagri breed of sheep and its crosses with Suffolk and Dorset. *Indian J. Anim. Sci.* **48**: 915–916.

Bhat, P.P. and Baruah, P. (1980). Note on the genetics of enzyme and serum protein polymorphism in three breeds of Indian goats. *Indian J. Anim. Sci.* **50**: 290–293.

Bhat, P.P. and Bhat P.N. (1979). Relationship of transferrin and haemoglobin types with pre-weaning growth in Muzaffarnagri (*Ovis aries*) and its crosses with suffolk and Dorset breeds. Proc. XIV International Congress of Genetic (Part-II), Moscow. C-31, pp. 422 (Abstr.).

Bhat, P.N, Bhat, P.P., Khan, B.U., Goswami, D.S. and Singh, B. (1981). Animal Genetic Resources in India. Bulletin No. 192, National Dairy Research Institute, Karnal.

Bhat, P.P., Bhat, P.N. and Trivedi, K.R. (1978). Relationship of transferrin and haemoglobin types with preweaning growth in Muzaffarnagari (*Ovis aries*) breed and its crosses with Suffolk and Dorset breeds. Proc. 14th Int. Cong. Genet. Moscow (Abstr.) pp. 422.

Bhat, P.P., Bhat, P.N. and Trivedi, K.R. (1979). Relationship of transferriun and haemoglobin types with preweaning growth in Muzaffarnagri and its crosses with suffolk and Dorset breeds. Proceedings of XIV International Genetic Congress. Moscow. 21-30 August, 1976. Abstract [*Anim. Breed. Abstr.*) **47(6)**: 2929].

Bhat, P.P., Khanna, N.D. and Bhat, P.N. (1973). Studies in Blood Group and Biochemical Polymorphism in Indian Sheep. *Scientific Report*. Animal Genetics Division, IVRI, Izatnagar.

Bhat, P.P., Khan, B.U., Santiago, T.C. and Sahni, K.L. (1981). Potassium and Haemoglobin polymorphism in Muzaffarnagari breed of sheep. *Indian J. Anim. Sci.* **51**: 1147–1151.

Bhuvankumar, C.K., Krishnamurthy, U.S. and Rathnasabapathy, V. (1974). Haemoglobin and transferrin polymorphism in Mecheri sheep. *Cherion* **3(1)**: 26–28.

Blackshaw, A.W. (1958). The effects of glycerol on the paravital staining of spermatozoa. *Aust. Vet. J.* **34**: 71–76.

Bogart, R. and Mayer, D.T. (1946). Environmental temperature and thyroid gland involvement in lowered fertility of ram. Res. Bull. Mo. Agric. Exp. Sta. No. 402: 43. (*Anim. Breeding Abstr.* **14**: 235).

Bohra, S.D.J. (1984). Crossbreeding for evolving new mutton breeds. Breed and breed cross differences in growth, efficiency of feed converson and carcass yield. Ph.D. Thesis, Kurukshetra University, Kurukshetra.

Bosc, M.J. (1972). The induction ad synchronisation of lambing with the aid of dexamethason. *J. Reprod. Fert.* **29**: 347–357.

Boujenane, I. and Brodford, G.E. (1991). Genetic effects on ewe productivity of crossing D'man and Sardi breeds of sheep. *J. Anim. Sci.* **69**: 525–530.

Boujenane, J., Khallouk, M. and Kerfal, M. (1988). Genetic parameters of reproductive performance of D'man ewes. Proc. Third World Congress on Sheep and Beef Cattle Breeding. 19-23 June, Paris Vol. **2**.

Brown, D.E., Cameron, C.W. and Ngere, L.O. (1972). Synchronisation of oestrus and early post-partum rebreeding of Nungua Blackhead sheep in Ghana, *Ghana J. agri. Sci.* **5**: 183–187.

Cardellino, R.C. (1986). Breeding programmes for multiple purpose breeds in Uruguay, Proc. 3rd World Cong. Genet. *Applied Livestock Prod.* **IX**: 650–656.

Carillo, A.M.D. and Macocha, C.E. (1991). Heritibility of litter size in sheep. *Animal Breed Abstr.* **59**: 1032.

Carter, H.B. and Clarke, W.H. (1957). The hair follicle group and skin population of Australian Merino sheep. *Aust. J. Agric.* **8**: 91–108.

Castillo Rojas, J., Hernandez Ledezma, J.J., Berruecos, J.M. and Lopez Angeles, J.J. (1977). Comportamiento reproductivo del borrego Tabasco mantenido en clima tropical. III. Puberted y duracion del estro (Reproductive performance of Tabasco lambs maintained in a tropical climate. III. Puberty and duration of oestrus), *Teenica Pecuaria en Mexico* **32**: 32–35.

Chang, H.C. (1945) . The sperm production of adult rams in relation to frequency of semen collection. *J. Agric. Sci.* (Camb), **35**: 245–46.

Charyulu, E.K. and Acharya, R.M. (1984). Heterosis in skin follicular and fleece characteristics. *Indian J. Anim. Sci.* **54**: 1003–1005.

Chaudhry, A.L. (1988). Interospection of carpet wool avaliability and production strategy in India. Paper presented at National Seminar on Carpet Wool Production and utilisation held at Central Sheep and Wool research Institute, Avikanagar, Oct. 24-26, 1988.

Chopra, S.C. and Acharya, R.M. (1971a). Genetic and Phenotypic parameters of body weight in Bikaneri sheep (Magra Strain). *Anim. Prod.* **13**: 343.

Chopra, S.C. and Acharya, R.M. (1971b). A note on non-genetic factors affecting body weights of Bikaneri sheep (Magra strain). *Anim. Prod.* **13**: 349–351.

Chopra, S.C. and Chopra, S.C. (1972). Factors affecting wool characteristics in Nali and Lohi sheep. *Indian J. Anim. Sci.* **42**: 363–369.

Compendium (1982). National Seminar on Sheep and Goat diseases. Held at CSWRI, Avikanagar, May 28-30, 1982.

Comstock, R.E., Green, W.W., Winters, L.M. and Nordskog, A.N. (1945). Studies of semen and semen production. *Tech. Bull. Minn. Agric. Exp. Sta.* No. 162: 55 *Animal Breeding Abstr.* **13**: 37.

Corroll, N., Bogart, R. and Price, D. (1968). Genetics of sheep transferrin as determined by disc electrophoresis. *Genetics* **60**: 208 (Abstr.).

Cundiff, D.V., Gregory, K.E. and Koch, R.M. (1982). Selection for increased survival from birth to weaning. Proc. 2nd World Congress Genet. *Livestock Prod.* **IX**: 310–337.

Cupps, P.T., MuGowan, B., Rohlmann, D.F., Reddon, A.R. and Weir, W.C. (1960). Seasonal changes in the semen of rams. *J. Anim. Sci.* **19**: 208–213.

Daflapurkar, D.K., Kaushik, S.N. and Kalpatal, B.G. (1979). Comparative studies on wool production and wool quality traits in Patanwadi and Deccani crosses. *Indian Vet. J.* **56**: 764–767.

Dangi, K.S., Boylan, W.J. and Rempel. W.E. (1988). Productivity of some pure breeds of sheep producing both purebred and crossbred lambs: Reproductive traits of ewes and lamb weights. *Indian J. Anim. Genet. Breed.* **10**: 1–8.

Dass, G.S. and Acharya, R.M. (1970). Growth of Bikaneri sheep. *J. Anim. Sci.* **31**: 1–4.

Dass, M.N. and Rajagopalan, V.R. (1956). Studies on the scope for improvement of Bellary sheep. *Indian J. Vet. Sci. and Anim. Husb.* **26**: 171–183.

Deshpande, B.R. (1970). Morphological changes in the semen picture and testicular tissue with artificially induced sub-fertility and sterility by scrotal insulation in rams. Ph. D. Thesis M.P.K.V. Rahuri (India).

Deshpande, A.D., Kashirsagar, S.G. and Tadpatrikar, N.S. (1976). Studies on sheep haemoglobin. *Mahavet* **2(4)**: 101–103, 105–106 (*Anim. Breed. Abstr.* **44**: 5717)

Dhanda, O.P. (1970). Studies on oestrus pattern, hormonal synchronisation of oestrus and subsequent fertility in Nali sheep. M.Sc. Thesis Haryana Agricultural University; Hisar.

Dickerson, G.E. (1973). Inbreeding and heterosis in animals. Proc. Anim. Breed. and Genet. Symp. in Honor of J.L. Lush. ASAS and ADSA, Champaign, Illinois.

Dickerson, G.E., Glimp, H.A. and Gregory, A. (1975). Genetic resources for efficient meat production in sheep. Preweaning viability and growth of Finn sheep and domestic crossbred lambs. *J. Anim. Sci.* **34**: 940.

Dodd, C.J. and Delahunty, J.R. (1983). Efficiency of selection for the overall index and its component traits in large flocks of sheep in New Zealand. *Proc. New Zealand Soc. Anim. Prod.* **4(3)**: 193–196.

Domigo, W.R. and Klyne, W. (1949). A photoelectric flame photometer. *Biochem. J.* **49**: 400–408.

Dregne, H., Kassa, M. and Rzanov, B. (1991) A new assessment of the world status of desertific-action. *Desertification Control Bulletin* **20**: 6–18.

Dun, R.B. (1956). Temporary infertility of rams associated with flooding. *Aust. Vet. J.* **32**: 1–3.

Dun, R.B., Ahmed, W. and Morrant, A.J. (1960). Annual reproductive rhythm in Merino sheep related to the choice of a mating time at Trangie, Central Western New South Wales. *Aust. J. agric. Res.* **11**: 805–826.

Dwarkanath, P.K. and Joshi, B.K. (1973). Haemoglobin types in Sonadi breed of sheep. *Univ. Udaipur Res. J.* **11**: 18–21.

Dziuk, P.J., Graham, E.F., Donker, J.D., Marion, G.B. and Petersen, W.E. (1954). Some observations in collection of semen from bulls, goats, boars and rams by electrical stimultion. *Vet. Med.* **48**: 455–58.

Easley, G.T. (1951). A report of artificial insemination: Factors affecting reproduction and semen studies in sheep of the highland pastures of Southern Peru. *J. Amer. Vet. Med. Ass.* **119**: 278–282.

El-Homosi, F.F. and El-Hafiz, G.A.A. (1982). Reproductive performance of Ossimi and Saidi sheep under two prepubertal planes of nutrition. *Assiut Vet. Med. J.* **10(19)**: 59–66.

El-Wishy, A.B. (1974). Some aspects of reproduction in fat tailed sheep in subtropics. IV. Puberty and sexual maturity. *Zeitschrift fur Tierzuchtung ung Zuchtungsbiologie*, **91**: 311–316.

Emmens, C.W. (1950). Fertility in the male. (Supplementary). Progresses in the Physiologty of Farm Animal. (*Ed.*) J. Hammond. London: Butter woths Scientific Publications. pp. 1047–1116.

Emmens, C.W. and Robinson, T.V. (1962). Artifical Inseminaiton. *In*: Sheep. The semen of Animals and Artficial Inseminaiton (*Ed.*). J.P. Maule Commonwealth Agricultural Bureaux. Farnham Royal. Bucks. England Engela. D.J. 1948/Fertility in rams. *Fmg. S. Afr.* **23**: 218–282 (*Anim. Breeding Abstr.* **26**: 1025).

Engela, D.J. (1948). Use of fodder trees and shrubs Qd. Deptt. preimary Industries Div. Plt. Ind. Advisory leaflat 1024.

Evans, J.V. (1957). The stability of the potassium concentrate in the erythrocytes of individual sheep compared with variability between different sheep. *J. Physiol.* **136**: 42–59.

Evans, J.V. and Mounib, M.S. (1957). A survey of potassium concentration in the red blood cells of sheep. *Nature* **176**: 177.

Evans, J.V., Harris, H. and Warren, F.L. (1958). The distribution of haemoglobin and blood potassium types in British breeds of sheep. *Proc. Roy. Soc. B.* **149**: 249–252.

Evans, J.V., King, J.W.B., Cohen, B.L., Harris, H. and Warren, F.L. (1956). Genetics of haemoglobin and blood potassium differences in sheep. *Nature*, Lond. **178**: 849–850.

Fahmy, M.H. (1986). Preliminary results on fertility, prolificacy, lamb production and carcass traits of Romanov sheep in Canada Proc. 3rd World Cong. *Genetics Applied Livestock Prod.* **IX**: 559–593.

Fahmy, M.H. (1987). The accumulation effect of Finnsheep breeding in crossbreeding schemes: Wool production and fleece characteristics. *Can. J. Anim. Sci.* **67**: 1–11.

Fahmy, M.H. (1989). Reproductive performance, growth and wool production of Romanov sheep in Canada. *Small Rum. Res.* **2**: 253–264.

Fall, A., Diop, Sandford, J., Wissoq Durkin, J. and Trail, J.C.M. (1982). Evaluation of the productivities of Djallonke sheep and N-Dama catle at the Centre de Recherches Zootechniques, Kolda, Senegal. Research Report No. 3. Intern. Livestock Centre for Africa, Addis Ababa, Ethiopia.

FAO (1989). Biotechnology for Livestock Production. Proccedings of the expert consultation on the application of Biotechnology in Livestock Production and Health in developing Countries, October 6-10, 1986, Rome, Italy. Division of Plenum Publishing Corporation, New York.

Ferrall, G.J.M. (1976). Phenotypic and genetic parameters of reproduction in ewes. *Anim. Breed. Abstr.* **23**: 295–304.

Fesus, L. (1972). Apparent disturbed segregation at the haemoglobin and transferrin loci in Hungarian Merino sheep. XII Europ. Conf. Anim. blood Grps. *Biochem. Polymorph. Bpest* **1970**: 567–573.

Flamarique, C.V.C. (1992). Efficiency of selection in sheep. *Anim. Breed. Abstr.* **60**: 5097.

Fletcher, I.C., Lishman, A.W., Thring, B. and Holmes, J.A. (1980). Plasma progesterone levels in lactating ewes after hormone-induced ovulation during the non-breeding season. *S. Afr. J. Anim. Sci.* **10**: 151–157.

Foilmom, S.N., Lunca, I., Bratescu and Otel, V. (1956). Cercetari asupra diluarii Si Conservarii Spermei de berbec. si taur. (The dilution and storage of ram and bull semen). *Anal. Inst. Cero. Zootec.* **14** : 231–241 (*Anim. Breeding Abstr.* **25**: 1943).

Forrest, P.A. and Bichard, M. (1974). Analysis of production record from a low land sheep flock. Phenotypic and genetic parameters for reproduction performance. *Anim. Prod.* **19**: 33–45.

Fraser, A.H.M. (1954). Influence of nutrition on wool growth. *Nutr. Abstr. Rev.* **4**: 9–13.

Gaillard, Y. (1979). Caracteristiques de la reproduction de la brebis Oudah (Reproductive characteristics of the Uda ewe), Revue d'Elevage et de Medecine Veterinaire des Pays Tropicaux. **32**: 285–290.

Galal, E.S.E., Bedier, N.Z., Younis, A.A. and Mansour, H. (1988). Estimates of genetic and phenotypic parameters for some traits and productivity indices in Barki ewes. Proc. 3rd World Congress sheep and cattle breeding. 19-23 June, 1988 Paris. Vol. **2**.

Ghanekar, V.M., Soman, B.V. and Bhatwadekar, S.D. (1972). A note on some vital and meat characteristics in Bannur sheep. *Indian J. Anim. Sci.* **42**: 584–586.

Ganesakale, D. (1975). Effect of season of mating on the economic traits in sheep. *Cheiron* **4**: 40–44.

Ghosh, P.K., Eyal, E. and Evans, J.V. (1965). The blood of desert sheep. Paper presented at the Australian Arid Zone Research Conf. Alice Springs, W.T. Northern Territory, Australia.

Gill, R.S. and Negi, S.S (1971). Requirement of maintenance digestible crude protein for adult sheep. *Indian J. Anim. Sci.* **41**: 980.

Gunn, R.M.C., Sanders, R.N. and Granger, W. (1942). Studies in fertility in sheep. 2. Seminal changes affecting fertility in rams. Bull. Coun. Sci. Industr. Res. Aust., No. 148–149 (*Anim. Breeding Abstr.* **11**: 236).

Gupta, D.C. and Pant, K.P. (1980). Growth of indigenous and crossbred lambs in hot semi-arid region. *Indian Vet. J.* **57**: 821–824.

Gupta, B.R., Reddy, K.K. and Munerathnam, D. (1987). Puberty and sexual maturity in Nellore sheep and its synthetics with Dorset. *Indian J. Anim. Sci.* **57**: 1013–1016.

Hafez, E.S.E., Baderldin, A.L. and Darwisch, Y.H. (1955). Seasonal variations in semen characteristics of sheep in the subtropics. *J. agric. Sci.* (Camb), **45**: 283–292.

Hazel, L.N. (1943). The genetic basis for constructing selection indexes. *Genetics* **28**: 476–490.

Hazel, L.N. and Lush, J.L. (1942). The efficiency of three methods of selection. *J. Heredity.* **33**: 393.

Hazra, A. and Patnayak, B.C. (1975). Evaluation of *Cenchrus ciliaris* pasture in terms of some blood biochemical parameters in ewes. CSWRI Annual Report 1975.

Henderson, C.R. (1951). Cornell University, Ithaca, New York (mimeograph).

Hill, J.R. (Jr.), Hurst, V. and Godleg, W.C. (1958). A comparision of reconstituted skim milk and egg yolk sodium citrate as extenders for ram semen. *Amer. J. Vet. Res.* **19**: 132–134.

Hohenboken, W. and Cochran, P.E. (1976). Heterosis of ewe lamb productivity. *J. Anim. Sci.* **42**: 819–823.

Honmode, J. (1970). Gestation period in Rambouillet, Chokla and Malpura ewes. *Indian J. Anim. Sci.* **40**: 229.

Honmode, J. (1971). Effect of megasterol, megasterol acetate and progesterone on oestrus and fertility in cyclic Malpura and Chokla ewes. *Indian J. Anim. Sci.* **41**: 972.

Honmode, J. (1972). Annual Report, Central Sheep Wool Research Institute, Avikanagar, Rajasthan.

Honmode, J., Bohra, S.D.J. and Patil, B.D. (1971). Oestrus synchromsation in anoestrus motegenous and exstifc ewes – A note. *Indian J. Anim. Sci.* **41**: 1066.

Honmode, J., Chatterjee, R. and Patil, B.D. (1971). Investigation into the effect of season, hormone and plane of nutrition on reproduction and characteristics of wool of six indigenous sheep breeds. Annual Report, Central Sheep and Woool Research Institute, Avikanagar Rajasthan, pp. 48.

Honmode, J., Patil, B.D. and Pachlag, S.V. (1971). Hormonal induction of superovulatory oestrus in early postpartum indigenous and crossbred ewes. *Indian J. Anim. Sci.* **41**: 977.

Honmode, J. Patil, B.D. and Tiwari, S.B. (1971). Reproductive performance of Chokla rams as influenced by differential feeding. *Indian Vet. J.* **48**: 1044–1047.

Honmode, J. and Tiwari, S.B. (1968). Artificial breeding of Rambouillet, Malpura, Chokla, R × C and R × M ewes during March/April. Annual Report, Central Sheep and Wool Research Institute, Avikanagar, Rajasthan. pp. 4.

Honmode, J. and Tiwari, S.B. (1973). Artificial breeding of exotic, indigenous and their crossbred ewes during March/April. Annual Report, Central Sheep and Woo Research Institute, Avikanagar, Rajasthan.

Honmode, J. and Tiwari, S.B. (1974). Effect of frequency of semen collection on quality and fertility of Malpura and Chokla rams. *Indian Vet. J.* **51**: 100–104.

Hulet, C.V., Voigtlander, H.P. (Jr.), Pope, A.L. and Caside, L.E. (1956). The nature of early season infertility in sheep. *J. Anim. Sci.* **15**: 607–616.

Indian Livestock Census (1987, 1994). Directorate of Economics and Statistics, Department of Agriculture and Cooperation, Ministry of Agriculture, Govt. of India.

IASRI (1997). Agricultural Research Data Book 1996, IASRI, New Delhi.

Indo-Soviet Seminar (1980). Breeding of fine wool sheep for arid and semi-arid areas. December 14-16, 1980, CSWRI, Avikanagar.

Idybsn, D. (1956). Effektivnostj ispolizovanija moloka v kacestve razbavitelja semen ni barana.

(The effectiveness of using milk as a diluent for ram semen), *Ovcevodstuv* **1956(4)**: 33–37 (*Anim. Breeding Abstr.* **25**: 231).

Jones, L.P. (1982). Economic aspects of developing breeding objectives: A specific example; breeding objectives for Merino sheep. Future developments in the Genetic improvement of Animals, Academic Press, Sydney, Australia, pp. 119–136.

Joshi, D.C. (1966). Studies on high altitutde Himalayan pastures (Bugiyals) of Uttar Pradesh. *Indian Vet. J.* **43**: 1019–1027.

Joshi, D.C. and Ludri, R.S. (1967). The chemical composition and nutritive value of Buffaloe (*Pennistum lacerdum* Griseb) and *Panicum kabulabula* grasses. *Indian Vet. J.* **44**: 800–804.

Joshi, C.G., Patel, M.M. and Shah, R.R. (1991). Efficiency of index selection in sheep. *Indian J. Anim. Sci.* **61**: 196–198.

Joshi, J.D. and Singh, G. (1968a). Studies on the ram semen diluents I. Effect of egg yolk citrate series diluents on livability and morphology of ram spermatozoa. *Indian J. Vet. Sci.* **38**: 574.

Joshi, J.D. and Singh, G. (1968b). Studies on the ram semen diluents III. Effect of skim-milk citrate series diluents on livability and morphology of ram spermatozoa. *Indian J. Vet. Sci.* **38**: 583.

Joshi, J.D. and Singh, G. (1968c). Studies on ram semen diluents II. Effect of tomato juice series diluents on the livability and morphology of ram spermatozoa. *Indian J. Vet. Sci.*, **38**: 591.

Kalla, S.D., Dwarkanath, P.K. and Singh, Madho (1970). Haemoglobin polymorphism in the sheep of north Rajasthan. *Indian J. Exp. Biol.* **8**: 60–61.

Kalla, S.D., Dwarkanath, P.K. and Singh, Madho (1971). Haemoglobin polymorphic studies in relation to wool quality in sheep of west Rajasthan. *Indian J. Anim. Sci.* **41**: 109–112.

Kalla, S.D. and Ghosh, P.K. (1975). Blood biochemical polymorphic traits in relation to wool production efficiency in Indian sheep. *J. Agric. Sci. Camb.* **84**: 149–152.

Kalla, S.D., Ghosh, P.K. and Taneja, G.C. (1972). Erythrocyte glutathione level in relation to blood potassium type in the Rajasthan desert sheep. *Anim. Blood. Grps. Biochem. Genet.* **3**: 121–123.

Kalra, S. (1976). Comparison of ewe productivity in various crossbred grades of Rambouillet × Chokla and purebred Rambouillet and Chokla. M.Sc. Thesis, Haryana Agric. Univ, Hisar.

Kammlade, W.G. (Sr.) and Kammalade, W.G. (Jr.) (1955). Sheep Science. J.B. Lippincott Company, New York.

Kandasamy, N. (1979). Haemoglobin and blood potassium polymorphism in Nilgiri sheep. *Indian J. Hered.* **11**: 117–124.

Kandasamy, N., Arora, C.L. and Gupta, D.C. (1982). Studies on wool quality attributes in Nali, Chokla and cross-bred lambs. *Indian Vet. J.* **59(2)**: 132–137.

Kandasamy, N., Gupta, D.C. and Arora, C.L. (1980). Variation of gestation length in crossbred ewes. *Indian J. Hered.* **12**: 57-62.

Kandasamy, N. and Pant, K.P. (1980). Gestational oestrus during pregnancy in sheep. *Indian Vet. J.* **57**: 434–35.

Kapoor, U.R., Aggarwala, On.N., Panchauri, P.C., Nath, K. and Narayan, S. (1972). The relationship between diet, the copper and sulphur content of wool and fibre characteristics. *J. Agric. Sci.* (Camb.) **79**: 109–114.

Karam, N.A. (1959). Selecting yearling Rabmani sheep. *J. Anim. Sci.* **18**: 1452–1459.

Kasseme, R., Owen, J.B., Fadel, J., Juha, H. and Whitanker, C.J. (1989). Aspects of fertility and lamb survival in Awassi sheep under semi arid condition. Res. and Dept. in Agric. **6**: 161.

Kaushish, S.K. (1989). Studies on reproduction in sheep. Relationship between gestation length, birth weight and weight at service. Paper presented at V. Annual Conf. SAPI. Munnar (Kerala) Dec. 27-30, 1989.

Kaushish, S.K. (1971). Studies on gestation length, blood constituents and process of parturition in sheep. M.Sc. Thesis, Haryana Agricultural University, Hisar.

Kaushish, S.K. and Arora, K.L. (1973). Effect of weight of dam at seirce on gestation length, parturition and weight of lamb born to Nali sheep. *Indian J. Anim. Prod.* **3** (**1**): 41–50.

Kaushish, S.K. and Arora, K.L. (1974a). Studies on reproduction in sheep. 3 Factors affecting gestation length. *Indian J. Anim. Prod.* **5** (**1–4**): 85–90.

Kaushish, S.K. and Arora, K.L. (1974b). Studies on reproduction in sheep. 4. process of parturition. *Indian J. Anim. Sci.* **44**: 667–671.

Kaushish, S.K. and Arora, K.L. (1975). Studies on relationship among weight of placenta, number of cotyledons and birth weight. *Indian J. Anim. Sci.* **45**: 248–251.

Kaushish, S.K. and Arora, K.L. (1983). Effect of weights and mesurements on placenta weight and number of cotyledons in Nali and Lohi sheep. *Intern. J. Trop. Agri.* **1**(**2**): 167–169.

Kaushish, S.K. and Arora, K.L. and Dhanda, O.P. (1973). Studies on reproduction in sheep 2. Time of parturition. *Indian J. Anim. Prod.* **4**(**1**): 56–59.

Kaushish, S.K. and Sahni, K.L. (1976). Effect of feeding animal protein (Egg + milk) and trace elements and provision of cooler climate on libido, semen quality and certain physiological reactions of Russian Merino rams during summer season. *Indian J. Anim. Sci.* **46**: 135–36.

Kaushish, S.K., Rawat, P.S. and Kalra, D.B. (1986). Studies on reproduction in sheep 6. Effect of gestation length and dams weight at service and lambing on preweaning body weight in exotic fine wool breeds. *Indian J. Anim. Reprod.* **7**(**1**): 92–94.

Kaushish, S.K. and Sahni, K.L. (1975). A note on the time of parturition in Russian Merino ewes. *Indian Vet. J.* **52**: 837–839.

Kaushik, S.N. and Singh, B.P. (1968). Studies of the extent and nature of variation in body weight in Rampur Bushair × Polwarth crossbred lambs. *J. Anim. Morphol. Physiol.* **15**: 78–83.

Keast, J.C. and Meeley, F.H.W. (1949). Some observations on artificial in semination os sheep. *Aust. Vet. J.* **25**: 281–287.

Kennedy, J.P. (1967). Genetic and phenotypic relationship between fertility and wool production in two year old Merino sheep. *Aust. J. Agric. Res.* **18**: 515–522.

Kerr, S.E. (1937). Studies on the inorganic composition of blood IV. The relationship of the potassium in the acid soluble fractions. *J. Biol. Chem.* **117**: 227–235.

Khan, B.U. and Bhat, P.N. (1981). Note on potassium Polymorphism in in Muzaffarnagri sheep. *Indian J. Anim. Sci.* **51**(**9**): 884–885.

Kishore, K., Gour, D., Rawt, P.S. and Arora, C.L. (1980). Note on gestation length in cross-bred sheep. *Indian J. Anim. Sci.* **50**: 565–567.

Kishore, K., Grour, D., Rawat, P.S. and Malik, R.C. (1982). A study of some reproductive traits of Avikalin and Avivastra ewes and factors affecting them. *Cheiron* **11**: 20–24.

Kleeman, D.O., Smith, D.H., Grimise, R.J. and Holloway, R.E. (1991). Comparative performance of South Australian Merino and Poll Dorset × Merino ewes mated to Suffolk rams. *Small Rum. Res.* **5**: 129–140.

Koger, M. (1951). Storage, dilution and use of ram semen in artificial breeding of sheep. *Bull. Max. Agric. Exp. Sta.* No. 366: 12 pp. (*Anim. Breeding Abstr.* **21**: 789).

Krishnamurthy, U.S. and Rathnasabapathy, V. (1977). Genetics of blood potassium in Nilgiri, Merino and their crossbred sheep. *Indian Vet. J.* **54**: 36–41.

Kothandapani, R., Krishnamurthy, U.S. and Rathnasabapathy, V. (1974). Haemoglobin polymorphism in Mandya sheep. *Cherion* **3**: 58–60.

Krishnan, T.S. (1939). Nutrition of sheep and wool production. *Indian. J. Vet. Sci.* **9**: 59–60.

Krishnappa, S.B. (1985). Investigations on the productivity of Nali and its crosses with Russian Merino and Corriedale Sheep. Ph.D. Thesis, Haryana Agric. University, Hisar.

Kumar, Satish, Bohra, S.D.J., Bapna, D.L., Singh, R.N., Nivsarkar, A.E. and Arora, C.L. (1988). Crossbreeding for mutton production in India. Reproductive performance of Malpura, Sonadi and their F_1 crosses with Dorset and Suffolk. *World Rev. Anim. Prod.* **24**(**3**): 20–24.

Long, T.E., Thomas, D.L. Fernando, R.L., Lewis, J.M., Garrigus, U.S. and Waldron, D.F. (1989). Estimation of individual and maternal heterosis, repeatability and heritability for ewe productivity and its components in Suffolk and Targhee Sheep. *J. Anim. Sci.* **67**: 1208–1211.

Labban, F. (1973). A study on Wool Characteristics of the Barbary Sheep. Minstry of Agriculture, Tripoli, Libyan Arab Republic.

Lardy, H.A. and Phillips, P.H. (1939). Preservation of spermatozoa. Proc. Amer. Soc. Anim. Prod., pp. 219–221.

Lasley, J.F., Easiloy, G.T. and McKenzie, F.F. (1942). A staining method for the differenation of live and dead spermatozoa. I. Applicability to the staining of ram spermatozoa. *Anat. Rec.* **82**: 167–174.

Macpherson, J.W. (1957). Goats milk as a semen diluent. *Can. J. Comp. Med Vet. Sci.* **21**: 161–162.

Mahalanobis, P.C. (1928). A statistical study at Chinese head measurements. *J. Asiatic Soc. Bengal* **25**: 301–377.

Maheshwari, (1971). Cited by Patnayak (1980c).

Maheshwari, S.R. (1977). A comparative study in energy and protein requirement of the native and crossbred lambs. Ph. D. Thesis, Panjab University, Chandigarh.

Mali, P.C., Patnayak, B.C. and Bedekar, A.R. (1983). Plane of nutrition of Marwari ewes in different seasons – grazing on natural range pasture of arid region. Annals-Arid Zone **12**: 323.

Malik, R.C. and Acharya, R.M. (1972). A note on factors affecting lamb survival in Indian sheep. *Anim. Prod.* **14**: 123–125.

Mann, T. (1964). The biochemistry of semen and of the male reproductive tract. Indian Edn. Methuen and Co. Let., London.

Mathur, C.S. (1960). Common fodder grasses native to the desert soil of Rajasthan and their feeding value. *Indian J. Vet. Sci.* **30**: 69–103.

Mathur, P.B. and Dubey, S.C. (1994). Sheep and Goat Diseases. ICAR, New Delhi.

Mathur, C.S., Kalla, S.D. and Mohan, Mani (1972). Observations on the sulphur content of wool from different parts of the body of Magra sheep. Wool and Woolens of India, October.

Maussen, B., Preisinger, R. and Kalm. E. (1988). Analysis of fertility status of North German sheep population. Proc. 3rd World Cong. sheep and beef cattle breeding. 19-23 June, 1988 Paris Vol.2.

Maqsood, M. (1951). Seasonal variations in the testis histology of the ram. *Vet. Rec.* **63**: 597–98.

Marston, M.R. (1955). Nutritional factors involved in wool production by Merino sheep. *Aust. J. Sci. Res. Series.* **1(3)**: 376–387.

Martemucci, G., Ballitti, E., Toteda, F. and Manchisi, A. (1983). Study of first signs of oestrus in ewe lambs according to genotype. Annalidella Facolta di Agraria, University di Bari, **31**: 185–200 (*Anim. Breeding Abstr.* **5**: 4979).

Mathur, A.K., Srivastava, R.S., Anil Joshi and Kalra, D.B. (1989). Pellet freezing of ram semen. *Indian J. Anim. Sci.*, **59**: 1529.

Mehta, B.S., Kandasamy, N. and Arora, C.L. (1986). Life time productivity of farmers breed exotic (Russion Merino, Rambouillet) ewes in semi-arid tropics. *Indian J. Anim. Sci.* **56**(1): 158–160.

Mehta, B.S., Kandasamy, N. and Arora, C.L. (1979). Adult body weight of indigenous and crossbred ewes under semi-arid conditions in India. *Trop. Anim. Hlth. Prod.* **11**: 227–230.

Meikle, H.E., Wickham, G.A., Rae, A.L., Dobbie, J.L. and Hichey, S.M. (1988). Follicle and fleece characteristics of Merinos, Romneys and Merino-Romney crossbreds. Proc. New Zealand Soc. Animal Prod. **48**: 195–200, c.f. *Anim. Breed. Abstr.* (1990) **58**: 594.

Meyer, H., Lohse, B. and Groning, M. (1967). A contribution of haemoglobin and potassium polymorphism in sheep. *Z. Tierzucht Zucht. Biol.* **83**: 34–37, c.f. *Anim. Breed. Abstr.* (1968) **36**: 249.

Miller, S.J. (1975). Personal communication from his data on A.I. in sheep in Australia.

Mohd. Yussuff, M.K., Dickerson, G.E. and Yang, L.D. (1992). Reproductive rate and genetic variation in composite and parental population experimental results in sheep. *J. Anim. Sci.* **70**: 673–688.

Mohan, Mani and Acharya, R.M. (1979). Heterosis in body weight and survival in sheep. *Indian J. Anim. Gen. Breed.* **1**: 92–98.

Moore, B.H. (1949). Characteristics of subtrates and media essential for metabolism and motility of ram, boar and stallion spermatozoa. Microfilm Abs., IX: 12–13 (*Anim. Breeding Abstr.* **19**: 30).

More, T., Pachlag, S.V. and Sahni, K.L. (1975). Incidence of blood potassium type and mortality in sheep under semi-arid conditions. *Indian Vet. J.* **52**: 687–690.

Morely, F.H.W. (1948). The effect of flystrike on the scrotum on subsequent fertility in rams. *Aust. Vet. J.* **24**: 94–95.

Morley, F.H.W. (1990a). The profitability of ewe selection. *Proc. Aust. Soc. Anim. Prod.* **18**: 312–315, c.f. *Anim. Breed. Abstr.* (1991) **59**: 5482.

Morley, F.H.W. (1990b). Fibre diameter and profits from selection of Merino ewes. Wool Technology and Sheep Breeding **38**: 101–103, c.f. *Anim. Breed. Abstr.* (1991) **59**: 8332.

Morris, C.A., Clarke, J.N. and Smeaton, J.E. (1983). Objectives for sheep improvement sheep Production. **I**: 143–167. *Eds.* Wichham, G.A. and McDonald, M.F. Ray Richards, Auckland.

Nadkarani, U.G. (1968). Discriminatory techniques in classing of fleeces. *J. Indian Soc. Agric. Stat.* **20(2)**: 22–28.

Nainar, A.M., Krishnamurthy, U.S. and Rathnasabapathy, V. (1977). Serum alkaline phosphatase and its association with R-O-i blood groups in Nilgiri sheep. *Cheiron* **6**: 136–141.

Narain, P. and Garg, L.K. (1975). A possible use of discriminant function and D^2-Statistic for comparing different grades of sheep in a crossbreeding programme. *Indian J. Anim. Sci.* **45**: 243–247.

Narayanaswamy, M. and Balaine, D.S. (1976). A note on the oestrus cycle in Bannur sheep. *Indian Vet. J.* **53**: 235–236.

Narayanaswamy, M., Balaine, D.S. and Chopra, S.C. (1975). A note on gestation length in Mandya sheep. *Indian J. Anim. Sci.* **45**: 915–917.

Naude, J.E. and Grant, J.L. (1979). An investigtation into an eight month reproduction cycle for sheep. *In*: Division of Livestock and Pastures Annual Report for the year ended 30 September, 1978, pp. 223–224.

NCA (1976). Report, Part VII: Animal Husbandry. Government of India, Ministry of Agriculture and Irrigation, New Delhi.

Nel, J.A., Mostert, L. and Steyn, M.G. (1960). Karakul breeding and research in South-West Africa with special reference to the Neudam Karakul stud. *Anim. Breed. Abstr.* **28**: 89–101.

Negi, P.R. and Bhat, Pran, P. (1980). Genetic variability of blood serum proteins and enzymes in Gaddi sheep and its half breds with Rambouillet and Russian Merino. *Indian J. Anim Sci.* **50(11)**: 1004–1008.

Nicholas, F.W. and Smith, C. (1983). Increase rate of genetic change in dairy cattle by embryo transfer and sampling. *Anim. Prod.* **36**: 341–353.

Niekerk, C.H. Van (1979). Limitations to female reproductive efficiency. In Sheep Breding. 2nd edn, pp. 303–313 (*Eds*) G.L. Tommes, D.E. Robertson, R.J. Lightfoot, Butterworths, London, UK.

Nivsarkar, A.E. and Acharya, R.M. (1982). Note on feed lot and carcass characteristics of Malpura sheep. *Indian J. Anim. Sci.* **52**: 1262–1264.

Nivsarkar, A.E., Arora, R.L. and Arora, C.L. (1987). Post-weaning performance of Russian Merino lambs under semi-arid tropical climate. *Indian J. Anim. Sci.* **57(5)**: 486–489.

Nivsarkar, A.E., Arora, C.L., Rawat, P.S. and Sanghi, V.N. (1983). Reproductive efficiency of Merino and Rambouillet sheep in the semi-arid conditions of Rajasthan. *Livestock Adviser* **VIII**: 39–45.

Nivsarkar, A.E., Arora, R.L., Arora, C.L., Rawat, P.S. and Kumar, Mahesh (1984a). Factors affecting pre-weaning body weights in Soviet Merino sheep. *Indian J. Anim. Sci.* **54(2)**: 211–213.

Nivsarkar, A.E., Arora, R.L., Arora, C.L., Rawat, P.S. and Kumar, Mahesh (1984b). Pre-weaning body weight of Rambouillet lambs. *Indian J. Anim. Sci.* **54(4)**: 385–387.

Oldeman, L.R., Hakkeling, R.T.A. and Sombroek, W. G. (1990). "World Map of the Status of Human-Induced Soil Degradation; Explanatory Note". The Global Assessment of Soil Degradation, ISRIC and UNEP in cooperation with the Winand Staring Centre, ISSS, FAO and ITC.

Ortavant, R. (1958). Le cycle spermatogenotique Chez le belier. Thesis, University of Pairs.

Pacho, H.N. (1973). Diagnostico de gestaction por laparotomia en ovejas (Pregnancy diagnosis by laparotomy in ewes), *Veterinaria* (Mexico) **4**: 161–165.

Patnayak, B.C. (1980a). Feeding of protected proteins to sheep. Proceeding of Symposium on protein utilization in ruminants. NDRI, Karnal.

Patnayak, B.C. (1980b). Livestock productivity on arid and semi-arid pasture. Lecture notes, ICAR Summer Institute on Management of Arid Zone of India (Vol. 3) CAZRI, Jodhpur.

Patnayak, B.C. (1980c). Problems and prospects in the nutrition of arid zone Livestock. *In*: "Arid Zone Research and Development" *Ed.* H.S. Mann.

Patnayak, B.C. (1981). Researches in India on nutrient requirements for sheep. 1st Proceeding of Society and National Seminar on Sheep and Goat Production and Utilization. *Eds.* Basuthakur, A.K. and Acharya, R.M.

Pattie, W.A. and Smith, M.D. (1964). A comparison of the production of F_1 and F_2 Border Leicester × Merino ewes. *Aust. J. Exp. Agric Anim. Husb.* **4**: 80–85.

Peters, H.F., Sken, S.B. and Hargrave, H.J. (1961). Genetic trends in performance of the Romnelet sheep during the period of breed development. *Can. J. Anim. Sci.* **41**: 126–133.

Plant, J.W. (1980). Pregnancy diagnosis in the ewe. *World Anim. Rev.* **36** : 44–47.

Poggenpoel, D.G. (1991). Efficiency of early selection of Merino rams on a selection index. *South African J. Anim. Sci.* **21**: 55–58, c.f. *Anim. Breed. Abstr.* **59**: 6811.

Poonia, J.S. (1982). Studies on preweaning growth and wool quality traits in nali lambs and its crosses with Carriedale and Russian Merino. MVSc. Thesis, Haryana Agric. Univ., Hisar.

Prasad, V.S.S., Arora, C.L., Singh, R.N. and Bohra, S.D.J. (1981). Assessment of carcass and mutton quality in native and cross-bred sheep. *Indian J. Anim. Sci.* **51(6)**: 638–642.

Proceeding of National Seminar on Carpet Wool Production and Manufacture (1983). Held at Bikaner. March 5-6, 1983, CSWRI, Avikanagar.

Proceeding of National Convention on Sheep (1983). Held at University of Agricultural Sciences, Bangalore, November 11-13, 1983.

Purohit, G.R., Ghosh, P.K. and Taneja, G.C. (1973). Effects of varying degree of water restriction on the distribution of body water in high ad low potassium type Marwari sheep. *J. Agric. Sci., Camb.* **80**: 177–180.

Purser, A.F. (1965). Repeatability and heritability of fertility in hill sheep. *Anim. Prod.* **7**: 75–82.

Purvis, I. (1990). Heterosis between strain of Australian stud Merino Sheep. Proc. 4th World Cong. Genet. Applied Liverstock Prod. XV: 220–223, c.f. *Anim. Breed. Abstr.* (1991) **59**: 330.

Raina, B.L. (1969). Studies on the blood groups of Indian Sheep. Ph.D. Thesis, Agra University Library, Agra.

Ramamurty, A. (1963). Breeding performance of Nilgiri, Romney Marsh and kuibishow rams. *Indian Vet. J.* **40**: 621.

Ranjhan, S.K., Singh, C.P., Nadger, S.R. and Talpatra, S.K. (1959). Studies on some high protein green feeds of Uttar Pradesh. 1. Chemical composition and nutritive value of the green feed kerr (*Carthamus tictorius* Linn.). *Indian Vet. J.* **36**: 138–144.

Rao, L.R., Purushotham, C. and Reddy, K.K. (1978). Studies on gestation period in Mandya and Nellore breeds of sheep. *Livestock Adv.* **3(6)**: 9–12.

Rastogi, R., Boylan, W.J. and Rempel, W.E. (1988). Reproductive performance of purebred sheep and their two and three breed crosses. *J. Anim. Sci.* **66**: 118–124.

Rastogi, R., Boylan, W.J. and Rempel, W.E. (1975). Lamb performance and combining ability of Columbia, Sufolk and Targhee breeds of sheep. *J. Anim. Sci.* **41**: 10–15.

Rastogi, R., Boylan, W.J., Rempel, W.E. and Windels, H.F. (1982). Specific three-breed crosses in sheep I. Reproduction and lamb survival. *Indian J. Anim. Gen. and Breed.* **4**: 14–19.

Rawal, C.V.S., Singh, R., Garg, R.C. and Luktuke, S.N. (1986). A study on post partum oestrus interval in Muzaffarnagri sheep. *Indian J. Anim. Reprod.* **7**: 95–98.

Reeve, E.C.R. and Robertson, F.W. (1953). Factors affecting multiple births in sheep. *Anim. Breed. Abstr.* **21**: 211.

Reid, J.T., Bensadown, A., Bull, L.S., Burton, J.M., Gleeson, P.A., Han, I.K., Joo, Y.D., Johnson, D.E., McManus, W.R., Paladines, O.L., Stroud, J.W., Tyrrell, H.F., Van Neikerk B.D.H., Welligton, G.H. and Wood, J.D. (1968). Changes in body composition and meat characteristics of cattle, pigs and sheep during growth. Proc. Cornell Meat Conf. 1968. New York, Oct, 1968. pp. 183–187.

Report (1994). Technical committe of Direction for improvement of Animal Husbandry and Dairying Statistics – Department of Animal Husbandry and Dairying, Ministry of Agriculture (GO1), New Delhi.

Richardson, C. (1972). Pregnancy diagnosis in the ewe: A review, *Vet. Record* **90**: 264–275.

Robertson, H.A. and Sarda, I.R. (1971). A very early pregnancy test for mammals: Its application to the ow, ewe and sow. *J. Endocrinol.*, **49**: 407–419.

Roux, P.J. le, Wyk, L.C. van (1977). Studies on reproduction in Karakul sheep: Effect of shortened length of gestation after flametha zone treatment on the quality of Karakul lamb pelts 1971–1975. *In* Agri. Res. 1976. pp. 86–87. Department of Agricultural Technical Services, Pretoria, South Africa.

Sahni, K.L. (1967). Studies on quality, preservation and fertility of semen of sheep and goat. Ph.D. Thesis Agra University, Agra.

Sahni, M.S., Bapna, D.L., Batra, H.S., Mohd, A. and Verma, S.K. (1991). Selection studies in Marwadi sheep: Performance studies. Annual Report (1990–91). Central Sheep and Wool Research Institute, Avikanagar, pp. 10–11.

Sahani, M.S., Dwivedi, V.K., Dubey, S.C. and Arora, C.L. (1980). Note on body weight from birth to first breeding and fleece weight in native and cross-bred (quarter-bred) sheep produced under field conditions in semi-arid areas of Rajasthan. *Indian J. Anim. Sci.* **50**(7): 580–581.

Sahani, M.S. and Pant, K.P. (1978). Breed differences in the duration of pregnancy in sheep. *Indian Vet. J.* **55**: 99–102.

Sahni, K.L. and Roy, A. (1967). A note on summer sterility in Rommney Marsh rams under tropical conditions. *Indian J. Vet. Sci.* **37**: 335–38.

Sahni, K.L. and Roy, A. (1969). Influence of season on semen quality of rams and effects of diluters and dilutions on *in vitro* preservation. *Indian J. Anim. Sci.* **39**: 1–14.

Sahni, K.L. and Roy, A. (1972). A note on seasonal variation in semen production of Corriedale and Corriedale × Bikaneri (Magra) rams under tropical condition. *Indian J. Anim. Sci.* **42**: 99.

Sahni, K.L. and Tiwari, S.B. (1972). Reproductive efficiency of young ewes of various sheep breeds. Annual Report, Central Sheep and Wool Research Institute, Avikanagar, Rajasthan.

Sahni, K.L. and Tiwari, S.B. (1974b). Effect of two ejaculation frequencies on semen characterisation of Rambuillet rams in different seasons. *Indian vet. J.* **52**: 252–255.

Sahni, K.L. and Tiwari, S.B. (1981). Reproductive performance of bucks and rams under hot arid and hot humid conditions. 1st Proceeding of the Society and National Seminar on Sheep and Goat Production and Utilization.

Salamon, S. and Robinson, T.J. (1962). Studies on the artificial insemination of Merino Sheep. I and II. *Aust. J. Agri. Res.* **13**.

Saxena, J.S., Kulshreshtha, S.K. and Upadhyaya, R.B. (1972). Studies on assessment of palatability and nutritive value of green panic (*Panicum maximum* var trichoglume). *Indian Vet. J.* **49**: 177–179.

Saxena, V.B. and Singh, G. Preservation of ram spermatozoa. I. Effects on motility, livability and morphology of ram spermatozoa in four semen diluents. *J. Anim. Morph. Physiol.* **14**: 114–120 (*Anim. Breeding Abstr.* **37**: 2662).

Schmalwasser, T., Konig, K.H., Al-Ali, A. and Al-Chikh, A. (1990). Improvement of reproductive performance in sheep by means of breeding. *Anim. Breed. Abstr.* **58**: 4426.

Seth, O.N. (1968). Influence of haemoglobin varient on the fertility in Bikaneri (Magra) sheep. *Curr. Sci.* **37**: 231–232.

Seth, O.N. (1973). A note on certain economic characters in relation to haemoglobin types in Bikaneri (Magra) sheep. *Indian J. Anim. Sci.* **43**: 549–552.

Seth, O.N., Pandey, M.P. and Roy, A. (1971). Effect of haemoglobin types on fertility in Bikaneri sheep. Workshop on blood Grps. and Biochemical polymorphism in Indian Fur Animals. Hyderabad, April 10-12, 1971.

Sharma, S.K., Bhat, P.P. and Bhat, P.N. (1976). Haemoglobin and transferrin polymorphism in Muzaffarnagri breed of sheep and its crosses with Corriedale. *Anim. blood Grps. Biochem. Genet.* **7**: 127–132.

Sharma, S.K., Bhat, Pran P., Bhat, P.N. and Garg, R.C. (1978). Genetic and non-genetic factors affecting pre-weaning body weights in Muzaffarnagri and Corriedale × Muzaffarnagri halfbreds. *Indian J. Anim. Sci.* **48(3)**: 177–180.

Sharma, R.P., Tiwari, S.B. and Roy, A. (1969). Effect of frequency of semen collection from ram on seminal attributes and fertillity. *Indian J. Anim. Sci.* **39**: 15.

Shrestha, J.N.B., Rempel, W.E., Boylan, W.J. and Miller, K.P. (1983). General, specific maternal and reciprocal effects for ewe productivity in crossing five breeds of sheep. *Can. J. Anim. Sci.* **63**: 497–509.

Shukla, D.D. and Bhattacharya, P. (1952). Seasonal variation in reaction time and semen quality. *Indian. J. Vet. Sci.* **22**: 109.

Shukla, K.S. and Ranjhan, S.K. (1969). Chemical composistion and nutritive value of Gokhru, *Indian Vet. J.* **46**: 715–718.

Sidwell, G.M. and Miller, L.R. (1971). Production in some purebreds of sheep and their crosses I. reproductive effciency in ewes. *J. Anim. Sci.* **32**: 1084–1089.

Sidwell, G.M. Everson, D.O. and Terrill, C.E. (1964). Lamb weights in some purebreds and crosses. *J. Anim. Sci.* **23**: 105–110.

Singh, R.N., Arora, C.L., Bohra, S.D.J., Bapna, D.L. and Nivsarkar, A.E. (1981). Note on variation in the pre- and post-weaning body weights in Malpura and Sonadi, and their crosses with Suffolk and Dorset. *Indian J. Anim. Sci.* **51(9)**: 877–880.

Singh, Harbans and Earl, N. Moore (1968). Livestock and Poultry Production. Prentice Hall of India Pvt. Ltd., New Delhi.

Singh, C.S. and Joshi, D.C. (1957). Goja (*Amaranthus spinosa* Linn.) a drought resistant evergreen useful feed for sheep. *Indian Vet. J.* **34**: 191–196.

Singh, Gurmej, Gupta, D.C., Bohra, S.D.J. and Arora, C.L. (1987). Genetic and non-genetic factors affecting lamb mortality and its causes under farm conditions in semi-arid tropics. *Indian J. Anim. Sci.* **57(8)**: 871–877.

Singh, Gurmej, Mehta, B.S., Sethi, I.C. and Arora, C.L. (1987). Genetic and non-genetic factors affecting growth traits of Nali and its crossbred lambs under semi-arid conditons. *Indian J. Anim. Sci.* **57(7)**: 728–734.

Singh, R.N., Nivsarkar, A.E., Bohra, S.D.J., Kumar, Mahesh and Arora, C.L. (1984). Performance of

Malpura and Sonadi, and their halfbreds with Dorset and Suffolk. *Indian J. Anim. Sci.* **54(11)**: 1084–1086.

Singh, M., Patnayak, B.C., Singh, N.P., Arya, D.L. and Sidiqui, S.S. (1977). Observations on the effect of feeding level on canary coloration of Indian wools. *Indian J. Anim. Sci.* **47**: 34–38.

Singh, M. and Rai, A.K. (1981). Adaptation of sheep to hot arid environment. 1st Proceeding of the Society and National Seminar on Sheep and Goat Production and Utilization.

Singh B. and Roy, A. (1963). Studies on certain aspects of sheep and goat husbandry. VI. Seasonal variation in the semen qualit of Corriedale and Bikaneri rams. *Indian J. Vet. Sci. and Anim. Husb.* **33**: 211–214.

Singh, G. and Saxena, V.B. (1965). Effect of various dilutors on preservability and morphology of ram spermatozoa. *Agri. Res.* **5**: 215.

Singh, L.B., Singh, Mohan and Dwarkanath, P.K. (1972a). Variation of haemoglobin polymorphism in Russian Merino located at two different semi-arid (Jaipur) and Himalayan range (Simla). *Indian J. Anim. Hlth.* **11**: 239–240.

Singh, L.B., Singh, Mohan and Dwarkanath, P.K. (1972b). Some observed benefits of haemoglobin type B in desert sheep of Rajasthan. *Curr. Sci.* **41**: 740–741.

Singh, L.B., Singh, Mohan and Dwarkanath, P.K. (1973). A note on the significance of haemoglobin and blood potassium types. *Indian J. Anim. Sci.* **43**: 890.

Singh, L.B., Singh, Mohan, Dwarkanath, P.K. and Lal, A. (1975). Adaptive significance of haemoglobin and blood potassium types. *Curr. Sci.* **44**: 474–475.

Singh, L.B., Singh, Mohan and Dwarkanath, P.K. (1976a). Haemoglobin and blood potassium type in some indigenous, exotic and crossbred sheep. *Indian J. Anim. Sci.* **46**: 345–350.

Singh, L.B., Singh, Mohan, Chaudhary, A.L. and Dwarkanath, P.K. (1976b). Effect of Hb and K type of some production traits in Magra sheep. *Indian J. Anim. Hlth.* **15**: 139–143.

Singh, N.P., Singh, M. and Patnayak, B.C. (1979). Effect of energy and protein levels on nutrient digestibility and wool production. *Indian J. Anim. Sci.* **49**: 277.

Sinha, N.K., Wani, G.M. and Sahni, K.L. (1980). Note on the incidence of gestational oestrus in Muzaffarnagri ewes. *Indian J. Anim. Sci.* **50**: 1009.

Skolovskaya, I.L., Drozdova, L.P., Goluisheva, M.G., Korotkov, A.I., Maximov, U.V. and Lebedeva, V.A. (1956). Improvement of diluents for semen of farm animals. *Proc. Lenin Acad. Agri. Sci.* **XXI**: 17–24 (*Vet. Bull.* **XXVII**: 937).

Srivastava, V.K. and Roy, A. (1969). The influence of season of birth and age on the carcass quality of Magra lambs. *Indian J. Anim. Sci.* **39**: 294–306.

Steinfeld, H., Gerber, P., Wassenaar, T., Castel, V., Rosales, M. and de Haan, C. (2006). Livestock's long shadow. Environmental issues and options. FAO/LEAD, Rome.

Talpatra, S.K. and Bansal, C.S. (1957). Studies on indigenous creepers: A new rainfed high protein phosphorous rich plant, *Indian J. Vet. Sci.* **27**: 45–60.

Taneja, G.C. (1968). Hairiness in wool in relation to blood potassium types in Marwari sheep. *Experientia* **24**: 696.

Taneja, G.C. (1970). Blood potassium types in sheep in relation to animal production in arid environment – A review. *Proc. Indian Acad. Sci. B* **36(5)**: 311–330.

Taneja, G.C. (1972). Genetic control and adaptive significance of potassium types in some indigenous and exotic breeds of sheep in India. *J. Genet.* **61**: 64–77.

Taneja, G.C. and Abichandani, R.K. (1967). Genetic basis of blood potassium concentration in sheep. *Indian J. Exp. Biol.* **5**: 226–228.

Taneja, G.C. and Ghosh, P.K. (1965). Variation in potassium concentration of blood in Marwari sheep. *Indian J.Exp. Biol.* **3**: 166.

Taneja, G.C. and Ghosh, P.K. (1967). Body weight and fleece weight in relation to blood potassium types in Marwari sheep (A preliminary note). *Indian Vet. J.* **44**: 402–407.

Taneja, G.C., Ghosh, P.K., Abichandani, R.K. and Goyal, D. (1971). Seasonal variation in blood composition in high and low potassium type Marwari sheep. *J. Agric. Sci. Camb.* **77**: 37–41.

Taneja, G.C., Narayan, N.L. and Ghosh, P.K. (1969). On the relationship between wool quality and the type and concentrate of blood potassium in sheep. *Experientia* **25**: 1200–1201.

Taparia, A.L. (1972). Breeding behaviour in Sonadi sheep. *Indian J. Anim. Sci.* **42**: 576–579.

Terrill, C.E. (1974). Sheep. *In*: Hafez, Reproduction in Farm Animals, (*Ed*). Hafez, E.S.E. 3rd edn. pp. 265–274. Lea and Febiger, Philadelphia, USA.

Terrill, C.E. (1958). Fifty years of progess in sheep breeding. *J. Anim. Sci.* **17**: 944–959.

Teixeira, F.J.L., Bessa, M.N., Souza, A.A. de, Freitas, J.P. de and Mendonca, H.L. (1980). Contributicao ao estudo da efciencia reproductiva de ovelhasdeslanades Morada Nova-Variedade Branca (Reproductive efficiency of White Morada Nova sheep). In Pesquisa em ovino no Brasil 1975–1979. pp. 55, Empresa Brasileira de Pesquisa Agropecuaria, Bage, Brazil.

Thapan, P.C., Bell, B.S. and Parker, C.F. (1970). Reproductive performance in Targhee sheep. *Indian J. Anim. Sci.* **40**: 223–228.

Thapan, P.C., Bell, B.S. and Parker, C.F. (1979). Index selection as a method of estimating breeding value in Targhee sheep. *Indian J. Anim. Prod.* **11**: 20–28.

Thwaites, C.J. (1981). Development of ultrasonic techniques for pregnancy diagnosis in the ewe. *Anim. Breeding Abstr.* **49** : 427–434.

Tiwari, S.B. and Honmode, J. (1968). Age of matrutiy of young rams of various breeds with special reference to semen production. Annual Report, Central Sheep and Wool Research Institute, Avikanagar, Rajasthan, pp. 4.

Tiwari, S.B. and Sahni, K.L. (1981). Puberty and reproductive performance in ram lambs of indigenous exctic and cross-bred sheep under semi-arid conditions. *Indian J. Anim. Sci.* **51**: 634–638.

Tiwari, S.B., Sharma, R.P. and Roy, A. (1969a). Process of parturition in sheep and goat and morphological characteristics of their placenta. *Indian Vet. J.* **46**: 576.

Tiwari, S.B., Srivastava, R.Ş. and Kalr, D.B. (1978). Effect of stage ofheat on conception rate. Annual report. Central Sheep and Wool Research Institute, Avikanagar, Rajasthan.

Tomar, S.S. (1979). Time of parturition in sheep. *Indian J. Anim. Res.* **13**: 68–70.

Trapp, M.J. and Slyter, A.L. (1983). Pregnancy diagnosis in the ewe. *J. Anim. Sci.* **57**: 1–5.

Trivedi, K.R. and Bhat, P.P. (1978). A note on transferrin polymorphism in Muzaffarnagari breed of sheep (*Ovis aries*) and its crosses with Suffolk and Dorset. *Indian J. Anim. Sci.* **48**: 318–320.

Turner, H.N., Hayman, R.H., Triffitt, L.K. and Prunster, R.W. (1962). Response to selection for multiple births in the Australian Merino. A progress report. *Anim. Prod.* **4**: 165–176.

Turner, H.N. and Kotch, J.H. (1961). Studies on the sodium-potassium balance in erythrocytes of Australian Merino sheep. *Aust. J. Biol. Sci.* **14**: 260–273.

Valencia, J., Barron, C., Ferandez-Baca, S. (1977). Pubertad en corderos Tabasco × Dorset (Puberty in Tabasco × Dorset lambs), *Veterinaria* (Mexico) **8**: 127–130.

Venkatghalam (1960). Chemical composition and nutritive value of buffalo grass (*Brachiaria mutica*). *Indian Vet. J.* **37**: 260–262.

Vesley, J.A. and Peters, H.F. (1964). The effects of breed and certain environmental factors on birth and weaning traits of range sheep. *J. Anim. Sci.* **44**: 215–219.

Vesley, J.A. and Peters, H.F. (1981). Lamb production from ewes of four breeds and their two, three and four breed crosses. *Can. J. Anim. Sci.* **61**: 271–277.

Volvani, R.(1953). Seasonal variations in spermatogenesis of some farm animals under the climatic conditions of Israel. *Bull. Res. Coun. Israel* **3**: 123–126 (*Anim. Breeding Abstr.* **22**: 1279).

Wani, G.M. (1981). Ultrasonic pregnancy diagnosis in sheep and goats – A review. *World Rev. Anim. Prod.* **17(4)**: 43–48.

Wani, G.M. and Sahni, K.L. (1981). Ultrasonic pregnancy diagnosis in ewe under tropical field conditions. *Indian J. Anim. Sci.* **51**: 194–197.

Waston, R.H. (1952). Seasonal variation in fertility in Marino sheep. *Pastoral Rev.* **62**: 324 (*Anim. Breeding Abstr.* **21**: 272).

Webster, G.M. and Haresign, W. (1981). A note on the use of dexamethasone to induce parturition in the ewe. *Anim. Prod.* **32**: 341–344.

Webster, W.M. (1952). Infertility in rams, Proc. N.Z. Aci. Anim. (*Anim. Breeding Abstr.* **20**: 1136).

Williams, S.M., Garrigus, U.S., Norton, H.W. and Nalbandov, A.V. (1956). The occurrence of estrus in pregnant ewes. *J. Anim. Sci.* **15**: 978–983.

Wilson, R.T. and Durkin, J.W. (1983). Livestock production in central Mali: Weight at first conception and ages at first and second parturitions in traditionally managed goats and Sheep. *J. Agric. Sci.* (Camb.) **100**: 625–628.

Yalcin, B.C. (1973) Studies on phenotypic and genetic parameters and on the improvement through selection of important fertility traits in Kenya Merino sheep. *Anim. Breed. Abstr.* **41**: 2635.

Yenikoye, A., Andre, D., Ravault, J.P. and Mariana, J.C. (1981). Etude de quelques caracteristiques de reproduction chez la brebis Peulth, du Niger (Some reproductive characteristics of the Fulani ewe is Niger). *Reprod. Nutr. Develop.* **21**: 937–951.

Yenikoye, A., Pelletier, J., Andre, D. and Mariana, J.C. (1982). Anomalies in ovarian function of Peulh ewes. *Theriogenology* **17**: 355–364.

Young, S.S.Y. (1961). The use of sire's and dam's records in animal selection. *Heredity* **16**: 91–102.

Young, S.S.Y., Turner, H.N. and Dolling, C.H.S. (1963). Selection for fertility in Australian Merino sheep. *Aust. J. Agric. Res.* **14**: 460–482.

Young, L.D. and Dickerson, G.E. (1991). Comparison of Booroola Merino and Finn sheep. Effects on productivity of mates and performance of crossbred lambs. *J. Anim. Sci.* **69**: 1899–1911.

Younis, A.A. and Afifi, E.A. (1979). On the occurrence of oestrus in pregnant Barki and Merino ewes. *J. Agric. Sci.* (Camb.) **92**: 505–506.

Younis, A.A. and El-Gaboory, I.A.H. (1978). On the diurnal variation in lambing and time for placeta expulsion in Awassi ewes. *J. Agri. Sci.* (Camb.) **91**: 757–760.

Younis, A.A., El-Gaboory, I.A., El-Tawil, E.A. and El-Shobokshy, A.S. (1978). Age at puberty and possibility of early breeding in Awassi ewes. *J. Agri. Sci.* (Camb.) **90**: 255–260.

Widda, W.F. (1954). Differences in cation concentrations in foetal and adult sheep erythrocytes. *J. Physiol.*(Lond.) **125**: 18–19.

INDEX

D

E

F

G

H

I

J

K

L

M

N

O

P

R

S

T

V

W

Y